Therapeutic Revolutions

Therapeutic Revolutions

Pharmaceuticals and Social Change in the Twentieth Century

EDITORS JEREMY A. GREENE,
FLURIN CONDRAU, AND
ELIZABETH SIEGEL WATKINS

The University of Chicago Press   Chicago and London

The University of Chicago Press, Chicago 60637
The University of Chicago Press, Ltd., London
© 2016 by The University of Chicago
All rights reserved. Published 2016.
Printed in the United States of America

25 24 23 22 21 20 19 18 17 16   1 2 3 4 5

ISBN-13: 978-0-226-39073-4 (cloth)
ISBN-13: 978-0-226-39087-1 (paper)
ISBN-13: 978-0-226-39090-1 (e-book)
DOI: 10.7208/chicago/9780226390901.001.0001

Library of Congress Cataloging-in-Publication Data
Names : Greene, Jeremy A., 1974– editor. | Condrau, Flurin, editor. |
    Watkins, Elizabeth Siegel, editor.
Title: Therapeutic revolutions : pharmaceuticals and social change
    in the twentieth century / Jeremy A. Greene, Flurin Condrau, and
    Elizabeth Siegel Watkins, editors.
Description: Chicago ; London : The University of Chicago Press,
    2016. | Includes index.
Identifers: LCCN 2016018233 | ISBN 9780226390734 (cloth :
    alk. paper) | ISBN 9780226390871 (pbk. : alk. paper) |
    ISBN 9780226390901 (e-book)
Subjects: LCSH: Pharmaceutical industry—Social aspects. |
    Therapeutics—Social aspects.
Classification: LCC HD9665.5.T46 2016 | DDC 303.48/3—dc23
    LC record available at https://lccn.loc.gov/2016018233

♾ This paper meets the requirements of ANSI/NISO Z39.48-1992
(Permanence of Paper).

Contents

Medicine Made Modern by Medicines

JEREMY A. GREENE, FLURIN CONDRAU, AND ELIZABETH SIEGEL WATKINS

Asked to compare the practice of medicine today to that of a hundred years ago, most people will respond with a story of therapeutic revolution: back then we had few effective remedies, now we have more (and more powerful) tools to fight disease. These narratives of medical modernity are often illustrated with specific pharmaceuticals: antibiotics that defeated infectious diseases, vaccines that prevented childhood diseases, antineoplastic drugs that fought cancers, cardiovascular drugs that helped stem the epidemic of heart disease, immunosuppressants that made complex organ transplants possible, psychotropic drugs that controlled the demons of psychosis and lifted the veil of depression. These stories have become familiar catechisms of the biomedical present. They suggest sudden and dramatic forms of social change that followed in the wake of a series of magic bullets discovered over the course of the twentieth century. In these versions of history, medicine was made modern—and effectual—by medicines.

Stories of glorious therapeutic revolutions can have strong appeal even in the face of later complications. The twenty-first century has already witnessed a number of high-profile pharmaceutical scandals, from the increased risk of suicide associated with selective serotonin uptake inhibitors (SSRIs), such as Prozac, Paxil, and Zoloft, to the increased rates of cardiovascular death associated with the painkiller Vioxx and the diabetes drug Avandia. These and other stories of lives damaged by therapeutic enthusiasm remind us that the massive rollout of new pharmacotherapeutics has the potential to harm as well as help. Likewise, the differential availability of antiretroviral cocktails in affluent versus poor countries has produced a heterogeneous map of HIV/AIDS mortality that shows just how unevenly these therapeutic revolutions are experienced across time and space. In recent decades, reports of a "crisis of innovation" in the multinational pharmaceutical sector have led many industry commentators to speculate as to whether a therapeutic revolution that began in the mid-twentieth century has ground to a halt in the twenty-first.[1]

Yet narratives of revolutionary change in biomedical therapeutics continue to have lasting explanatory power. This book engages the concept of therapeutic revolution developed in 1977 by Charles E. Rosenberg, who used the term to describe a fundamental shift between the beginning and the end of the nineteenth century in lay and professional understandings of efficacy, or what makes a medicine work.[2] For Rosenberg, therapeutic efficacy was both historically contingent and locally specific; it mattered where, how, by whom, for whom, and within what cultural and cognitive framework a medical intervention was employed. This concept has since been extended and amended to account for changes in medicine in the twentieth and twenty-first centuries as well. The aim of this volume is to open this contextual notion of therapeutic revolution to analysis and debate. We examine the collective memory of a powerful twentieth-century pharmacotherapeutic revolution as a key and largely unexamined narrative in popular, professional, and scholarly histories of biomedicine. This is not to deny the utility of revolution as an analytic implement in the historian's tool kit. Rather, the essays collected in this volume seek to challenge the linearity of this historical narrative, provide thicker descriptions of the process of therapeutic transformation, and explore the complex relationships between medicines and social change. Working on three continents and touching upon the lived experiences of patients and physicians, consumers and providers, marketers and regulators, and many other actors and agents, the contributors to this volume cumulatively reveal the tensions between universal claims of therapeutic knowledge and the specificity of local sites in which they are put into practice. Collectively we ask: What is revolutionary about therapeutics?

Revolutionary Narratives in Science and Medicine

Shortly after Rosenberg developed his analysis of nineteenth-century therapeutic revolutions, the German historian Reinhart Koselleck published an influential series of essays on the history of the concept of revolution itself.[3] Koselleck argued that the term meant different things to different actors at different moments in time—especially to those with a stake in the production and circulation of narratives of social change. The concept of the industrial revolution, for example, was initially proposed by industrialists themselves, as a means to naturalize some of the calamitous social change that accompanied the development of the spinning jenny, the steam engine, and the assembly line. As Kosseleck argued, to understand why Friedrich List depicted the social transformation taking place in eighteenth-century Germany as an industrial revolution, one needed also to understand that List was a Prussian railroad entrepreneur, economist, and politician with a specific set of interests in the work that the narrative of "industrial revolution" did for apportioning blame and responsibility for social inequity to a historical process rather than to individual actors.[4] When the British historian Arnold Toynbee published his influential history of the industrial revolution in the 1880s, the master narrative he generated for this era was not politically neutral and had lingering ramifications in both scholarly and popular imaginaries.[5,6] These master narratives have staying power, and can function to obscure rather than explain the broader social, political, and cultural dimensions of historical change.

Revolutionary narratives have long interfered with more nuanced interpretations of the history of science. The memorable first sentence of Stephen Shapin's 1996 book *The Scientific Revolution* ("There was no such thing as the Scientific Revolution and this is a book about it") acknowledged that, in the six decades of scholarship on the Scientific Revolution since Alexander Koyre first coined the term in the 1930s, the perspective of elite scientists had been privileged over more careful social and cultural accounts of how knowledge was produced, circulated, and consumed. Even before Shapin's account, Roy Porter had argued in 1986 that the concept of revolution was not particularly useful as an analytical category, but was perhaps more useful as a reflection of the interests of those who used it.[7]

We begin with the premise that to talk about a therapeutic revolution is to talk about a particular actor's narrative of the past rather than an objectively evident event. Over the course of the twentieth century, many physicians, pharmaceutical executives, public health officials, and patient activists have found it important to distinguish a scientific, rational present from a traditional, irrational past. Stories about therapeutic revolutions help these self-identified biomedical moderns separate their own present from whatever precursors they label as premodern. In order to understand the social utility of these master narratives of modernity and progress we must look for more complex and subtle stories of health, disease, and therapeutic change in the twentieth century.[8]

Stories that serve to sever modern (scientific) medicine from its premodern (traditional) past long predate the twentieth century. One can trace a long tradition of these narratives. Well before Galen's extensive writings on medicine in the second century, would-be therapeutic reformers had created narratives of modernity that differentiated the elegance of new diagnostic and therapeutic approaches from the alleged superstition and ignorance of their predecessors. The historiography of how medicine became scientific remains one of the most common popular narratives by which physicians and the public understand the history of medicine as necessarily progressive and the medicine of the present as necessarily modern.

However, not all stories of change and modernity require revolution as their principal narrative device. As the title of and introduction to William Osler's popular lectures on *The Evolution of Modern Medicine* (1913) suggested, would-be moderns could also take "an aeroplane flight over the progress of medicine through the ages," and chart a story of modernization that emphasized continuity instead of change. Fielding Garrison noted in his preface to the first edition of Osler's text that "the slow, painful character of the evolution of medicine from the fearsome, suspicious mental complex of primitive man, with his amulets, healing gods, and disease demons, to the ideal of a clear-eyed rationalism is traced with faith and a serene sense of continuity."[9] This curious relationship between revolution and evolution as metaphors for how medicine becomes modern is the subject of David Jones's essay in this collection. As change and continuity are two of the historian's principal analytics, it is often surprisingly difficult to disentangle one from the other.

Whether medicine has changed by revolution or evolution, a substantial corpus of medical historiography since Osler has continued to emphasize the role of science in making medicine modern, especially

4

in popular histories, from Paul deKruif's *Microbe Hunters* in 1926 to Paul Starr's *The Social Transformation of American Medicine* in 1982 to Siddhartha Mukherjee's *The Emperor of All Maladies* in 2010.[10] Even as professional historians of medicine have come to challenge the Whiggishness of such books, many scholarly synthetic accounts continue to perpetuate the notion that medicine was made modern by adopting new ways of knowing.

In these accounts, the advent of twentieth-century pharmaceuticals is but one of many aspects of medical modernity. These tellings and re-tellings of how medicine become modern frequently rely on narratives of revolution to depict pivotal and disruptive moments in the history of biomedicine. A short list of these disruptions in the nineteenth century would include the anatomo-pathological revolution of the Paris School, the anesthetic revolution of ether and chloroform, the antisep-tic revolution of Semmelweis and Lister, the bacteriological revolution of Pasteur and Koch, and the laboratory revolution in the life sciences. It is worth considering each of these in turn, as they laid the founda-tion for expectations and interpretations of therapeutic revolutions in the twentieth century.

The anatomo-pathological revolution has been the subject of many historical treatments, of which Michel Foucault's *The Birth of the Clinic* is perhaps the most revolutionary. In his dramatic reimagining of the medical gaze at the Paris Clinic in the early nineteenth century, Fou-cault described a transition from premodernity to modernity in French medicine that was as abrupt as the parallel transition from the *ancien regime* to the First Republic in French government. In the nearly con-temporaneous book *Medicine at the Paris Hospital*, Erwin Ackerknecht located these same historical actors within a greater continuity of thought and action, but did not downplay the powerful transforma-tions of either the French Revolution or the rise of the pathological an-atomical gaze. In neither Foucault's seismic account nor Ackerknecht's more gradualist telling, however, could the revolution acted out at the Paris Clinic be called a therapeutic revolution. With the exception, per-haps, of the conflict between Broussais and Louis over the therapeutic role of bloodletting, Paris clinicians became celebrated for their diag-nostic skills and were far less revolutionary in their approaches to treat-ments, outcomes, or the perspectives of patients themselves.[11]

In similar fashion, the developments of anesthesia and antisepsis—commonly narrated as the anesthetic revolution and the antiseptic revolution—have been told and retold as a set of milestones in medical science that potentiated new surgical practices.[12] In *Medicine and Soci-*

ety in America (1960), Richard Shryock reflected on the 1846 discovery of ether as the first breakthrough medical discovery to cross the Atlantic from west to east; two decades later, A. J. Youngson would pinpoint the introduction of chloroform and aseptic surgery as the precise moment of *The Scientific Revolution in Victorian Medicine.* Even though more nuanced works, such as Martin Pernick's *A Calculus of Suffering,* have located the rapid changes wrought by anesthesia and antisepsis in relation to a more complicated culture of surgery, consumer response, and doctor-patient interactions in general, the overwhelming narration of this episode remains one of intellectual discovery and rapid dissemination—a scientific revolution in miniature.[13]

For Roy Porter, however, only one revolution in medicine met the criteria he had laid out in 1986 for describing the scientific revolution: the bacteriological revolution. Two decades later, Michael Worboys argued that labeling the transformations in bacteriology as a revolution blurred rather than explained what was going on in that field as well as in the wider realms of biology and medicine.[14] According to Worboys, an evolutionary model of gradual change better explains how bacteriology began to reshape the medical landscape, not least because of the many different readings of and practices in that discipline in its various contexts. Worboys's critique of Porter's bacteriological revolution recapitulated earlier debates over the scientific revolution. In this case, a revisionist history has disputed earlier interpretations that a novel discipline, bacteriology, had created a new way of knowing the world of disease, and in so doing, had overturned an *ancien regime* of epistemology.

All of these central revolutionary narratives retell revolutions within medicine as a subgenre of scientific revolution. Leading bacteriologists (especially Louis Pasteur and Robert Koch) have been the most prominent subjects of studies engaging with the revolutionary nature of laboratory changes and medical science, with physiologists and pathologists (especially Claude Bernard and Rudolf Virchow) close behind.[15] Celebratory accounts that depicted these figures as revolutionaries helped subsequent generations of physicians to congratulate themselves about their own modernity and social legitimacy. They have since been followed by newer tales for the twentieth and twenty-first centuries celebrating the heroes of evidence-based revolutions, genomic revolutions, and neuroscience revolutions.

Each of these narratives captures important facets of changes in medical knowledge and practice. But we should be careful not to assume that stories physicians tell about themselves reflect the only way of understanding the practices and meanings that constitute modern medicine. Anesthesia and antisepsis made abdominal surgery newly possible, but the popularization of elective surgeries for the gallbladder owes as much to the managerial and business practices of Charles and Elton Mayo as it does to the scientific demonstrations of Morton or Lister. The rise of germ theory may have showcased the role of the laboratory in modern medicine for audiences of physicians in the 1880s, but as Nancy Tomes has shown in her history of the advertising of antibacterial goods in the mass market of the early twentieth century, it made medicine modern to the general public via the marketplace.[16] Likewise, Jonathan Liebenau has shown that the American pharmaceutical industry's first investment in medical laboratories provided an aura of scientific marketing rather than any true commitment to the research and development of innovative drug products. As the twentieth century progressed, the drug industry dedicated significant effort to producing and popularizing mass-media propaganda about the benefits of modern medicine to the everyday citizen, from the series of *Your Doctor and You* advertisements by Parke-Davis that ran in popular magazines in the 1920s and 1930s to the *Great Moments in Medicine* oil paintings commissioned by that same company and made available in boxed sets of reproductions for physicians to hang in their offices in the 1950s.[17] These narratives of medical modernity were far from disinterested.

Revolutionary Therapeutics: The Symbolic Power of Pharmaceuticals

One of the more obvious stakeholders invested in narratives of a twentieth-century therapeutic revolution was the pharmaceutical industry. Just as Antoine Lavoisier was happy to coin the phrase "chemical revolution" to describe the transformation that his own work embodied in the eighteenth century, the powerful narrative of the chemical revolution in psychiatry was put to use in the promotion of antipsychotic drugs in the early 1960s (see figure 0.1 and chapter 3). Physicians and patients were also invested in producing and consuming stories of revolutionary change in medicine. In the 1970s, the prominent academic physician Walsh McDermott (discussed in chapters 6 and 7) invoked narratives of therapeutic revolutions to defend the value of biomedicine in an atmosphere of increasing criticism. His contemporary, the physician-author Lewis Thomas, also took this

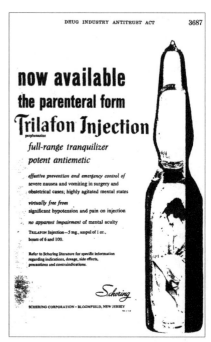

FIGURE 0.1 This advertisement for Schering's tranquilizer Trilafon suggested that if anti-psychotics had been available in the eighteenth century, they might have altered the course of the French Revolution. Advertisement (October 1956) reproduced in *Drug Industry Antitrust Act: Hearings before the Subcommittee on Antitrust and Monopoly of the Committee on the Judiciary, United States Senate, Eighty-Seventh Congress* (Washington: Government Printing Office, 1962), 3686–3687.

approach in *The Youngest Science: Notes of a Medicine-Watcher*, a 1983 memoir of sorts that marched through the decades of the twentieth century differentiating the therapeutically modern present of the (retired) author from the primitive therapeutics of his early twentieth-century medical training.[18]

The practice of medicine in 1980 looked quite a bit different than it had in 1930; it should not be surprising that physicians and historians would have sought to describe this era of transformation, just as others had done for the nineteenth century.[19] From the vantage point of the early twenty-first century, historian John Lesch looked back at the development of the sulfa drugs as a signal event in a chronology of a twentieth-century therapeutic revolution. Noting that he was not the first historian to employ the term, he defined "therapeutic revolution" as shorthand for an aggregate of events, including rapid expansion of pharmaceutical research, development, and production, that issued a steady and, in quantitative terms, unprecedented flow of new medicines onto the market and into medical practice, a flow that has continued from the late 1930s to our own day and that has transformed the practice of medicine.[20]

Lesch did not coin the term "therapeutic revolution" in 2007, nor did Rosenberg coin it in 1977. Rather, the term appears to have developed over time by historical actors as they described their enthusiasm and anxiety in a world of expanding pharmacotherapeutic possibility.

In 1933, *Fortune* magazine celebrated hormone research as "the most important field in medical research" and noted six products that were "of definite and immediate utility to the man-in-the-street and the physician-in-the-office": thyroid extracts, pituitary hormones, adrenalin, cortin, estrin, and insulin. The article described these developments as "authentic miracles" flowing from the laboratories of research scientists.[21] While mindful of the long history of unfulfilled promises made by quacks and other healers, the authors took pains to differentiate the applications of modern laboratory-based scientific research from snake oil and other alleged miracle cures. A decade later, the science writer Waldemar Kaempffert wrote of "the coming revolution in medicine" in *The American Mercury*. "A revolution is under way," he gushed, with sulfa drugs as "the harbinger of a new medical day."[22] In addition to recognizing the "sober, patient, cautious, academic gentlemen" conducting endocrine research, *Fortune* was also fascinated by the burgeoning American pharmaceutical industry in which many medical scientists were employed. In 1940 the magazine profiled Abbott Laboratories as an exemplary modern pharmaceutical firm: in spite of the lingering Great Depression, Abbott's sales and profits had doubled since 1935.[23] Part of the success of drug companies was attributed to their investment in research and development, as the pharmaceutical industry significantly expanded its research operations in the interwar years.[24]

Of course, many other social and technological developments, well beyond the pharmaceutical industry, contributed to the modernization of medicine in the later twentieth century, including the standardization of hospital practice and medical education, the expansion in federal funding of basic and clinical research, the routinization of clinical trials as a form of knowledge production, and the establishment of evidence-based medical practice. Therapeutic revolutions also took place beyond the field of pharmacology: in surgical procedures, diagnostic tests, medical devices, psychodynamic therapies, and occupational and physical therapies. Contemporaries appreciated the significance of surgeons, for example, and wrote about them in popular articles such as "Miracle Men," a tribute to military medicine during World War II.[25] In 1976, upon signing into law the bill that gave the US Food and Drug Administration (FDA) authority to regulate medical devices, President Gerald Ford proclaimed, "Medical and diagnostic devices have produced a therapeutic revolution."[26]

Yet of the many new forms of therapeutics developed by the expanding biomedical enterprise, the contributors to this volume have decided to focus specifically on pharmaceutical agents, as objects that have done substantial and symbolic work to transform the image and substance of medicine in the twentieth and twenty-first centuries. We have chosen to shine our spotlight on drugs because their status as everyday consumer goods—common things in our lives—affords historians and ethnographers the opportunity to examine the development and use of the concept of therapeutic revolution from a variety of standpoints.

A Guide to This Volume

The chapters of this volume highlight ways in which histories of therapeutic revolutions have been mobilized for specific purposes by specific historical actors, many of whom have come from within the field of medicine itself. In addition to the medical profession and the pharmaceutical industry, other organizations, from governmental bodies to civic groups, have found uses for the notion of therapeutic revolution. In areas such as the management of psychosis and tuberculosis control, narratives of transformative pharmaceuticals have served to justify public and private divestment from institutional forms of care. In turn, advocacy groups with foci as varied as population control, international consumerism, and the politics of living with mental illness have predicated their calls for action on the same master narrative of therapeutic revolution. Collectively, the chapters of this volume explore how actors with entirely orthogonal political positions on policies of pharmaceutical pricing, regulation, and intellectual property law nonetheless found that narratives of the twentieth-century therapeutic revolution were critical to their respective political platforms.[27]

The first chapters revisit three of the most common narratives linking therapeutic change with social change: the antibiotic revolution, the relationship of the birth control pill to contraceptive and feminist revolutions, and the chemical revolution in psychiatry. In chapter 1, Scott H. Podolsky and Anne Kveim Lie map out the strategies by which leading academic physicians in the field of infectious diseases used the history of the antibiotic revolution to argue for more resources and attention to their field, creating a series of doomsday "post-antibiotic" futures that portrayed the medical world as continuously under threat of reverting to a dangerous premodern state. In chapter 2, Elizabeth Siegel Watkins narrates how the birth control pill extended the reach of pharmacy beyond the treatment of disease, bringing about substantive and relatively sudden changes in both physician practice and patient experience in the realm of family planning, which resulted in large-scale transformations in contraceptive behaviors. In chapter 3, Nicolas Henckes provides a revisionist history of the "chemical revolution" in psychiatry, positing that the concept of therapeutic revolution might be better understood as a marketing strategy for psychotropic drugs that met with great success within the medical profession and ultimately within the public perception and historiography of medicine itself. He demonstrates that this marketing strategy was not containable: what started as localized changes in therapeutic practices within mental hospitals soon developed a ripple effect that expanded beyond the walls of the asylums, beyond the intellectual circles of the psychiatric profession, and out to the fabric of societies in which the lives of the mentally ill—both the treated and the untreated—were incorporated.

The next two chapters investigate how claims about a mid–twentieth-century therapeutic revolution have coincided with a robust debate about the relevant yardsticks by which such change ought to be measured. In chapter 4, Nils Kessel and Christian Bonah make use of an unusual set of historical sources—the archive of data from the market research giant IMS Health—to reexamine histories of therapeutic change from the standpoint of pharmaceutical consumption. In their close examination of therapeutic behavior in West Germany in the 1960s and 1970s, the narrative of therapeutic revolution appears as more of a wishful marketing slogan of pharmaceutical marketers, spreading narratives of therapeutic revolutions that were not necessarily borne out in their own market research data. In chapter 5, Janina Kehr and Flurin Condrau explore how twentieth-century histories of therapeutic revolutions have hinged in part on the emergence of formal protocols of proof, especially in the statistical methods that underscore therapeutic efficacy.[28] Their joint history and ethnography of treatment for tuberculosis (TB) examines the paradoxical ways in which clinical trials data demonstrating the efficacy of streptomycin and other TB drugs at mid-century had the unanticipated consequence of transforming tuberculosis from a subject of cutting-edge biomedical research into a disease that physicians considered "boring" or "neglected," until it reemerged in the late twentieth century as a new set of more "exciting" biomedical objects: multidrug resistant tuberculosis (MDRTB) and extremely drug-resistant tuberculosis (XDRTB).

The efficacy and economics of modern medicine form a third area of focus for this volume. In the realm of international development, as Jeremy Greene describes in chapter 6, patchy access to the whole armamentarium of modern medicines differentially shaped the health profiles and policies of many nations in the global South and drove decisions for both American policy makers and pharmaceutical manufacturers. The economics of modernization in developing countries has for the past five decades been interwoven with the economics of pharmaceutical manufacturing, marketing, and distribution. In chapter 7, Paul Farmer, Matthew Basilico, and Luke Messac reanimate a lingering debate over the value of biomedical interventions in public health that has taken place in the fields of economics, epidemiology, and public health since the provocations of the British physician Thomas McKeown over the "role of medicine" in the early 1960s. Reengaging this debate over the contested revolutionary status of modern medicines, the authors bring to bear new data—economic, epidemiological, and ethnographic—to articulate a fresh perspective on the role of medicine in global health in the twenty-first century.

The next two chapters examine in close detail these claims of history and geography of therapeutic change in two sub-Saharan African nations with very different positions in the global circulation of pharmaceuticals. In chapter 8, Julie Livingston traces a comparison between the practice of biomedicine and the efficacy of its therapeutic interventions, as well as the effects on health care provider practices and patients' lived experiences, in the complex delivery of cancer care in Botswana. Livingston's textured ethnography of care in an oncology clinic calls attention to the nonuniversality of biomedical therapeutics and the spatial and temporal lacunae that separate biomedical practices in local contexts. Modern medicines circulate too easily, and not easily enough. This problem of the perilous forms by which modern medicines do or do not circulate is taken up in the Nigerian context by Kristin Peterson in chapter 9. In her ethnography of the Idumota drug market in Lagos, Peterson explores how the production and circulation of drugs in Nigeria—and the fates of Nigerians employed in the drug trade and as consumers of drug products—have been inextricably bound up with global financial markets and the profit motives of multinational corporations.

The volume concludes with a pair of reflections on the meanings of therapeutic revolutions in contemporary biomedicine. In chapter 10, David S. Jones discusses the role of revolution and evolution as distinct metaphors that physicians and scientists use to explain therapeutic change over time. Jones reviews a century's worth of medical literature to trace the use of both concepts by a series of historical actors engaged with processes of biomedical innovation, with particular focus on the development of a broad armamentarium of cardiovascular therapeutics over the twentieth century. Finally, in chapter 11, Charles E. Rosenberg revisits the concept of therapeutic revolutions as a durable preoccupation in the stories we tell ourselves regarding the modernity of our medicine. Reconsidering the same topic from a remove of nearly four decades, Rosenberg suggests how and why therapeutics have become such important things to think *with*, for historians, providers, and patients alike.

Thinking about revolutions as rhetorical tools and analytical devices requires awareness of both continuity and change, as well as attention to context and contingency in asking *how* medicine changes, for whom, where, and with what consequences. The chapters in this volume explore these questions in a variety of geographical, chronological, and institutional settings, through the methods of both ethnography and history, and on a range of scales from local to global.

This volume probes our common-sense assumptions about what makes medicine modern. Taken together, these chapters offer many intersections and common themes. Each chapter investigates a specific claim about the modernity of medicines through careful scrutiny of historical materials or ethnographic research. Each chapter examines the work done by a specific narrative of therapeutic revolution. Each chapter asks what new form of modernity, what new way of life was understood to be brought into being in tandem with revolutionary new therapeutics. As a whole, this volume explores the many layers of transformative power of modern pharmaceuticals—the vaunted therapeutic revolution of the twentieth century—over a wide range of therapeutic and diagnostic areas, from acute illness to chronic disease management and from psychoactive drugs to contraception. At the end, the reader is left to consider how past narratives have brought us to a present in which the therapeutic future can be pictured as bright, bleak, or somewhere in between.

NOTES

1. David Healy, *Let Them Eat Prozac: The Unhealthy Relationship between the Pharmaceutical Industry and Depression* (New York: NYU Press, 2006); Terence Neilan, "Merck Pulls Vioxx Painkiller From Market, and Stock Plunges," *New York Times*, 30 September 2004; Gardiner Harris, "F.D.A. to Restrict Avandia, Citing Heart Risk," *New York Times*, 23 September 2010; UNAIDS, *The Gap Report* (2014), http://www.unaids.org/en/media/unaids/contentassets/documents/unaidspublication/2014/UNAIDS_Gap_report _en.pdf; United States Government Accountability Office, *New Drug Development: Science, Business, Regulatory, and Intellectual Property Issues Cited as Hampering Drug Development Efforts*, GAO-07-49 (November 2006). For criticism of the "innovation crisis," see Donald W. Light and Joel R. Lexchin, "Pharmaceutical Research and Development: What Do We Get for All That Money?" *BMJ* 2012; 345:e4348.
2. Charles E. Rosenberg, "The Therapeutic Revolution: Medicine, Meaning and Social Change in Nineteenth-Century America," *Perspectives in Biology and Medicine* 1977, 20(4): 485–506. John Harley Warner used empirical evidence from two hospitals to trace therapeutic usage over the nineteenth century in *The Therapeutic Perspective: Medical Practice, Knowledge, and Identity in America, 1820–1885* (Cambridge, MA: Harvard University Press, 1986), but this methodology was not taken up by other scholars.
3. Reinhart Koselleck, *Futures Past: On the Semantics of Historical Time*, translated by Keith Tribe (New York: Columbia University Press, 2004, originally published in German in 1979).
4. Ibid.
5. Arnold Toynbee, *The Industrial Revolution of the Eighteenth Century in England; Popular Addresses, Notes, and Other Fragments by the Late Arnold Toynbee, Together with a Reminiscence by Lord Milner* (London: Longmans Green, 1884); Arnold Toynbee, *Progress and Poverty: A Criticism of Mr. Henry George being Two Lectures Delivered in St. Andrew's Hall, Newman Street* (London: Kegan Paul, Trench and Co., 1883).
6. Nicholas F. R. Crafts, *British Economic Growth during the Industrial Revolution* (Oxford: Clarendon Press, 1985); Maxine Berg and Pat Hudson, "Rehabilitating the Industrial Revolution," *Economic History Review* 45 (1992): 24–50.
7. Stephen Shapin, *The Scientific Revolution* (Chicago: University of Chicago Press, 1996); Alexandre Koyre, *From the Closed World to the Infinite Universe* (Baltimore: Johns Hopkins University Press, 1957); Thomas Kuhn, *The Structure of Scientific Revolutions* (Chicago: University of Chicago Press,

1962); Roy Porter, "The Scientific Revolution: A Spoke in the Wheel?" in Roy Porter and Mikulas Teich, editors, *Revolution in History* (Cambridge: Cambridge University Press, 1986).

8. The concept of modernity itself has been subject to multiple usages and interpretations. Here we are interested in modernity as a historical trope for social, cultural, economic, political, and intellectual trends of rationalization, professionalization, secularization, bureaucratization, and industrialization, as well as prioritization of and belief in individual freedom, equality, and perfectability. While these trends developed at different times in different places, this volume focuses on modernity as it was conceived, shaped, and revised during the long twentieth century.

9. William Osler, *The Evolution of Modern Medicine: A Series of Lectures Delivered at Yale University on the Silliman Foundation in April, 1913*. See preface by Fielding H. Garrison. The observation that a narrative of modernity should be informed by the continuous presence of the premodern in the contemporary world can be found throughout the volume. In Osler's words: "It has been a slow and gradual growth, and not until within the past century has science organized knowledge—so searched out the secrets of Nature, as to control her powers, limit her scope and transform her energies. The victory is so recent that the mental attitude of the race is not yet adapted to the change. A large proportion of our fellow creatures still regard nature as a playground for demons and spirits to be exercised or invoked." Quote on 5–6.

10. Lewis Thomas, *The Youngest Science: Notes of a Medicine-Watcher* (New York: Viking, 1983); Paul Starr, *The Social Transformation of American Medicine* (New York: Basic Books, 1982); Siddhartha Mukherjee, *The Emperor of All Maladies: A Biography of Cancer* (New York: Scribner, 2010).

11. The work of Pierre C. A. Louis, perhaps, standing as an important exception. Michel Foucault, *The Birth of the Clinic: An Archaeology of Medical Perception* (London: Routledge, 1973); Erwin Ackerknecht, *Medicine at the Paris Hospital, 1794–1848* (Baltimore: Johns Hopkins University Press, 1967). See also Gunter Risse, *Mending Bodies, Saving Souls: A History of Hospitals* (Oxford, UK: Oxford University Press, 1999); Caroline Hannaway and Anne LeBerge, editors, *Constructing Paris Medicine* (Amsterdam: Editions Rodopi B. V., 1998); Erwin Ackerknecht, "Broussais, or a Forgotten Medical Revolution," *Bulletin of the History of Medicine* 27 (1953): 320–43.

12. Stephanie Snow, *Blessed Days of Anaesthesia: How Anaesthetics Changed the World* (Oxford, UK: Oxford University Press, 2008).

13. Richard Shryock, *Medicine and Society in America, 1660–1860* (New York: Cornell University Press, 1960); A. J. Youngson, *The Scientific Revolution in Victorian Medicine* (New York: Holmes and Meier, 1979); Martin Pernick, *A Calculus of Suffering: Pain, Professionalism, and Anaesthesia in Nineteenth-Century America* (New York: Columbia University Press, 1985).

14. Roy Porter, *The Greatest Benefit to Mankind: A Medical History of Humanity*

(New York: HarperCollins, 1997); Andrew Mendelsohn, "Cultures of Bacteriology: Foundation and Transformation of a Science in France and Germany 1870–1914," PhD dissertation, Princeton University, 1996; Michael Worboys, "Was There a Bacteriological Revolution in Late Nineteenth-Century Medicine?" *Studies in the History and Philosophy of Biological and Biomedical Sciences* 38 (2007): 20–42.

15. Not all biographic approaches need to focus on the scientist as revolutionary. Bruno Latour's *Pasteurization of France* (Cambridge, MA: Harvard University Press, 1993) argues implicitly against the revolutionary nature of Louis Pasteur's innovations, while Christoph Gradmann's *Laboratory Disease: Robert Koch's Medical Bacteriology* (Baltimore: Johns Hopkins University Press, 2009) locates Robert Koch's bacteriology in a historical context of Prussian medical sciences and politics. Both scholarly works contain an interesting ambiguity: they are concerned with key individuals who have been regarded as singlehandedly changing the face of medicine, yet their research suggests that the contributions of both men had more to do with historical circumstance than with genius.

16. Nancy Tomes, *The Gospel of Germs: Men, Women, and the Microbe in American Life* (Cambridge, MA: Harvard University Press, 1999).

17. Jeremy Greene and David Herzberg, "Hidden in Plain Sight: Marketing Prescription Drugs to Consumers in the Twentieth Century," *American Journal of Public Health* 100 (2010): 793–803; Jonathan M. Metzl and Joel D. Howell, "Great Moments: Authenticity, Ideology, and the Telling of Medical 'History,'" *Literature and Medicine* 25 (2006): 502–521; Jacalyn Duffin and Alison Li, "Great Moments: Parke, Davis and Co., and the Creation of Medical Art," *Isis* 86 (1995): 1–29.

18. Walsh McDermott, "Historical Perspective" *Pharmaceuticals for Developing Countries* (Washington: Institute of Medicine, 1979); Ivan Illich, *Medical Nemesis: The Expropriation of Health* (New York: Pantheon Press, 1975); Thomas McKeown, *The Role of Medicine: Dream, Mirage, or Nemesis* (Oxford, UK: Blackwell, 1976); Archibald L. Cochrane, "Medicine's Contribution from the 1930s to the 1970s: A Critical Review," in G. Telling-Smith and N. Wells, editors, *Medicines for the Year 2000* (London: Office of Health Economics, 1979); Lewis Thomas, *The Youngest Science: Notes of a Medicine-Watcher* (New York: Viking, 1983).

19. W. F. Bynum, *Science and the Practice of Medicine in the Nineteenth Century* (Cambridge: Cambridge University Press, 1994), xi.

20. John E. Lesch, *The First Miracle Drugs: How the Sulfa Drugs Transformed Medicine* (Oxford: Oxford University Press, 2007), 4.

21. "The Endocrine Glands," *Fortune* 8 (November 1933): 76–94.

22. Waldemar Kaempffert, "The Coming Revolution in Medicine," *American Mercury* (July 1943): 66–71, quotes on 66 and 71. As noted earlier, the concept of revolutionary change in therapeutics was not a twentieth-century invention. Observers of nineteenth-century medicine often predicted a

revolution on the horizon. For example, in 1888 Austin Flint, a preeminent physician and heart researcher, wrote in an article in *The Forum* titled "A Possible Revolution in Medicine": "The science and practice of medicine and surgery are undergoing a revolution of such magnitude and importance that its limits can hardly be conceived." Three years later, he titled his follow-up article, about Koch's method of treatment for tuberculosis, "The Revolution in Medicine." See Austin Flint, "The Revolution in Medicine," *The Forum* (January 1891): 527–536, quote on 527.

23. "Abbott Laboratories," *Fortune* 22 (August 1940): 63–68, 102–110.
24. John Parascandola, "Industrial Research Comes of Age: The American Pharmaceutical Industry, 1920–1940," *Pharmacy in History* 27 (1985): 13–14.
25. Peter Bowman, "Miracle Men," *The New Masses* (12 May 1942): 10–11.
26. Harold M. Scheck, Jr., "F.D.A. to Control Medical Devices," *New York Times,* 29 May 1976, 9.
27. For a book-length study of this argument, see Stephen Epstein, *Inclusion : The Politics of Difference in Medical Research* (Chicago: University of Chicago Press, 2007).
28. Harry M. Marks, *The Progress of Experiment. Science and Therapeutic Reform in the United States, 1900–1990* (Cambridge: Cambridge University Press, 1997).

Futures and Their Uses: Antibiotics and Therapeutic Revolutions

SCOTT H. PODOLSKY AND ANNE KVEIM LIE

In 1960, two of the leading apostles of the antimicrobial era concluded their book *The Antibiotic Saga* with a broad look to the future:

Can we expect more? In the last fifteen years we have grown to expect a great deal from medicine. . . . Diseases that in the old days were a death warrant are now cured promptly with the new "miracle drugs." . . . Although we are nowhere near the end, progress is rapid and certain and the time will come, and within this century, when we will be looking for a disease for a newly discovered drug to cure![1]

Over a half century later, as antibiotic resistance threatened this particular vision, Dame Sally Davies, the British government's chief medical officer, depicted a very different antibiotic future. "Antimicrobial resistance is a ticking time-bomb not only for the UK but also for the world," she noted in 2013. "We need to work with everyone to ensure the apocalyptic scenario of widespread antimicrobial resistance does not become a reality." Within twenty years, she warned, even minor surgery could lead to death through untreatable infection, in a health system reminiscent of that of the nineteenth century.[2]

These could not be more disparate depictions of the future. The first is an acknowledgment of a therapeutic revolution. The second is a dystopic warning of the end of this revolution and all the components of medicine (from chemotherapy to organ transplantation) and society that have accompanied it. Each is historically situated, revealing the aspirations, fears, and limitations of its era. And neither quote appeared in isolation. The post–World War II era was awash in paeans to the antimicrobial "miracle drugs." The twenty-first century is similarly awash in bitter prophecies concerning their imminent demise. Yet this evolution has not been a straightforward diachronic process from utopia to dystopia. Contrary to what is often claimed, warnings regarding the end of antibiotics can be traced almost to the start of the antibiotic era. Furthermore, current jeremiads are not necessarily the same as prior ones, and they have been and may be mobilized for very different ends and even in direct opposition to one another.

Futures are not often studied within medical history. Indeed, past futures tend to be forgotten. As Peter Burke has pointed out, the notion of the future was placed on the historian's agenda only relatively recently, when it was pioneered by the German historian Reinhart Koselleck in the latter half of twentieth century.[3] In 1964 Koselleck, together with his colleague Reinhart Wittram, invented the concept of a "past future." By that they meant a future that was not the future of the present, but the future as it was conceived at some time in the past.[4] In several articles and essays, Koselleck has argued that past and future cannot be reduced to dimensions to be viewed from the perspective of the present, but should be acknowledged as historical subjects in their own right.[5]

Koselleck's project has recently been taken up by a number of scholars in the field of science and technology studies (STS), who see the production of futures as a key mode of how science operates.[6] Researchers in this field of work accordingly shift the analytic angle from "looking into the future to looking at the future, or how the future is mobilized in real time to marshal resources, coordinate activities and manage uncertainty."[7] This literature engages with the future as an object of critique in its own right.[8] Promises and potentials are seen to be "fundamentally generative" in the production of artifacts and knowledge; they can be *performative* in the sense that they not only describe future technologies but also help bring them into being. Expectations can help innovators, scientists, and public health officials mobilize support and funding for emerging artifacts, but can also encourage change in practice and policy. Interestingly, several of the key actors currently involved in depicting a post-antibiotic era have recently reflected on such implications of their discourse.[9]

These insights from the sociology and anthropology of science strengthen Koselleck's challenge to focus on the temporal layers in our historical material. However, as historians we are particularly interested in studying how these future visions have *changed* over time. The sociology of expectations has focused largely on contemporary aspects, and has largely neglected to study how such different expectations can be in conflict both diachronically and synchronically.[10] More generally, we ask what can we learn from the *history* of expectations and from the rise and fall and intersection of particular futures past. We argue that expectations of the future tell us a great deal about both the scientific and cultural contexts of their origins.[11]

In particular, the revolutions claimed in the name of antibiotics and the crises envisioned regarding their usage and enduring utility provide instructive analytical lenses. We do not intend here to focus on the veracity of past or current utopias or dystopias, or whether they have been hyperbolic or unnecessarily glum. Instead, we intend to bracket off the *telos* and look at historical futures as a structure of analysis. We attempt to unpack the degree to which such visions have been revelatory, performative, heterogeneous, and at times conflicting.

We begin with a review of the major transformation in medicine ushered in by the advent of the antimicrobial wonder drugs and an examination of the futures claimed in their name. We then proceed to examine two intersecting strands of therapeutic reform (the first beginning in the 1950s, the second not fully taking off until the 1980s) justified by alternate antibiotic dystopias, before proceeding to more general reflections on antibiotics and the nature of therapeutic "revolutions," and on the performative aspects of expectations in medicine. We focus on the American and British contexts, but many of the lessons learned can be generalized more broadly. Through such analysis, we hope to shed light on therapeutic futures, past and present.

The Antibiotic Revolution

The advent of the sulfa drugs and antibiotics during the 1930s, 1940s, and 1950s was characterized both at the time and thereafter—by clinician-scientists, the media, and the pharmaceutical industry alike —as ushering in a "therapeutic revolution."[12] Antibiotics appeared at a moment particularly prone to describing scientific development in revolutionary terms. After all, the phrase "scientific revolution" had only recently entered common use after Alexandre Koyre gave it broader visibility in 1939.

The infectious disease expert Lawrence Garrod remarked in 1950 of the sulfonamides in the midst of a sober "refresher course" on the use of penicillin: "They had brought about the greatest therapeutic revolution of modern times, and completely transformed the outlook in conditions which formerly had a high mortality."[13] In less sober fashion, Paul de Kruif, the well-known author of *Microbe Hunters*, reported in 1948: "A fantastic breed of 'doctors' can now rescue millions of sick human beings who only seven years ago were sure to die. . . . Pasteur once prophesied that it would sometime be within man's power to wipe microbic maladies from the face of the earth. It seems now that Pasteur's apparently wild prediction may come true."[14] Industry depicted such a victory as both an inspiring portrait of medical modernity and a cheery portent of the future, as in Lederle's early 1950s advertisement for its broad-spectrum antibiotic Aureomycin, headlined "Thank heavens it's only pneumonia!" The ad went on to boast: "This teamwork between laboratory and clinical research workers is typical of American preoccupation with working to do things *better*, for *more* people, by *everybody. Through research, they live who would have died!*"[15]

Medical therapeutics, in the context of the sulfa drugs and antibiotics, had completed a radical break with the past. As popular writer Fred Reinfeld concluded his 1957 book *Miracle Drugs and the New Age of Medicine*: "We have seen, first, how ineffectually disease was treated in the pre-scientific age of medicine. Then we saw the effects of Pasteur's germ theory of medicine, and the extraordinary impetus it gave to vaccines, chemotherapy, and antibiotics. Finally, we saw what profound and revolutionary changes have been brought about by the introduction of the miracle drugs. So far-reaching has this change been that more than half of our drugs date back only as far as the 1930's."[16] And this revolution was still ongoing. Selman Waksman, the microbiologist who first isolated streptomycin (and who coined the very term "antibiotic"), in reflecting in 1960 on the history and future of antibiotic research, foresaw the developments of antibiotics to eliminate tuberculosis, completely control childhood diseases, and treat both viruses and cancer.[17]

As historian John Lesch has related, medical practice itself was portrayed as undergoing a fundamental shift in the process—and not always to the economic benefit of the physician. As one sulfa-drug–administering clinician quipped in the *Journal of the American Medical Association*:

The Doctor now sits in his big office chair,
With his feet on the top of the table;
His brow is all furrowed with deep lines of care,
(Just picture him there, if you're able)—
His office is filled, but with big empty chairs,
And his office force all now are idle,
While the sound of a snore from the office next door,
Makes a pain in his head and his middle. . . .
So with nothing to do, and the prospect so blue
We can only sit down and discuss
NOT what have we done with this product,
BUT—WHAT HAS IT DONE WITH US?[18]

As an article in *The American Mercury* echoed a decade later, upon the advent of the broad-spectrum antibiotics: "Entire categories of diseases which used to keep doctors solvent year in and year out have been reduced to an easily managed and, from the M.D.'s angle, unprofitable estate. It speaks well for the ethics of the medical community that, by and large, its enthusiasm for miracle medicaments remains high notwithstanding."[19] Nonetheless, despite such fears about the expendability of the clinician, they were hardly realized. As historian Robert Bud has described, antibiotics came to stand as the "brand" of a powerful modern medicine wielded by increasingly powerful clinicians in the midst of medicine's golden age.[20]

Such powerful medicines, moreover, were expected to be studied and provided by an increasingly well-funded biomedical enterprise and an increasingly research-driven pharmaceutical industry. As industry provided such wonder drugs as corticosteroids, major and minor tranquilizers, and antihypertensives, the future of biomedicine itself appeared to be tied to ongoing investment in both the public and private spheres, and could be used to further advocate for and justify such investment. More broadly, antibiotics could stand as symbols of modernity and national pride.[21] Globally, they could be seen as key components of post–World War II reconstruction—through the attempts of the United Nations Relief and Rehabilitation Agency (UNNRA) and the World Health Organization (WHO) to foster penicillin production worldwide—or as centerpieces of such vertical programs as the WHO's effort to eradicate yaws using penicillin.[22]

For some, the advent of powerful antimicrobials constituted an opportunity to provide treatment to all people suffering from infectious disease around the globe, and even to extinguish such diseases. To others, however, antibiotics had the ability to threaten the future world order. In a collection devoted to "the impact of antibiotics on medicine and society," one contributor's concerns about the future impact of antibiotics on worldwide population growth drew on popular neo-Malthusianism: "It is important to face the fact that today the antibiotics are keeping alive hundreds of thousands of persons who are a drain on the community. . . . Modern medicine is keeping alive those who will never 'pull their weight in the boat.' Modern medicine may soon keep alive also the hordes of starving people in the Orient. What will happen when the old terrible epidemics of plague, cholera, and dysentery no longer close millions of hungry mouths?"[23]

Reinhart Koselleck has argued that the concept of revolution has become an iconic mode of historical thought for the modern period because it not only opens up a space for the new but also actively fosters, encourages, and promotes it.[24] Moreover, revolution is a concept that has intrinsic temporal aspects, pointing not only towards contemporary affairs but also backwards and forwards.[25] Calling the advent of antibiotics a therapeutic revolution not only implied the announcement of a profound break with (outdated, old-fashioned) past treatment regimes, but posited medical science as a distinctively modern, truly groundbreaking activity.

In other words, the "antibiotic revolution" contained within itself a description of a future of progress. For the pharmaceutical industry, the concept of the therapeutic revolution also served other rhetorical purposes. The promise of an ongoing revolution, fueled in the postwar years by a steady stream of new antibiotics entering the market, could mobilize support and create allies in the continuous development of lifesaving drugs. A common feature of these pronouncements of revolution was that they incorporated both a dramatization of the new, in emphasizing a radical break with the past, and a faith in a technological fix to whatever hurdles (e.g., antibiotic resistance) might arise.

However, such futures were also offset by contemporaneous counterprojections and frank dystopias. The therapeutic futures associated with antibiotics and a powerful medical profession abetted by a powerful pharmaceutical industry were not without their perceived future threats and hazards. We next describe two particularly important lineages of dystopic critique—important as models for the mobilization of expectations, as well as for their enduring relevance to today's efforts to forestall the "end of antibiotics." The first dystopia concerns the "end" of rational therapeutics and the processes by which drugs enter the marketplace and physicians are educated about them; the second concerns the dilemma of antibiotic resistance and the very "end" of antibiotics themselves.

Taking the Miracle out of the Miracle Drugs

A darker view of the antibiotic future emerged amidst concerns regarding the increasing intimacy between medicine and marketing, science, and commerce. And the reforms set in motion by this future focused on constraining the introduction and marketing of novel antibiotics.

As of 1948, penicillin and streptomycin accounted for nearly the entirety of US antibiotic production. But these weren't patented medicines, and in December of that year, Lederle offered its first patented antibiotic, Aureomycin, for interstate sale, to be quickly followed by Parke-Davis' Chloromycetin and Pfizer's Terramycin (the first "broad-spectrum" antibiotics). In 1948, US antibiotic output had totaled 240,000 pounds; by 1956 it had surpassed three million pounds, reflecting a combination of real medical need and a dramatic escalation in pharmaceutical promotion.[26]

By the 1950s it had become apparent that such widespread antibiotic usage could be associated with side effects, superinfections, excessive cost, diagnostic sloppiness, and resistant bugs. An emerging cohort of infectious disease experts and therapeutic reformers, led by Harvard's Maxwell "Max" Finland and the University of Illinois's Harry Dowling, perceived a regulatory vacuum in the absence of action taken by the agencies—namely, the American Medical Association (AMA) and the Food and Drug Administration (FDA)—traditionally responsible for examining, if not curtailing, such exuberant prescription practices. The engagement of such would-be antibiotic reformers was catalyzed especially by the advent of "fixed-dose combinations" of antibiotics (i.e., set dosages of more than one antibiotic in a single pill), promoted from the mid-1950s onward on the basis of still broader microbial coverage, potential synergy in their attack on microbes, and thus utility for the practicing physician without easy access to laboratory testing. Reformers warned against such nonspecific "shotgun" therapies, claiming that antibiotics in combination could be antagonistic to one another and that combinations were, in the best case, bacterial strain-dependent and hence not amenable to fixed-dose preparations in the first place.[27]

Nevertheless, the pharmaceutical industry jumped at the opportunity to market them, introducing at least sixty fixed-dose combination antibiotics on the US market by 1956. In contrast to the conservative stance of Finland, Dowling, and their colleagues in academia, Henry Welch, the head of the FDA's Division of Antibiotics (the government agency tasked with regulating such market entry) provided de facto government support for such therapeutic exuberance. Throughout the 1950s, Welch ran an annual international symposium in the nation's capital and published the proceedings under the title *Antibiotics Annual*. At his fourth annual symposium on antibiotics in October of 1956, Welch remarked of the fixed-dose combination antibiotics:

These presentations and others indicate a distinct trend toward combined therapy, not an old fashioned "shotgun" approach, but a calculated rational method of attacking the problem of resistant organisms. It is quite possible that we are now in a third era of antibiotic therapy; the first being the era of the narrow-spectrum antibiotics, penicillin and streptomycin; the second, the era of broad-spectrum therapy; the third being an era of combined therapy where combinations of chemotherapeutic agents, particularly synergistic ones, will be customarily used.[28]

In this setting—with drugs approved by the FDA on scant evidence and extensively promoted on the basis of "testimonials" (basically extended case series)—the anxieties of such leaders of academic medicine as Finland, Dowling, and the University of Iowa's William Bean (one of the nation's leading experts on nutrition and vitamins) were raised to crisis levels. Each of these academicians had maintained a long-standing skepticism about the enthusiastic adoption of new drugs and the increasing role of the pharmaceutical industry in promoting such uncritical enthusiasm.[29] Fixed-dose antibiotics and their marketing heralded a dystopic future in which bluster would supersede substance, bringing down the entire and increasingly interrelated edifice of medicine and the pharmaceutical industry. Welch's "third era" utopia would become their dystopia.

In his 1955 article "Vitamania, Witchcraft, and Polypharmacy," Bean delivered an unfavorable interpretation of the therapeutic tea leaves:

The omens which indicate the direction of therapy today . . . point to a reversion to the polypharmacy which condoned combinations of puppy dog fat, powdered unicorn's horns, dried mosquito wings, and spider webs, obtained from a graveyard in the dark of the moon, and brewed by witches as panacea for real and imagined ills of every kind.[30]

Bean thus warned of a future characterized not by progress, but by a reversion to a medieval approach to therapy.

Dowling went further, anticipating a future of disillusionment, caused by physicians' increasing eagerness to promote new and useless medications, eventually causing nothing less than the ultimate discrediting of the medical profession. In a speech before the AMA in 1957 with the provocative title "Twixt the Cup and the Lip," he reported:

The techniques that had been used so successfully in the advertising of soaps . . . tooth pastes . . . cigarettes, automobiles, and whiskey could be used as successfully to advertise drugs to doctors. . . . With the inevitable disillusionment that comes with the failure of each useless modification to make any advance, the pharmaceutical industry will lose its prestige and with this will lose its financial backing. It will fall, and the medical profession will be dragged down with it.[31]

Three years later, Finland repeated these concerns with the caveat that the greatest threat to therapeutic autonomy stemmed not from government incursion, but from pharmaceutical marketing. He lamented:

Perhaps the greatest objection to the use of fixed combinations of antibiotics, and in fact, in prescribing any of the mixtures of drugs now being marketed by the various pharmaceutical firms, is that they have removed the physician from his important status as an educated and rational individual who acquires his own knowledge, experience, and skill and applies them to the choice of therapy as required for his patient.[32]

This was a concern about the fate of rational therapeutics writ large. As Dowling and Finland (along with seven other infectious disease experts aligned in their camp) noted elsewhere, "If this trend is not checked now, the practicing physician will soon be confronted with such a bewildering array of antibiotic combinations supported by multicolored promotional material . . . that rational chemotherapy will give way to chaos."[33] The term "rational" had taken on skeptical as well as moral overtones, to be juxtaposed with the influences of commerce, ignorance, intellectual lassitude, and fear.

As scholars within the sociology of expectations remind us, every future is predicated on another to be avoided, and in parallel with positive promises and hopes of future developments, fears and concerns about future risk are also significant features of the dynamics of expectations.[34] In our analysis we observe this phenomenon in reverse:

in parallel with fears and concerns about the future risk of the end of the medical profession, positive promises and hopes of future developments coexisted. Finland and his group of reformers attempted to counter the case series–based "testimonials" upon which fixed-dose combination antibiotics were justified with calls for "controlled clinical studies" that would place pharmaceuticals under appropriate scrutiny to buttress a more rational therapeutics.[35]

The activities of the academic medical reformers ultimately intersected with the concerns of reformers in the media, at the FDA, and in Congress, and found a public stage in Senator Estes Kefauver's hearings on the pharmaceutical industry, beginning in 1959.[36] Earlier that year, John Lear had published a widely read and cited muckraking article in the *Saturday Review*, "Taking the Miracle out of the Miracle Drugs," which railed against antibiotic marketing and misuse. Reflecting the concerns of Finland and Dowling and warning of an impending disaster unless remedies were taken, he nevertheless emphasized that "there is time to avert catastrophe by reversing the trend."[37]

Lear's first article influenced the direction of Kefauver's hearings, as did his later works exposing a potential conflict of interest involving FDA antibiotics chief Henry Welch.[38] Perhaps the most shocking moments of the Kefauver hearings were devoted entirely to Welch, as it became apparent that he had received 7.5 percent of all advertising revenue in his publications, along with 50 percent of all reprint sales, earning $287,000 from such activities during the 1950s. Welch's very announcement that the fixed-dose combination antibiotics (especially Pfizer's Sigmamycin) ushered in a "third era" of antibiotic therapy was found to have had been written by a Pfizer employee.[39] The projected coming of a third era of antibiotics and the promise of the pharmaceutical industry as the ideal promoter of this new era were contested and ultimately refuted at the Senate hearings.

Welch was forced to resign, and the day after the hearings, FDA Commissioner George Larrick endorsed "a proposal that the new drug section of the Food, Drug, and Cosmetic Act require a showing of efficacy as well as a showing of safety."[40] The Kefauver-Harris Drug Amendments would be passed in the fall of 1962, mandating proof of efficacy via "well-controlled" studies prior to new drug approval. A key outcome of the amendments would be the Drug Efficacy Study and Implementation process, or DESI, entailing the review of medications approved between 1938 and 1962 and the removal from the market of those not proven efficacious. By the end of the 1960s, every fixed-dose combination antibiotic would be purged from the American market, representing the newfound power of the FDA to unmake pharmaceutical markets in the name of a rational therapeutics. Industry opposition emerged at this point, as Upjohn took the case for its fixed-dose combination antibiotic, Panalba, all the way to the Supreme Court. But the judiciary found in favor of the FDA, ending a critical era of therapeutic reform and laying the foundation for the FDA's regulatory structure of new drug approval that persists to the present.[41]

Thus, skeptical academic reformers had set forth a particular medical-pharmaceutical dystopia, engendering a series of reforms that profoundly shaped both the existing antibiotic market and the process of FDA drug regulation more broadly. And yet, while a particular therapeutic dystopia with respect to "irrational" drugs was seemingly averted, another appeared in its place. The marketing and prescribing of inappropriate drugs was replaced by the irrational overprescribing of appropriate drugs.

In other words, with respect to antibiotics, there had clearly been limitations to a reform movement that formulated its politics around a specific dystopic vision. By the early 1970s, HEW Deputy Assistant Secretary for Health Henry Simmons and Johns Hopkins's Paul Stolley pointed to a 30-percent increase in national antibiotic usage between 1967 and 1971, as compared to a 5-percent increase in the US population, asking, "Have we reached the point where the enormous use of antibiotics is producing as much harm as good?"[42] Simmons and Stolley's commentary, published in *JAMA*, was ironically entitled "This is Medical Progress?" By the end of the 1970s, Calvin Kunin (a leading infectious disease researcher, former fellow under Max Finland, and former head of one of the key DESI panels ruling in favor of eliminating the existing fixed-dose combination antibiotics) lamented in a similar vein: "A decade ago some of us working in this field believed that we had scored a major victory when the Food and Drug Administration removed fixed-dose combination antibiotics from the market. . . . This was no victory, but abject defeat; these drugs were almost immediately replaced by [other, more expensive drugs]. . . . I do not mean to imply that removing irrational drugs was a mistake in itself, but we expected too much for our efforts."[43] The overuse of these new drugs, combined with aggressive "educational" programs by the pharmaceutical industry, had continued to shape medical practice. Such concerns grew ever more relevant in the context of increasing antibiotic resistance.

Antibiotic Resistance and the Collision of Dystopias

At the same time that Finland and Dowling were projecting their dystopian futures of irrational medicine, a parallel set of dangerous antibiotic futures were beginning to be mobilized by a larger community of researchers initially concerned with the prospect of staphylococcal resistance and soon extending to more generalized antibiotic resistance. In contrast to the futures of irrational drugs described above, the projected futures of antibiotic resistance opened a larger window for potential responses, some of which in ensuing decades would complement the project for rational medication use grounded in tightly controlled clinical trials, and others which would conflict with it.

By the 1950s, René Dubos (discoverer of two of the first identified antibiotics, tyrothricin and gramicidin, at the Rockefeller Institute in the 1930s) had already claimed that any dream of permanent victory over microbes was delusional and based on overly static notions of nature and mankind.[44] Dubos may have accepted such a process as inevitable, but others attached a moral valence to the process. As the well known British general practitioner Lindsey Batten warned during the "Discussion on the Use and Abuse of Antibiotics" at the Royal Society of Medicine in 1954: "Those deadly staphylococci . . . are not pirates or privateers accidentally encountered, they are detachments of an army. They are also portents. . . . We should study the balance of Nature in field and hedgerow, nose and throat and gut before we seriously disturb it. Again, we may come to the end of antibiotics. We may run clean out of effective ammunition and then how the bacteria and moulds will lord it."[45]

A 1963 publication by the Japanese bacteriologist Tsutomu Watanabe, "Infective Heredity of Multiple Drug Resistance in Bacteria," further propelled such fears.[46] Shigella, a cause of bacillary dysentery, had been found to be increasingly resistant to multiple antibiotics in Japan; by 1959, researchers had discovered that such drug resistance could be passed *across* bacterial species by extrachromosomal plasmids. From a genetic standpoint, this development represented a fascinating example of "infective heredity" as posited during the previous decade by scientists such as Joshua and Esther Lederberg. However, it was a quick and alarming jump from notions of infective heredity to "infectious drug resistance." "Superbugs" could be envisioned and named.[47] In 1966, a *New England Journal of Medicine* editorial entitled "Infectious Drug Resistance" concluded, "Unless drastic measures are taken very soon, physicians may find themselves back in the preantibiotic Middle Ages in the treatment of infectious diseases."[48]

By the late 1960s, as the decade of Rachel Carson's *Silent Spring* was leading to the formation of the Environmental Protective Agency, the use of antibiotics in animal husbandry underwent increased scrutiny. A parallel concern attributed "environmental pollution with resistant microbes" to the prescribing habits of clinicians.[49] By the late 1970s, the image of microbial pollution had given way, in the wake of Pennsylvania's Three Mile Island nuclear meltdown in 1978 and the mass demonstrations against nuclear power in Europe and North America, to "an image of fallout akin to that from a leaking nuclear reactor."[50]

E. S. "Andy" Andersen, director of the Enteric Reference Laboratory in Colindale in North London, managed to escalate public concern in the United Kingdom about the use of drugs in agriculture into a fear of untreatable epidemics among humans in the future. This mobilized fear of impending disaster was reflected in the 1969 Swann report in Britain, which restricted the use of antibiotics in animal husbandry.[51] However, no such restriction on antibiotics in animal husbandry was forthcoming in the United States, as debate over evidentiary standards and the consequences of such restriction for agribusiness forestalled definitive measures for decades.[52] Tufts University's Stuart Levy mobilized attention to antibiotic resistance through the formation of the Alliance for the Prudent Use of Antibiotics in 1981.[53] But no real policy changes concerning antibiotic resistance occurred at the federal level, due to the backlash against the withdrawal of the fixed-dose combination antibiotics in the 1970s and the onset of Reagan-era public health retrenchment in the 1980s (including the reduction in the number of personnel at the National Center for Infectious Diseases at the Centers for Disease Control [CDC] by fifteen percent from 1985 to 1988).[54]

Much would change, though, once the Nobel Prize–winning microbial geneticist Joshua Lederberg began to support Levy's efforts in the late 1980s and early 1990s, mobilizing attention and resources around the crisis of "emerging infections." The notion of emerging infections contains its own sense of temporality and impending crisis.[55] The immediate stimulus to Lederberg's engagement with what would come to be considered emerging infections stemmed from his tenure as president of the Rockefeller University in New York City during the height of the AIDS epidemic.[56] In October 1987, Lederberg was asked by François Mitterand and Elie Wiesel to present at an explicitly future-oriented conference of Nobel Prize winners, "Facing the 21st Century: Threats and Promises." Lederberg's presentation was entitled "Medical

Science, Infectious Disease, and the Unity of Humankind." Describing an interwoven world of evolving humans and microbes, he derided "premature complacency" about infectious diseases in the wake of the wonder drugs, noting, with respect to AIDS, that "we will face similar catastrophes again."[57]

Lederberg—further influenced by Rockefeller University virologist Stephen Morse—pushed the Institute of Medicine (IOM) to convene a "multidisciplinary committee" on "emerging microbial threats" in 1991, resulting in the landmark report *Emerging Infections: Microbial Threats to Health in the United States*.[58] While the IOM report focused chiefly on viral pathogens, it helped give rise to a host of mutually reinforcing relationship building, media and government reports, and resource mobilization regarding antibiotic resistance as well. In a variety of publications, such as *Science*'s "The Crisis in Antibiotic Resistance"; *Newsweek*'s "The End of Antibiotics"; and Laurie Garrett's chapter in *The Coming Plague*, "The Revenge of the Germs," a scientific morality play depicted mutating bacteria responding to Darwinian selection pressures, a host of blameworthy actors responsible for generating such pressures, and an increasingly pessimistic view of the "arms race" between microbes and mankind, with the potential to culminate in "medical disaster" or a "post-antibiotic era."[59]

Four key policy recommendations could be found in the governmental and institutional reports on the problem of antibiotic resistance emanating from the United States throughout the 1990s.[60] The first was a call for increasing—and increasingly coordinated—surveillance, grounded in improved and standardized laboratory infrastructure. The second was an appeal for decreasing antibiotic usage, mandating interventions and studies at the levels of physicians, patients, and the animal husbandry industry. The third was a plea for increased research funding from the National Institute of Allergy and Infectious Diseases (NIAID), the CDC, and other agencies to study antibiotic resistance. And the dystopias mobilized were indeed generative. In the United States, the annual NIAID research funding for antibiotic resistance grew to $300 million by 2009, and the CDC would form the public-facing National Campaign for Appropriate Antibiotic Use in the Community in 1995 (renamed the Get Smart: Know When Antibiotics Work program in 2003). In Europe, the establishment of a continent-wide antibiotic resistance surveillance system was accompanied by both dramatically escalated research funding and individual national antibiotic stewardship programs.[61]

The fourth recommendation, which persists to the present, was a call for FDA regulatory changes to encourage industry—perceived to have turned its attention from antibiotics to more profitable drugs chronically prescribed for chronic diseases—to develop novel antibiotics, especially for drug-resistant bacteria. Here we see the collision of the two reform lineages reflected in the convergence of two dystopias: one, a fear of irrational drug development and marketing, and the other, a fear of losing the arms race between bugs and drugs. The role of the Infectious Diseases Society of America (IDSA) in this development has been telling. The IDSA formed in 1963, in the wake of the discrediting of Henry Welch and his conferences. Its first president was Max Finland, and its third was Harry Dowling. By 2003 the IDSA had entered the fray over antibiotic resistance with its own "Task Force on Ensuring the Future Availability of Anti-Infective Therapies." Although the IDSA also supported antibiotic stewardship programs, a focus on the drying up of the pharmaceutical pipeline for new antibiotics came to dominate its official pronouncements. In July 2004 the society released its report *Bad Bugs, No Drugs*, advocating a host of incentives for industry to reenter the arms race with microbes. These incentives—predicated on the "crisis" of antibiotic resistance—included tax credits for research and development, liability protection, wild-card patent extensions (in which a company's discovery of a novel antibiotic would be rewarded by the extension of a patent period for another existing drug of the company's choice), and a foreshortening of the investigative requirements to bring a novel antibiotic to market in the first place.

The irony of history that FDA antibiotic regulations—indeed, new drug regulations more broadly—had been fundamentally revised in the 1960s in response to antibiotic evaluations perceived to be overly loose and industry-friendly was not lost on former IDSA president Calvin Kunin as he watched an organization founded by Finland and Dowling recommend private-sector solutions. "Industry must take the lead to ensure success," the IDSA reported in *Bad Bugs, No Drugs*. "Industry decision-making is not perfect from a public health perspective, but the focus on financial incentives has made industry successful in the past, and new incentives can lead to future successes."[62] By 2010, the IDSA's recommendations had crystallized into a call for a "10 by '20 program" (the introduction to the market of ten new antibiotics by the year 2020) as the society lobbied both the FDA and Congress to incentivize novel industry antibiotic development.[63]

The advocacy of the IDSA and other bodies in the early twenty-first century appears to have had an impact at the FDA and elsewhere. FDA policy on antibiotic resistance shifted in 2012 with the passage of the GAIN (Generating Antibiotics Incentives Now) Act as part of FDASIA (the Food and Drug Administration Safety and Innovation Act), whereby "qualified infectious disease products" receive five years of extended market protection, and still more recently with discussions surrounding—and federal bills entailing—the proposed Limited Population Antibacterial Drug (LPAD) Approval Mechanism, in which antibiotics geared towards highly resistant organisms "would be studied in substantially smaller, more rapid, and less expensive clinical trials."[64] The downside to the LPAD process would be less precise estimates of both efficacy and safety, and an expectation that through antibiotic stewardship programs, education, and labeling alone (as opposed to more stringent regulatory measures), clinicians will avoid using approved remedies in more generalized or inappropriate situations.

In the present-day framing of policies to forestall an impending "post-antibiotic era," we thus see the collision of two reform efforts: one attempting to moderate the role of overenthusiasm and industry influence on the therapeutic rationality of the clinician, and the other relying on industry to save clinicians and their patients from a world bereft of effective antibiotics. These are not mutually exclusive categories. One can conceive of measures to incentivize industry to modulate the marketing of novel antibiotics—for example, the uncoupling of volume from profits.[65] Yet the very conduct of the randomized controlled clinical trial—whose role was cemented in the 1960s by the Kefauver-Harris amendments and the DESI process in the context of seemingly poor antibiotic studies—and its role in governing the therapeutic marketplace now appear at the point of impact of this collision. The role of the FDA in regulating antibiotics serves not only as a critical focus of contemporary antibiotic discussion, but as a fascinating case example of how evidence, institutional evolution, and articulated futures have played out in the ongoing formulation of the manner in which antibiotics are ideally developed and administered.

Revolutions and Futures

Was there really such a thing as an antibiotic revolution, or was this just another mobilization of a historical narrative by actors with economic and political interest in framing temporality? As Robert Bud and John Lesch have argued, it is hard to deny the dramatic transformation in the practice of medicine wrought by the advent of anti-infective miracle drugs. Dreaded diseases like puerperal fever and endocarditis became exemplars for conquered microbial foes, almost universally portrayed in such language of combat and conquest.[66] Furthermore, the advent of antibiotics has enabled the development of chemotherapy, organ transplantation, critical care medicine, and a wide range of medical technologies over the past seven decades. In this sense, they underpin medical technological modernity.

However, there are three important modifications to be made to this claim. First, the profound change in practice was geographically limited to one part of the globe. Despite initial optimism (and even Malthusian fears regarding the worldwide implementation of antibiotics), little concerted or consistent attention was directed to the global delivery of antibiotics, as is described elsewhere in this volume (see especially chapters 5, 6, and 7), and as the case of tuberculosis so glaringly illustrates.[67] Second, the sulfa drugs and antibiotics were revolutionary in no small measure because they were promoted as such, bringing into focus the rhetorical, performative impact of the term "revolutionary" itself. Third, if there was a revolution, its utopian heralding was balanced from the beginning by counterrevolutions or dystopias. Concerned individuals warned that this particular revolution might come to an end, and that it contained within it the seeds of its own undoing. At one level, William Bean, Harry Dowling, and Max Finland saw the marketing apparatus that accompanied the introduction of the antibiotics as the harbinger of an era of style over substance, necessitating educational or regulatory changes to forestall such a dystopia. At another level, the prospect of widespread antibiotic resistance generated concern about the "end" of the antibiotic revolution, necessitating other changes to forestall that particular dystopia. And through it all, there have been those, like René Dubos, who have denied the very possibility of a permanent (or static) "revolution," given that change itself is woven into the fabric of nature and mankind. Embedded in these last two notions is a return to the early modern meaning of the word "revolution": a cyclical process (described elsewhere in this volume), captured, for instance, in the warning of a return to a "pre-antibiotic Middle Ages."[68]

If the concept of an antimicrobial revolution continues to perform any work today, it does so primarily by way of announcing the limits or the end of that revolution. The sense is not that we are out of the revolutionary phase, and back to a form of, to use Kuhn's expression, "normal science," but rather that the revolution has ended, or that aspects of the apparent revolution may have been illusory from the beginning.

Indeed, if the antibiotic revolution stood at the head of a larger post–World War II therapeutic revolution, then the antibiotic dystopias described here—and certainly the predicted "end" of antibiotics—have served to focus reflection and anxiety on how the wonder drugs have been used, misused, and regulated, and on their relative benefits and costs, in all senses of those words.

The dystopias we confront today are not exactly the same as those of prior decades. Background assumptions about temporality or historicity, notably a progressive teleology and a belief in a brighter future, patterned the way in which twentieth-century contemporaries envisaged their opportunities to inaugurate a new age of medicine. In the 1950s, the context for the warnings of a future without antibiotics was an unwavering belief in continued economic growth and unbounded prosperity. History was conceived as a process with a clear direction, in which the future represented the realization of progress. Time was endowed with a historical quality: transformed into a dynamic force, it itself became an historical actor, the engine of a history still to be completed.[69] These warnings were thus coupled to a clear program of action and a hope for improvement and progress if the right measures were taken.

The German sociologist Hartmut Rosa argues that the twenty-first century, to the contrary, has been confronted with a process of "detemporalization."[70] After the weakening of the widespread belief in progress and hope for a future teleologically oriented towards improvement, the experience of time in which the unfolding of a society's history and an individual's life history appear to be both directed and controllable has been lost. Living in a period of constant acceleration, Rosa argues, has led to a loss of the sense of *directed* time. Along with other rising threats like climate change and global terrorism, the growth of antibiotic resistance has led to prevailing images of an inchoate and indeterminate future. This new experience of time has shaped both how society views its uncertain collective future (and acted in accordance with these future visions) and how individuals have imagined the direction of their lives.

As a result of rapidly altering background conditions—for instance, in economical and technoscientific developments—the twenty-first century has seen an accelerative pressure on political systems to deliver collective, binding solutions rapidly. Meanwhile, the time frames available for important political decisions have continually diminished, leading to the risk that temporary and provisional solutions will take the place of more strategic and long-term political designs. These have been some of the challenges confronting the WHO, for example, as it has intensified its efforts to confront antibiotic resistance.[71]

One important component of the antibiotic revolution has been the set of *futures* claimed in its name. As we have seen, these futures have been mobilized in very different ways by different actors. In the context of antibiotics, both positive expectations of ongoing therapeutic revolution and fears of the end of such a revolution have been used to marshal resources, coordinate activities, and manage uncertainty. Not only have different futures—from visions of a world rid of infectious disease to predictions of medicine made impotent by the loss of its most important weapons—been present at different chronological times; they have also coexisted at the same time. Furthermore, images of antibiotic crisis have been mobilized towards changing purposes and with downstream effects that at times are at odds with those initially imagined.

At the time of this writing, the antibiotic future—whether utopian or dystopian—has more often than not become an interconnected, global future, mandating attention to antibiotic development, regulation, and education worldwide and to the structural changes in health care and social systems that are necessary to permit the rational and ongoing distribution of effective antibiotics. Antibiotic futures and their economic, moral, political, and public health components continue to be mobilized on behalf of such once—and future?—miracle drugs. It is important to engage in the work these future visions do in a world of evolving microbes, patients, and histories.

NOTES

1. Henry Welch and Félix Martí-Ibañez, *The Antibiotic Saga* (New York: Medical Encyclopedia, 1960), 140–143.
2. Sally C. Davies, *Annual Report of the Chief Medical Officer, Volume Two, 2011: Infections and the Rise of Antimicrobial Resistance* (London: Department of Health, 2013 [first published online in March 2013]).
3. Peter Burke, "Reflections on the Cultural History of Time," *Viator* 35 (2004): 617–626, 620.
4. Lucian Hölscher, "Mysteries of Historical Order: Ruptures, Simultaneity and the Relationship of the Past, the Present and the Future," in *Breaking Up Time: Negotiating the Borders between Present, Past and Future*, ed. Chris Lorenz and Berber Bevernage (Göttingen: Vandenhoeck & Ruprecht, 2013), 149.
5. See Rheinart Koselleck, *Vergangene Zukunft: Zur Semantik geschichtlicher*

Zeiten (Frankfurt: Suhrkamp, 1985); Lucian Hölscher, *Die Entdeckung der Zukunft* (Frankfurt am Main: Fischer Taschenbuch Verlag,1999); Ernst Schulin, "Die Zukunft im historisch-politischen Denkens des zwanzigsten Jahrhunderts," in *Geschichte und Zukunft: Funf Vorträge*, ed. Heniz Löwe (Berlin: Duncker & Humblot, 1978), 91–111; Ulrich Raulff, *Der unischtbare Augenblick: Zeitkonzepte in der Geschichte* (Göttingen: Wallstein Verlag, 1999).

6. Mads Borup, Nik Brown, Kornelia Konrad, and Harro Van Lente, "The Sociology of Expectations in Science and Technology," *Technology Analysis & Strategic Management* 18 (2006): 285–298; Nik Brown and Mike Michael, "A Sociology of Expectations: Retrospecting Prospects and Prospecting Retrospects," *Technology Analysis and Strategic Management* 15 (2003): 3–18; Cynthia Selin, "The Sociology of the Future: Tracing Stories of Technology and Time," *Sociology Compass* 2 (2008): 1878–1895; *Contested Futures: A Sociology of Prospective Techno-Science*, eds. Nik Brown, Brian Rappert, and Andrew Webster (Aldershot, UK: Ashgate, 2000); Paul Rabinow and Talia Dan-Cohen, *A Machine to Make a Future: Biotech Chronicles* (Princeton, NJ: Princeton University Press, 2005).

7. Brown and Michael, "A Sociology of Expectations," 4.

8. Mike Michael, "Futures of the Present: From Performativity to Prehension," in *Contested Futures*, 21–39.

9. Brigitte Nerlich [with a Reply from Richard James], "'The Post-Antibiotic Apocalypse' and the 'War on Superbugs': Catastrophe Discourse in Microbiology, its Rhetorical Form and Political Function," *Public Understanding of Science* 18 (2009): 574–590; Tom Fowler, David Walker, and Sally C. Davies, "The Risk/Benefit of Predicting a Post-Antibiotic Era: Is the Alarm Working?" *Annals of the New York Academy of Sciences* 1323 (2014): 1–10.

10. Brown, Rappert and Webster, "Introducing Contested Futures: From Looking into the Future to Looking at the Future," in *Contested Futures*, 3–20; Nehrlich [with James], "'The Post-Antibiotic Apocalypse' and the 'War on Superbugs.'"

11. Frank W. Geels and Wim A. Smit, "Lessons from Failed Technology Futures: Pitholes in the Road to the Future," in *Contested Futures*, 129–155.

12. Peter Temin, *Taking Your Medicine: Drug Regulation in the United States* (Cambridge, MA: Harvard University Press, 1980), 58–87.

13. Lawrence P. Garrod, "Chemotherapy II: Indications for the Use of Penicillin," *British Medical Journal* 2 (1950): 617.

14. Paul de Kruif, "They Used to Have to Die," *Reader's Digest* (November 1948): 29, 32.

15. Lederle advertisement for Aureomycin, early 1950s, SHP's possession; see also Scott H. Podolsky, *Pneumonia before Antibiotics: Therapeutic Evolution and Evaluation in Twentieth-Century America* (Baltimore: Johns Hopkins University Press, 2006), 135.

16. Fred Reinfeld, *Miracle Drugs and the New Age of Medicine* (New York: Sterling Publishing, Inc., 1957), 111–112.

17. Selman Waksman, "Antibiotics—20 Years Later," *Bulletin of the New York Academy of Medicine* 37 (1961): 202–212; Similarly, wrote Reinfeld, "in the years to come we shall witness changes even more spectacular, and even more beneficial to mankind, as seemingly fantastic cures become commonplace." *Miracle Drugs and the New Age of Medicine*, 112.

18. G. M. Maxwell, "Sulphanilamide and the Doctor," *JAMA* 112 (25 March 1939): advertising section, 28. Quoted in John E. Lesch, *The First Miracle Drugs: How the Sulfa Drugs Transformed Medicine* (New York: Oxford University Press, 2007), 272.

19. Noah Fabricant and Terry Hillel, "Antibiotics Hit the Doctor's Wallet," *American Mercury* (December 1950): 744.

20. Robert Bud, *Penicillin: Triumph and Tragedy* (New York: Oxford University Press, 2007). On medicine's "golden age," see Allan M. Brandt and Martha Gardner, "The Golden Age of Medicine?" in *Companion to Medicine in the Twentieth Century*, eds. Roger Cooter and John V. Pickstone (New York: Routledge, 2003), 21–39.

21. Bud, *Penicillin*, 54–74.

22. Ibid., 84–103.

23. Walter C. Alvarez, "The Antibiotics and the Clinical Practice of Medicine," in *The Impact of the Antibiotics on Medicine and Society*, ed. Iago Galdston (New York: International Universities Press, 1958), 186.

24. Rheinart Koselleck, "Crisis," *Journal of the History of Ideas* 67 (2006): 357–400.

25. Rheinart Koselleck, "Historische Kriterien des neuzeitlichen Revolutionsbegriffs," in *Vergangene Zukunft*, 67–68.

26. *Economic Report on Antibiotics Manufacture* (Washington: US Government Printing Office, 1958), 67.

27. Scott H. Podolsky, "Antibiotics and the Social History of the Controlled Clinical Trial, 1950–1970," *Journal of the History of Medicine and Allied Sciences* 65 (2010): 327–367; idem, *The Antibiotic Era: Reform, Resistance, and the Pursuit of a Rational Therapeutics* (Baltimore: Johns Hopkins University Press, 2015), 43–51.

28. Henry Welch, "Opening Remarks," *Antibiotics Annual* (1956–1957): 2.

29. Each, it should be noted, had likewise trained at the Thorndike Memorial Laboratory at Boston City Hospital.

30. William B. Bean, "Vitamania, Polypharmacy, and Witchcraft," *Archives of Internal Medicine* 96 (1955): 141. On Bean, see Scott H. Podolsky and Jeremy A. Greene, "Are the Medical Humanities for Sale? Lessons from a Historical Debate," *Journal of the Medical Humanities* (2014), doi: 10.1007/s 10912–014–9301–9.

31. Harry F. Dowling, "Twixt the Cup and the Lip," *JAMA* 165 (1957): 659.

32. Maxwell Finland, "Antibacterial Agents: Uses and Abuses in Treatment and Prophylaxis," *Rhode Island Medical Journal* 43 (1960): 503.

33. Harry F. Dowling et al., "The Clinical Use of Antibiotics in Combination," *Archives of Internal Medicine* 99 (1957): 537.

34. Borup, Brown, Konrad, and Van Lente, "Sociology of Expectations," 285–298.

35. Podolsky, "Antibiotics and the Social History of the Controlled Clinical Trial, 1950–1970," 327–67; idem, *Antibiotic Era*, 51–72.

36. Richard McFadyen, "The FDA's Regulation and Control of Antibiotics in the 1950s: The Henry Welch Scandal, Félix Martí-Ibañez, and Charles Pfizer and Co.," *Bulletin of the History of Medicine* 53 (1979): 159–169.

37. John Lear, "Taking the Miracle out of the Miracle Drugs," *Saturday Review* 42 (3 January 1959): 35.

38. John Lear, "The Certification of Antibiotics," *Saturday Review* 42 (7 February 1959): 43–48; Richard Harris, *The Real Voice* (New York: MacMillan, 1964), 18–20.

39. *Administered Prices in the Drug Industry* [Kefauver hearings], 1959–1960, US Congress, Senate Committee on the Judiciary, Subcommittee on Antitrust and Monopoly, Part 22, 11968–69.

40. Ibid., 12128–12129.

41. Daniel Carpenter, *Reputation and Power: Organizational Image and Pharmaceutical Regulation at the FDA* (Princeton, NJ: Princeton University Press, 2010), 345–357; Podolsky, *Antibiotic Era*, 94–111.

42. Henry E. Simmons and Paul D. Stolley, "This is Medical Progress? Trends and Consequences of Antibiotic Use in the United States," *JAMA* 227 (1974): 1027–1028. On progress, see Rheinart Koselleck, "Fortschritt und Niedergang: Nachtrag zur Geschichte zweier Begriffe," in *Niedergang: Studien zu einem geschichtlichen Thema*, ed. Reinhart Koselleck and Paul Widmer (Stuttgart: Klett-Cotta, 1980), 214–230.

43. Calvin M. Kunin, "Impact of Infections and Antibiotic Use on Medical Care," *Annals of Internal Medicine* 89 part 1 (1978): 717.

44. René Dubos, *Mirage of Health: Utopias, Progress, and Biological Change* (New York: Harper & Brothers, 1959).

45. Lindsey W. Batten, in "Discussion on the Use and Abuse of Antibiotics," *Proceedings of the Royal Society of Medicine* 48 (1955): 360.

46. Tsutomu Watanabe, "Infective Heredity of Multiple Drug Resistance in Bacteria," *Bacteriological Reviews* 27 (1963): 87–115.

47. John A. Osmundsen, "Are Germs Winning the War against People?" *Look* (18 October 1966): 140–141.

48. "Infectious Drug Resistance," *NEJM* 275 (1966): 277.

49. Richard Gleckman and Morton A. Madoff, "Environmental Pollution with Resistant Microbes," *NEJM* 281 (1969): 677–678.

50. Calvin M. Kunin, "Antibiotic Accountability," *NEJM* 301 (1979): 380.

51. Bud, *Penicillin*, 163–191.

52. Mark R. Finlay, "Reframing the History of Agricultural Antibiotics in

the Postwar World: An International and Comparative Perspective," talk given at the 2011 European Science Foundation DRUGS Research Network Programme on "Beyond the Magic Bullet: Reframing the History of Antibiotics," University of Oslo; Claas Kirchhelle, "Pyrrhic Progress. Antibiotics in Western Food Production (1949–2013)," PhD thesis, Oxford University, 2015.

53. Stuart B. Levy, *The Antibiotic Paradox: How the Misuse of Antibiotics Destroys their Curative Powers*, 2nd edition (Cambridge, MA: Perseus Publishing, 2002), 304–10.

54. Tim Beardsley, "NIH Retreat from Controversy," *Nature* 319 (1986): 611; Stuart B. Levy, "Antibiotic Resistance 1992–2002: A Decade's Journey," in *The Resistance Phenomenon in Microbes and Infectious Disease Vectors: Implications for Human Health and Strategies for Containment* [workshop summary], eds. Stacey L. Knobler, Stanley M. Lemon, Marjan Nafjafi, and Tom Burroughs (Washington: National Academies Press, 2003), 37; Ruth L. Berkelman and Phyllis Freeman, "Emerging Infections and the CDC Response," in *Emerging Illnesses and Society: Negotiating the Public Health Agenda*, eds. Randall M. Packard, Peter J. Brown, Ruth L. Berkelman, and Howard Frumkin (Baltimore: Johns Hopkins University Press, 2004), 356–357; Podolsky, *Antibiotic Era*, 166–167.

55. Laurie Garrett, *The Coming Plague: Newly Emerging Diseases in a World out of Balance* (New York: Farrar, Straus, and Giroux, 1994).

56. Bud, *Penicillin*, 196–198; Podolsky, *Antibiotic Era*, 167–168.

57. Joshua Lederberg, "Medical Science, Infectious Disease, and the Unity of Humankind," *JAMA* 260 (1988): 684–85.

58. Joshua Lederberg, Robert E. Shope, and Stanley C. Oaks, Jr., eds., *Emerging Infections: Microbial Threats to Health in the United States* (Washington: National Academy Press, 1992); Nicholas B. King, "The Scale Politics of Emerging Diseases," *Osiris* 19 (2004): 62–76.

59. See, e.g., Harold C. Neu, "The Crisis in Antibiotic Resistance," *Science* 257 (1992): 1064–1072; Sharon Begley, "The End of Antibiotics?" *Newsweek* (7 March 1994): 63; Geoffrey Cowley, John F. Lauerman, Karen Springen, Mary Hager, and Pat Wingert, "Too Much of a Good Thing [internal story capsule]," *Newsweek* (28 March 1994): 50–51.

60. US Congress, Office of Technology Assessment, *Impacts of Antibiotic-Resistant Bacteria* (Washington: US Government Printing Office, 1995); American Society for Microbiology, *Report of the ASM Task Force on Antibiotic Resistance* (1995), produced as a supplement to *Antimicrobial Agents and Chemotherapy* and now available at the ASM website at http://www .asm.org/images/PSAB/ar-report.pdf.

61. Bud, *Penicillin*, 201–206; Moysis Lelekis and Panos Gargalianos, "The Influence of National Policies on Antibiotic Prescribing," in *Antibiotic Policies: Theory and Practice*, eds. Ian M. Gould and Jos W. M. van der Meer (New York: Kluwer Academic / Plenum, 2005), 545–566.

62. Infectious Diseases Society of America, *Bad Bugs, No Drugs: As Antibiotic Discovery Stagnates . . . a Public Health Crisis Brews* (Alexandria, VA: Infectious Diseases Society of America, 2004), 22; Calvin M. Kunin, "Why Did it Take the Infectious Diseases Society of America So Long to Address the Problem of Antibiotic Resistance?" *Clinical Infectious Diseases* 46 (2008): 1791–1792.

63. "The 10 × '20 Initiative: Pursuing a Global Commitment to Develop 10 New Antibacterial Drugs by 2020," *Clinical Infectious Diseases* 50 (2010): 1081–1083.

64. "Docket No. FDA-20120N-1248, Statement of the Infectious Diseases Society of America (IDSA)," available at IDSA website at http://www.idsociety .org/uploadedFiles/IDSA/Policy_and_Advocacy/Current_Topics_and _Issues/Advancing_Product_Research_and_Development/Bad_Bugs_No _Drugs/Statements/IDSA%20LPAD%20Statement%20to%20FDA.March %201%202013.pdf. The text of the Antibiotic Development to Advance Patient Treatment (ADAPT) Act can be seen at https://www.congress.gov/ bill/113th-congress/house-bill/3742/text. The text of the Promise for Antibiotics and Therapeutics for Health (PATH) Act can be seen at https://www .congress.gov/bill/114th-congress/senate-bill/185/text.

65. Kevin Outterson and Aaron S. Kesselheim, "Improving Antibiotic Markets for Long Term Sustainability," *Yale Journal of Health and Policy* 1010 (2011): 103–167; Kevin Outterson, John H. Powers, Gregory W. Daniel, and Mark B. McClellan, "Repairing the Broken Market for Antibiotic Innovation," *Health Affairs* 34 (2015): 277–285.

66. Lesch, *The First Miracle Drugs,* 269–276. Robert Bud has enumerated nine books published between 1943 and 1953 "with titles similar to *The Conquest of Disease.*" See Bud, *Penicillin,* 101. For a slightly earlier example, see F. Sherwood Taylor (with foreword by Henry E. Sigerist), *The Conquest of Bacteria: From Salvarsan to Sulphapyridine* (New York: Philosophical Library and Alliance Book Corporation, 1942). For other historical assessments of the antibiotic revolution, see, e.g., Harry F. Dowling, *Fighting Infection: Conquests of the Twentieth Century* (Cambridge, MA: Harvard University Press, 1977); Gladys L. Hobby, *Penicillin: Meeting the Challenge* (New Haven: Yale University Press, 1985); David Greenwood, *Antimicrobial Drugs: Chronicle of a Twentieth-Century Medical Triumph* (New York: Oxford University Press, 2008).

67. Salmaan Keshavjee and Paul E. Farmer, "Tuberculosis, Drug Resistance, and the History of Modern Medicine," *NEJM* 367 (2012): 931–936. Relatedly, for much of the history described here, almost no one challenged the limits of the technological fix itself in the face of broader structural constraints. See Allan M. Brandt, *No Magic Bullet: A Social History of Venereal Disease in the United States Since 1880* (New York: Oxford University Press, 1985).

68. "Infectious drug resistance," 277.

69. Rheinart Koselleck, *Vergangene Zukunft.*
70. Hartmut Rosa, *Social Acceleration: A New Theory of Modernity* (New York: Columbia University Press, 2013).
71. See, e.g., World Health Organization, *Antimicrobial Resistance: Global Report on Surveillance* (Geneva: World Health Organization, 2014).

Reconceiving the Pill: From Revolutionary Therapeutic to Lifestyle Drug

ELIZABETH SIEGEL WATKINS

When Margaret Sanger conjured up the vision of a birth control pill in 1946, she tapped into the budding zeitgeist of a mid-twentieth-century therapeutic revolution that was based on the promise and potential of a new generation of so-called wonder drugs.[1] By the time the first oral contraceptive was developed and approved for sale in 1960, dozens of pharmaceuticals had come onto the market for the treatment of a wide variety of conditions: infection and inflammation, hypertension and cancer, depression and anxiety. Available by prescription only, these drugs were powerful new weapons in the physician's arsenal. The ability to cure and prevent disease, or at least to allay its symptoms, greatly increased the authority and prestige of physicians and led to what has nostalgically been called the "golden age of American medicine."[2]

In this context, the birth control pill contributed to growing narratives of therapeutic revolution by extending the reach of pharmacy beyond the treatment or prevention of disease or illness. Oral contraceptives prevent pregnancy, not normally considered to be a state of disease or illness. Unlike the other methods available in 1960, the pill offered near-perfect effectiveness and the advantage of separating contraception in both time and place from the act of sexual intercourse. Over the course of a decade, this pharmaceutical innovation effected a contraceptive revolution in the United States and in many other countries by changing the ways people thought about, discussed, and used birth control.

Although much ink was spilled in the 1960s either crediting or blaming the pill for fomenting sexual revolution, it is clear from the historical record that the pill played only a supporting role as one of many factors contributing to the liberalization and democratization of sexual behaviors and attitudes. It played a similarly auxiliary part in the revolutionary appeal of the second-wave feminist activism that swept the United States in the late 1960s and 1970s. The ability to plan whether and when to have children enabled women to pursue a wider range of education and employment opportunities. The pill, in concert with a host of other social, cultural, and political forces, helped to make women's lives in the 1980s look very different from those of their mothers in the 1950s.

By the 1990s, the pill had become part of the birth control establishment, prescribed and used more often than any other method of reversible contraception. Well after its revolutionary heyday it still served as the standard to which newer methods were compared. Manufacturers of oral contraceptives shifted the focus of their marketing strategies away from the primary indication of family planning to emphasize instead the secondary effects of relieving discomforts resulting from the menstrual cycle, such as pimples, irritability, and monthly bleeding. This chapter explores the shift in the conceptualization of the pill from life-changing to life-enhancing, and from revolutionary to commonplace, as well as the implications of this shift for the trajectories of women, birth control, and pharmaceutical consumerism.

Inventing the Pill: A Modernist Project

The development of the oral contraceptive in the 1950s that initiated the contraceptive revolution in the 1960s was not instigated by any sort of popular front. There was no groundswell of dissatisfaction with existing birth control methods, no call for action from ordinary women and men, and certainly no interest from the medical profession's rank and file. Instead, the pill was conceived privately as a collaboration among four individuals: Margaret Sanger, Gregory Pincus, Katherine McCormick, and John Rock.[3]

By the early 1950s, Margaret Sanger had been pushed out of the movement she had almost single-handedly begun forty years earlier. The American Birth Control League had changed its name to the more centrist and family-friendly Planned Parenthood Federation of America and had hired a predominantly male leadership team, and the aging birth control pioneer struggled to find a role for herself.[4] After meeting the biologist Gregory Pincus through a mutual friend in 1951, Sanger— the woman who had helped to legalize the diaphragm in America— turned her attention to a novel technological solution to the age-old problem of fertility control.

Pincus was an expert in mammalian sexual physiology who earned national attention in 1934 for successfully fertilizing rabbit eggs in vitro. However, his fame turned to notoriety a few years later when a popular national magazine sensationalized his experiments in parthenogenesis. In 1937 his employer, Harvard University, denied him tenure, possibly because of the perceived indecency of his research program, political machinations within the biology department, anti-Semitism, or some combination of the three.[5] Unable to find an academic home, Pincus teamed up with Hudson Hoagland in 1944 to start an independent nonprofit research institute, the Worcester Foundation for Experimental Biology (WFEB). He evolved into a scientific entrepreneur, negotiating grants and contracts from philanthropic organizations, government agencies, and drug companies to fund the research activities of the labs at WFEB.

Pincus was eager to begin the scientific pursuit of hormonal contraception, but he lacked the funding to support this work. Both the federal government and pharmaceutical manufacturers saw birth control as too controversial an area for major investment, and nonprofits such as Planned Parenthood did not have the resources to make a significant financial contribution. Enter the benefactor Katherine Dexter McCormick.[6] Wealthy by both birth and marriage, McCormick—one of the first two women to graduate from MIT in 1904—was struck by tragedy when her husband was diagnosed with schizophrenia. For the forty years of her husband's illness, she poured her energy and considerable resources into two areas: women's suffrage (until 1920) and then research on schizophrenia. After her husband died in 1947, she ceased her support of this research. In casting about for a new cause, she renewed her acquaintance with Sanger, who introduced her to Pincus in 1953. At the age of seventy-eight, McCormick—with the help of her seventy-four-year-old friend Sanger—took on the financial underwriting of the contraceptive revolution.

For Sanger and McCormick, as for many of their contemporaries, science and technology held the key to a better future for humanity. Sanger had long believed in the potential of scientists to develop a solution to the specific problem of a better contraceptive. Since the mid-1910s, she had exhorted audiences to "force open the doors of the laboratories where our chemists will give the women of the twentieth century reliable and scientific means of contraception hitherto unknown."[7] She firmly believed that the social problem of birth control could be solved by the application of science, so long as time and funds were allocated to the effort. All sorts of seemingly miraculous advances in science and technology had taken place during her lifetime, and Sanger fully expected that her cause would also benefit from twentieth-century scientific progress.

Sanger also promoted the medical profession as the only "proper authority" for dispensing contraceptive information and devices. During the campaign to legalize birth control in the 1920s and 1930s, she figured that Congress was more likely to sanction contraceptives if they could only be obtained from licensed physicians. But she did not champion the medical profession for purely pragmatic political purposes. Her trust in organized medicine paralleled her own personal belief in science as a social good, and she conceived of a birth control pill as the next in a long line of pharmaceutical innovations to come out of American laboratories and to be prescribed by doctors.

McCormick shared this conviction in the potential of science, technology, and medicine to do good. In addition, she remained committed to her belief in women's rights. She had spent her young adulthood on the front lines of the women's suffrage movement, and she believed that the right to reproductive control was as important as the right to vote. For both McCormick and Sanger, birth control belonged in the hands of women. Unlike a later generation of feminists, these two saw woman-controlled contraception not as a burden, but as a blessing. A technological solution that separated contraception from the act of sex would be a crowning achievement of science in service to women and society. The pill project was truly modernist in its orientation and intent: the application of scientific (rational) knowledge would produce a socially useful goal, namely, the emancipation of women from the burden of unpredictable and unwanted pregnancies.[8]

The project advanced at the Worcester Foundation, bankrolled by McCormick. After animal experiments had identified the most promising steroidal candidates for hormonal control of conception, the next step was to test the products in women. As a PhD scientist, Pincus did not have the qualifications to work with human subjects, so he had to enlist the help of a physician. For the clinical trials of progesterone and synthetic progestins, he called on John Rock, an obstetrician-gynecologist at Harvard Medical School who specialized in problems of infertility.[9] A preeminent figure in reproductive medical research and clinical practice, Rock was also an observant Roman Catholic. His participation in the development of oral contraceptives in the 1950s foreshadowed and helped contribute to the widespread acceptance of the pill among the Catholic laity in America over the next two decades.

The human studies were conducted first on small groups of women at Rock's Fertility and Endocrine Clinic in Boston and at the Worcester State Hospital, and then in large-scale field trials in Puerto Rico and Haiti. Pincus and Rock used the synthetic progestin norethynodrel in combination with a small amount of synthetic estrogen, ethinyl estra-diol, manufactured by Searle under the brand name Enovid. Enovid had received FDA approval for gynecological disorders in 1957 and was thus on the radar of some physicians, as well as business analysts who tracked the pharmaceutical industry, but neither group anticipated its approval for contraceptive purposes anytime soon.

Physicians in the 1950s did not consider birth control counseling to be within their purview. A 1957 study of doctors' attitudes and prac-tices regarding contraception found that 70 percent of non-Catholic physicians (and 83 percent of Catholics) thought that family planning should be a supplemental medical service at the request of the patient, as opposed to a regular procedure offered by the physician. Half of the doctors in the study said that they never introduced the subject of birth control during their medical examinations of premarital patients; the same proportion never, or hardly ever, brought up the subject with their postpartum patients.[10] However, the reticence of physicians to discuss family planning with their patients did not mean that women were kept ignorant of birth control; instead, they consulted other sources for contraceptive information and advice. When Enovid won FDA approval for use as a contraceptive in 1960, women learned of its availability from newspapers and popular magazines and went to their physicians to ask for it. Since the pill could not be dispensed without a doctor's prescription, women's requests forced physicians to engage directly with the issue of birth control.

The Revolution Bred by the Pill

It is hard to overstate the rapid and pervasive influence of the birth control pill on women's contraceptive practices and the contraceptive marketplace in the 1960s and 1970s. In 1955, more than half of the American women who used birth control relied on either condoms (27 percent) or a diaphragm (25 percent). Ten years later, those figures had changed dramatically. In 1965, 27 percent of American women reported use of the pill, 18 percent used condoms, and just 10 percent relied on a diaphragm. Several factors contributed to this exponential rise in pill use during its first five years on the market. Physicians enthusiastically prescribed the pill because it was easy to do so, because it increased their jurisdiction in family planning, and because it provided them with a financial incentive, since patients were required to revisit doctors' offices to obtain renewals. Planned Parenthood encouraged its affiliates to provide the pill to clients, reaching many women who might not have had access to private physicians. And women encountered ample publicity about the pill in newspapers and popular magazines, prompting them to request oral contraceptives from doctors in clinics and private practices. By 1970 some nine million American women took the pill each day, with another ten million users worldwide. Three years later, oral contraceptives were used by 36 percent of white married women in the United States.[11] Only 13.5 percent reported using condoms, and a mere 3.4 percent used a diaphragm.[12]

Clearly, a revolution in contraception took place between 1960 and 1975 in the United States, if we define that revolution in terms of the methods most commonly used by white married women (these are the only data available because studies included only white married women during this period). It is important to point out that this change was in the *kind* of method used. Most married women were already using some form of birth control: 70 percent in 1955 and 81 percent in 1960. Similarly, attitudes toward contraception had evolved into acceptance decades earlier for most Americans. Only Roman Catholics expressed ambivalence about the use of birth control, but their attitudes changed rapidly in the era of the pill. A Gallup Poll taken in 1962 revealed that 56 percent of Catholics agreed that birth control information should be available to anyone who wanted it. When pollsters asked the same question two years later, 78 percent of Catholics concurred.[13] By the mid-1970s, any disparity in attitudes toward or use of the pill between US Protestants and Catholics had completely disappeared. For Catholic

Americans, the pill brought about a transformation in their obeisance to the central Roman Catholic tenet on procreation.

Coincident with the first decade of the pill was a liberalization of laws and policies about contraception. In 1960, thirty states had statutes prohibiting or restricting the sale and advertisement of contraceptives. In 1965, the US Supreme Court ruled in *Griswold v. Connecticut* that married couples had a right to privacy and, by extension, the right to purchase and use birth control.[14] This ruling superseded most proscriptive state laws, but Massachusetts continued to prohibit the distribution of contraceptives to unmarried individuals. In 1972, the Supreme Court extended the right of privacy in matters of birth control to the unmarried population in the *Eisenstadt v. Baird* decision on the grounds that the Massachusetts law violated the Equal Protection Clause of the Fourteenth Amendment to the Constitution.[15]

Lyndon B. Johnson became the first president to endorse fertility control when he proclaimed in his 1965 State of the Union Address: "I will seek new ways to use our knowledge to help deal with the explosion in world population and the growing scarcity in world resources."[16] Johnson's sanction of international population control brought about a concomitant interest in domestic family planning. Government officials realized that the United States should practice what it preached on the subject of birth control, so the federal government, mainly through the Department of Health, Education, and Welfare and the Office of Economic Opportunity's antipoverty program, became involved in sponsoring family planning programs around the country. At the same time, more and more states included family planning in their health services; from 1959 to 1966, the number of states operating contraceptive clinics rose from seven to thirty-five.[17] The pill was not the cornerstone of all population control programs; family planners, especially those targeting underdeveloped countries, were equally enthusiastic about the intrauterine device (IUD). However, the simple iconography of the pill dominated media attention. As *Time* magazine reported in 1967, "The open debate, covered matter-of-factly by the press, was further proof of the worldwide turnabout in attitudes toward birth control since the advent of oral contraceptives . . . In the past few weeks, newspapers and magazines have been filled with news of family planning, population control and the pill."[18]

Although physicians were the only ones authorized to write prescriptions for oral contraceptives, they did not wield complete control over the use of the pill. Indeed, the pill altered the relationships between doctors and patients in unanticipated ways. Traditionally, when a person felt ill and went to the doctor's office, he or she relied on the doctor to diagnose the condition and to determine the course of therapy. Oral contraception changed that script. Women knew exactly what the problem was (they wanted to prevent pregnancy) and how to treat it (by taking the pill); all they needed from the physician was a written prescription to take to the pharmacy to obtain the pills. Women who went to their physicians with specific requests for oral contraceptives no longer passively received medical care, but transformed into active participants. If a doctor refused to comply, the patient could, and frequently would, find another, more willing provider. In the early 1960s, women who requested oral contraceptives from their physicians helped to shift the balance of power in the traditional doctor-patient relationship. Later in the decade, when research studies called the safety of the pill into question, women felt confident enough to doubt their physicians' judgment and to demand full disclosure so that they could make their own informed decisions about whether or not to take it.

The popularity of the pill in the 1960s owed much to its efficacy and its discreetness. Its rate of effectiveness—98 to 99 percent—was far greater than that obtainable with any other method available at mid-century. This highly reliable contraceptive gave women, for the first time in history, the ability to choose whether and when to have children. It also provided, for the first time, a way to separate the act of contraception from the act of sexual intercourse, allowing women secure and total control over their own fertility, without the knowledge, participation, or approval of their sexual partners.

Do the changes brought about by the birth control pill add up to a therapeutic revolution? Revolution is variously defined as an instance of revolving, a forcible overthrow of a government or social order in favor of a new system, or—the apposite definition for this discussion—a dramatic and wide-reaching change in the way something works or is organized or in people's ideas about it.[19] "Therapeutic" can mean of or relating to the healing of disease, administered or applied for reasons of health, or having a good effect on the mind or body and contributing to a sense of well-being.[20] While the first definition does not apply, as pregnancy is not a disease to be healed, the other two are consistent with the reasons for which women chose (and choose) to use contraception. The application or administration of birth control to prevent pregnancy—or, looked at from a different angle, to maintain a non-pregnant state of health—certainly contributes to a woman's sense of well-being. And while many normative debates have asked whether the net effect of contraception is good or not, it clearly has the desired effect on the body, namely, the maintenance of the nonpregnant state. Did the advent of the pill lead to a dramatic and wide-reaching change in the way birth control works and in people's ideas about birth control? Absolutely. Moreover, the pill extended the reach of medico-pharmaceutical intervention beyond the treatment or prevention of disease, expanding the scope of medical practice to incorporate birth control.

Contemporaries clearly recognized the revolutionary potential of the pill. In 1968, the popular anthropologist Ashley Montagu pronounced its invention to be as important as the discovery of fire and the development of toolmaking, hunting, agriculture, urbanism, scientific medicine, and nuclear energy.[21] Individual women understood the pill's revolutionary nature at a very personal level. Said one satisfied user to her doctor in the late 1960s, "For the first time in eighteen years of married life I can put my feet up for an hour and read a magazine. . . . If you refuse to give me the pill, I'll go get it from someone else."[22]

Journalists also promoted the dual promise of oral contraception as the newest pharmaceutical technology to come out of the scientific laboratory. They touted its convenience for individuals and its potential for population control. Population control was depicted positively; journalists at the time made no mention of the possibility of either coercion or eugenics. The presentation of individual convenience, however, was tempered by the possibility of negative medical side effects, and newspapers and magazines devoted considerable space to these problems. However, much of the popular media coverage of the pill was infused with a faith in science to continue to make improvements in its formulation. Articles in mass market magazines demonstrated this modernist optimism. A piece in *Reader's Digest* proclaimed, "Almost surely, better pills with fewer side effects are on the way."[23] Gregory Pincus wrote in one of *Ladies' Home Journal*'s monthly "Tell Me, Doctor" columns: "With the application of the method of science to the problem of human fertility, the development of adequate methods is inevitable."[24] In addition, popular articles on oral contraception stressed the importance of the role of the physician. The pill could be obtained only by prescription; *Good Housekeeping* readers were told that its use was "a medical judgment to be made by physicians."[25] These articles implied that, while the drugs might produce side effects, scientists were working to correct these flaws and physicians had the wisdom to prescribe oral contraceptives prudently. In the early 1960s, newspapers and popular periodicals depicted the pill in particular and science in general as holding tremendous promise and potential, and women leaped to avail themselves of the latest innovation of medical science.

As the decade progressed, journalists debated the broader social impact of the pill, especially as it might influence morality, marriage, and family life. As these writers conveyed the impression of social upheaval to their readers, they pointed to the pill as an influential factor. The role played by the pill in contributing to women's sexual liberation or promiscuity depended on the editorial slant of the article; either way, the popular press implicated oral contraceptives to some extent in America's sexual revolution. Ultimately, few took seriously the conservative warnings that the pill would lead young women down a dangerous descent into promiscuity. It is not known how many single women were using oral contraceptives in the 1960s (because the first demographic study of contraceptive practices among unmarried women was not conducted until 1971), but to the growing number of young women attending college, joining the workforce, living on their own, and delaying marriage and motherhood, the pill was perceived as a revolutionary option at their disposal. By the late 1960s, the pill had become the icon of the sexual revolution; one particularly striking pictorial representation was a photo on the cover of *Time* magazine in 1967 of dozens of birth control pills fashioned into the shape of the scientific symbol for "female."[26]

At the same time, the thinking of 1960s-era population control advocates was characterized by the assumption that a technological solution to preventing pregnancy, as opposed to more sweeping educational and social reforms, could stimulate economic development abroad and reduce poverty at home. Although this overly simplistic model faded during the 1970s, birth control in general, and the oral contraceptive as a specific method, earned acknowledgement as a basic health care need for women of reproductive age, whether it was provided by physicians or obtained through other channels. In 1977 the World Health Organization (WHO) made this recognition explicit by including the combination progestin-estrogen oral contraceptive pill on its first list of "essential medicines." According to WHO, the pill was an essential medicine, defined as one that met "the priority health care needs of the population" because of its "public health relevance, evidence on efficacy and safety, and comparative cost-effectiveness."[27]

The reputation of the pill as being revolutionary in birth control was magnified by its ancillary role as something revolutionary in society. As discussed above, oral contraceptives transformed the personal lives of a significant proportion of Roman Catholic women by giving them a means to control their fertility. For these women, taking a daily pill seemed more palatable, and less sacrilegious, than using a barrier contraceptive at the time of intercourse. As more and more women ignored the papal proscription of artificial birth control, they contributed to the growing distance between the Catholic laity and hierarchy in the United States.

Furthermore, the pill played a part in the evolution of women's liberation, regardless of their religion. By providing women with the easy-to-use and highly effective means to delay and space childbearing or to avoid it entirely, the pill opened up new possibilities for them to work, study, and participate more fully in public life. At the same time, the pill was only one of several interacting developments and trends that produced a shift in American culture from restraint to openness in sex and sexuality. As I have explained elsewhere,[28] the pill did not single-handedly bring about a sexual revolution in the 1960s, but it did contribute to changing sexual mores, attitudes, and practices. When it came to sex, Americans in the 1970s thought, spoke, and behaved in ways that differed greatly from those just two decades earlier.

Bumps on the Revolutionary Road

The revolutionary impact and impression of the birth control pill was complicated in the late 1960s by research evidence that associated oral contraceptive use with increased risk of potentially fatal blood clots. These findings led to highly publicized Senate hearings about the safety of the pill in 1970, and eventually resulted in a patient package insert (the first of its kind) to warn consumers of the adverse health effect of this prescription drug.[29]

Concerns about the safety of the pill came to a head with the publication of journalist Barbara Seaman's book *The Doctors' Case against the Pill* in 1969, and Senator Gaylord Nelson's hearings on oral contraceptives as part of his committee's investigation of the drug industry in 1970. Seaman and Nelson represented a growing ambivalence toward science and medicine in American society. On the one hand, they questioned the merit and safety of the pill, a product of medical science and technology, and criticized scientists and physicians for their incursion into family planning. On the other hand, they based their critiques on evidence from scientists and physicians. They found themselves on opposing sides of the debate over the pill because of the way the Senate hearings were handled. Seaman was not invited to testify, nor were any users of the pill allowed to provide their perspective. To protest these omissions, women from the group DC Women's Liberation organized demonstrations at the Senate hearings. A group of twenty women arrived early each day and strategically placed themselves at the end and in the middle of every other row of seats. They came prepared with questions to interrupt the hearings, and with bail money tucked inside their boots.

Ironically, both the feminists and Senator Nelson agreed on most issues concerning the pill. They believed that the FDA had allowed the drug companies to market the pill without adequate tests of its long-term safety. To illustrate the problem of insufficient testing, both likened the millions of women who used the pill to unsuspecting guinea pigs in a massive experiment. In spite of the medical controversy over the safety of the pill, neither the feminists nor Nelson advocated a ban on the oral contraceptives; both realized that such a proposal was unrealistic and impractical and could lead to a black market for the pills which would be even more dangerous to women. Instead, both Nelson and the DC Women's Liberation group argued that women needed access to all available information on the pill so that they could make intelligent decisions about birth control. They agreed that the lack of informed consent stemmed from the problem of poor communication between doctors and patients. At this point, the opinions of the senator and the feminists diverged. For Nelson, the issue of informed consent could be solved by improving the channels of communication among the manufacturers, the FDA, physicians, and patients. The more radical feminists saw the pill as the tip of the iceberg of much larger problems in women's health care; they doubted that these problems could be solved within the context of the contemporary system of male-dominated medicine.

The outcome of the Senate hearings satisfied none of the interested parties. The FDA mandated a patient package insert for oral contraceptives to warn patients about the serious side effect of abnormal blood clotting. Consumers and feminists objected to the brevity of the insert: the one-hundred-word label described the availability of an information booklet, which the patient could request from her physician. The longer booklet was written by the American Medical Association in conjunction with the FDA and the American College of Obstetricians and Gynecologists; the onus fell on the patient to ask her doctor to give her the booklet. Physicians and manufacturers objected to any patient package labeling; they wanted control of all medical information to remain in the hands of medical professionals, and to be dispensed as physicians saw fit.

The percentage of contraceptive users taking the pill dropped during the 1970s as concerned women switched to nonhormonal methods or chose permanent sterilization. The pill's rate of use among married women fell from 36 percent in 1973 to 20 percent in 1982. During the same period, the rate of sterilization by tubal ligation among married women more than doubled, and more married men underwent vasectomies, so that by 1988 almost half of married couples relied on either male or female sterilization to prevent pregnancy.[30] A contemporary writing in *Family Planning Perspectives* observed, "The pill has fallen from its pedestal to take its place among the other contraceptives, each with flaws and assets."[31] Adverse health effects and the ensuing negative publicity toppled the pill from its vaunted status as a wonder drug. However, despite its decline in popularity and market share, it still remained the most popular *temporary* contraceptive among American women through the end of the twentieth century and into the twenty-first.[32]

The Revolution Comes to a Halt

In spite of the tumult over its safety, the pill had become so thoroughly incorporated into the fabric of American life after a few decades that it was no longer perceived as revolutionary in either a therapeutic or social sense. By 1990, 80 percent of all American women born after 1945 had used the pill at some time in their lives.[33] Part of the reason for the pill's enduring popularity was its lack of major competitors in the birth control marketplace. After the advent of the pill in 1960 and of IUDs a few years later, it took almost three decades for the next wave of contraceptive innovation to come to market in the United States. Research and development on the implant (Norplant) and the shot (Depo-Provera) began in the United States in the 1960s, but these methods did not receive regulatory approval until the early 1990s. They and the other methods that followed (the skin patch and the vaginal ring) were not radically inventive; they simply provided different delivery systems for synthetic hormones, the ones used in birth control pills, to enter the bloodstream.

The pill remained the standard on whose technology newer methods were based, and by which these were judged by contraceptive users. With the long delay in getting new methods to market and with those methods' relatively low level of innovation, there was a growing sense among birth control researchers, policy makers, and providers that the contraceptive revolution begun by the pill had ended. A 1988 article in *Family Planning Perspectives* asked, "Whatever Happened to the Contraceptive Revolution?"[34] A series of meetings held by the National Academy of Science resulted in a 1990 book called *Developing New Contraceptives: Obstacles and Opportunities*, which focused more on obstacles than on opportunities.[35]

In 1995, the editor of *Family Planning Perspectives*, Michael Klitsch, summarized the stasis in contraceptive research in an article titled "Still Waiting for the Contraceptive Revolution."[36] Klitsch reviewed the factors that others had identified as contributing to the death of the revolution. First, he noted the chilling effect of product liability costs, resulting from both individual and class action lawsuits against contraceptive manufacturers. Second, he pointed to the financial burden of increased government regulation of contraceptive products, as the FDA required more stringent testing of experimental methods in animals and humans. Third, he enumerated several reasons for changes in public opinions about contraceptives. As discussed above, the enthusiasm over the pill in the early 1960s had waned in the aftermath of the safety debate, intensified by media coverage of the adverse health effects. Klitsch attributed greater public scrutiny of and skepticism toward contraceptives in the 1970s (especially those pharmaceuticals and devices available by prescription only) to the growing influence of the consumer movement and the women's movement. The political dimension of American attitudes toward contraception was further complicated in the 1980s by the emergence of HIV and AIDS, against which only barrier methods offered any protection, and by the increasingly acrimonious abortion debate, which swept contraceptives into its maelstrom. Collectively, these factors thwarted enthusiasm for exploring novel approaches to preventing pregnancy. Finally, Klitsch reported that the pharmaceutical industry saw limited opportunities for growth (and profits) in the contraceptive sector of developed countries, because that market was already saturated with existing products. Companies feared that new contraceptives would not attract enough new users to be profitable, or that they might eat into the profits of their products already on the market. The safer bet was to stick with current product lines.

Thus, the solution for drug companies already in the oral contraceptive business was simply to tinker with existing product formulations. In addition, more companies entered the pill market after 1984, when the Drug Price Competition and Patent Term Restoration Act, also known as the Hatch-Waxman Act, set up the modern system of generic drug approval and regulation, allowing generic versions of existing formulations to be marketed and sold. The total number of different birth control pills available in the United States increased dramatically, and by 2007, there were more than 90 brand-name and generic oral contraceptive products on the market in the United States.[37] Physicians, pharmacists, and women could choose pills based on price, since the action of these contraceptives, or their therapeutic equivalents, was essentially the same.[38] Brand-name manufacturers had to find a way to make their products stand out from the generic crowd.

One of the tactics used by manufacturers to promote their products was to rebrand oral contraceptives as drugs to remedy acne, to suppress monthly menstruation, or to treat a condition called PMDD, premenstrual dysphoric disorder. In 1992, Ortho-McNeil petitioned the FDA for approval of Ortho Tri-Cyclen for the treatment of acne. While physicians had been prescribing oral contraceptives off-label for acne treatment since the 1960s, Ortho Tri-Cyclen was the first to seek and receive formal FDA approval for this indication. In 2003, Duramed, a subsidiary of Barr Pharmaceuticals, received FDA approval to market Seasonale as the first extended-cycle oral contraceptive, which reduced the number of bleeding periods from twelve to four per year.[39] As in the case of acne, the knowledge that taking oral contraceptives continually would eliminate monthly periods was not new. In fact, the first advertisement for Enovid in 1960 promoted it not for contraceptive purposes, but rather to postpone menstruation "for convenience, for peace of mind, for full efficiency on critical occasions."[40] What Barr did in 2003 was to formalize this indication. Three years later, Bayer won approval to market its Yaz brand as a treatment for PMDD and for acne. Bayer's birth control pill has been extraordinarily successful. In 2009, Yaz was the bestselling oral contraceptive on the American market. Moreover, it was the twenty-first bestseller among *all* prescription drugs in terms of number of prescriptions filled (almost ten million), and it ranked fiftieth overall in terms of retail sales ($700 million).[41]

These new uses for oral contraceptives—acne treatment, menstrual suppression, and PMDD therapy—and the aggressive marketing campaigns undertaken by their manufacturers suggest a shift in the pharmaceutical industry toward marketing birth control pills as lifestyle drugs. Variously defined as medications taken for cosmetic, enhancing, or recreational purposes, lifestyle drugs include products such as weight-loss tablets, impotence therapies, and hair restorers. It would be inaccurate, and anachronistic, to describe the birth control pill as the first lifestyle drug, as some historians have suggested.[42] Although the pill may have been the first prescription medication meant to be taken by healthy people, categorizing it first and foremost as a lifestyle drug diminishes its significance in meeting a critical and basic health need for both individuals and populations—namely, the need for a reliable and effective method for preventing pregnancy. Indeed, oral contraceptives confer a significant health benefit (avoidance of pregnancy) on their users that cannot simply be considered cosmetic, enhancing, recreational, or discretionary. However, the secondary effects for which some brands are promoted do fit the lifestyle characterization. Acne, monthly bleeding, and periodic moodiness are common conditions that constitute inconveniences, unpleasantness, and varying degrees of suffering, but they are rarely life-threatening or wholly debilitating. Moreover, the newest brands of birth control pills are not being marketed solely for the primary indication of family planning. For example, ads for Seasonale, and its more recent iteration, Seasonique, promote freedom from menstruation, not freedom from pregnancy.[43] Yaz's slogan, "Beyond birth control," implies that its real purpose is to deal with the miseries attendant upon menstruation, such as headaches, irritability, and pimples.[44] Pharmaceutical manufacturers are not selling contraception per se as a lifestyle option; rather, they pitch menstruation (or acne or moodiness) as an annoying condition to be ameliorated by their products. The emphasis on secondary effects instead of the primary indication in advertisements represents an attempt to differentiate products in a crowded field, since no one brand can claim superior efficacy in the prevention of pregnancy. When the contraceptive aspect takes a back seat, the pill appears as a lifestyle drug in its marketing materials.[45]

Legacies of the Revolution

The transition in the pill's social status—from a radically innovative drug that upended therapeutic and social conventions to a time-honored member of the pharmacopeia, considered so basic that it is marketed for its secondary effects—offers an interesting perspective for delineating the contours of this particular therapeutic revolution. The pill was one of the numerous so-called wonder drugs that traveled from scientific laboratories via pharmaceutical companies into physicians' armamentaria in the mid-twentieth century. The initial slope of the pill's revolutionary gradient was steep. Within its first decade, the pill had revolutionized more in the world of medicine than simply the therapeutics of birth control. In the early 1960s it expanded the therapeutic purview of physicians to incorporate the writing of prescriptions for healthy women of reproductive age. It also gave women greater agency in demanding the treatment they wanted to prevent pregnancy. At the end of the 1960s, the debate over the safety of the pill appeared to threaten the very foundations of medical practice by calling into question status quo relationships among patients, doctors, drug manufacturers, and government regulators. It gave a voice to feminist health activists and a platform for them to expound their critique of paternalistic medicine and their vision for more patient-centered health care. And this therapeutic revolution extended beyond medicine to contribute to social and cultural changes associated with sexual liberalization and second wave feminism.

Of course the pill did not introduce the notion of birth control; women have adopted technologies to control their fertility for millennia. Women in the United States reduced the national fertility rate from 7 children per woman in 1800 to 3.5 in 1900, using condoms, pessaries, doucheing, withdrawal, and abstinence as methods. What the pill introduced in the second half of the twentieth century was a new cultural norm in which women could expect to plan and schedule childbearing. The pill could effect such a sweeping shift because it differed from preexisting methods in its near-perfect effectiveness and thus its greater reliability.

This new cultural norm was part of a broader revolution in the roles and status of women in society that took shape after 1960. The biological responsibility of childbearing was not the only impediment to women's liberation; long-standing social mores, cultural habits, laws, and economic arrangements also contributed to the fettering of women. American women had begun to address these inequities as early as 1848, at the Seneca Falls Convention, but so long as biology was seen as destiny, women struggled to gain full equality with men. The pace at which laws, policies, and conventions were successfully challenged and overturned in the 1960s and 1970s was accelerated by women's control over their fertility. Wrote one observer in 1982: "Without the advent of the pill and women's response to the freedom it *promised*, our present age would clearly be very different, and so would our vision of the future."[46] As historian Elaine Tyler May neatly summarized in 2010: "Today, women no longer need to choose between having a family and a career. At the pill's 50th anniversary, that alone is well worth celebrating."[47]

Just as contemporaries recognized the immediate and potential revolutionary impact of the pill in the 1960s, reflections on the occasions of the pill's semicentennial also acknowledged its historic transformative effects. But it was clear to those taking retrospective stock that the revolution had run its course. From the vantage point of 2010, the pill had unleashed great changes in contraception and in women's lives compared to fifty years earlier, but the pace of change had slowed or stalled as the decades progressed. The success of the pill did not lead to further truly innovative contraceptive development, and concerns over its safety tempered initial enthusiasm about it as a method. By the 1970s, the pill was no longer lauded as a revolutionary therapeutic. Instead, it had settled into the workaday toolbox of modern medicine, alongside antibiotics, cortisone, and other mid-century wonder drugs, as a useful but somewhat flawed pharmaceutical product.

The decline in contraceptive research and development contributed, perhaps somewhat ironically, to the enduring popularity of the pill both in the United States and worldwide, because of the dearth of desirable alternatives. In addition, manufacturers developed oral contraceptive formulations in the 1970s and 1980s with smaller amounts of synthetic hormones and diminished side effects. Concerns about an increased risk of breast cancer were mitigated by research that demonstrated the pill's protective effect against ovarian cancer. With hundreds of millions of woman-years of experience by the end of the twentieth century, scientific, medical, and popular opinion cohered into a consensus that the pill was benign and beneficial. While no longer a revolutionary force, the pill rebounded from its nadir in the 1970s to its abiding position as reliable contraceptive workhorse.

As a revolutionary innovation in the 1960s, the pill brought birth control into the public eye as no method had done before. The lasting effect of that shift, in historian James Reed's words, "from private vice to public virtue" is that today more than 99 percent of American women between the ages of fifteen and forty-four who have ever had sexual intercourse have tried at least one contraceptive method; of those women, 82 percent used the pill.[48] The pill continues to have major, if not revolutionary, influence worldwide. In 2009 the pill, as compared to all other contraceptives, had the widest geographic distribution around the world and the greatest proportion of countries in which at least 30 percent of contraceptive users chose it as a method of birth control. Overall, more than 100 million women relied on it as their primary method.[49]

The pill also helped to normalize daily pharmaceutical consumption. For women in most countries in North America, Western Europe, and Oceania, oral contraceptives are still only available by prescription. This restriction is not the case in other parts of the world. Of 147 countries surveyed in 2012, almost 70 percent allowed the pill to be obtained without the explicit permission of a doctor.[50] With or without a physician's prescription, most oral contraceptives are taken for the purpose of preventing pregnancy, although, as we have seen, secondary indications to prevent menstruation, acne, and premenstrual dystrophic disorder may account for some proportion of that total. In that simple act of swallowing a tablet each morning or before going to bed, a hundred million individuals participate in the interwoven histories of birth control, women's liberation, and pharmaceutical consumerism that represent the legacy of the pill's therapeutic revolution.

NOTES

1. *Ellen Chesler, Woman of Valor: Margaret Sanger and the Birth Control Movement in America* (New York: Simon and Schuster, 1992), 407.
2. See, for example, John C. Burnham, "American Medicine's Golden Age: What Happened to It?" *Science* 215 (19 March 1982): 1474–1479.
3. See Elizabeth Siegel Watkins, *On the Pill: A Social History of Oral Contraceptives, 1950–1970* (Baltimore: Johns Hopkins University Press, 1998), chapter 1.
4. Chesler, *Woman of Valor.*
5. James Reed, *The Birth Control Movement and American Society: From Private Vice to Public Virtue* (Princeton, NJ: Princeton University Press, 1984), 324–325.
6. For the only book-length biography of Katherine Dexter McCormick, see Armond Fields, *Katherine Dexter McCormick: Pioneer for Women's Rights* (New York: Praeger, 2003).
7. Quoted in Chesler, *Woman of Valor.* 146.
8. For a helpful explication of modernity and modernism, see David Harvey, *The Condition of Postmodernity: An Enquiry into the Origins of Cultural Change* (Cambridge, MA: Blackwell, 1990), chapter 1, esp. 12, 31, 35.
9. For an excellent biography of John Rock, see Margaret Marsh and Wanda Ronner, *The Fertility Doctor: John Rock and the Reproductive Revolution* (Baltimore: Johns Hopkins University Press, 2008).
10. Mary Jean Cornish, Florence A. Ruderman, and Sydney S. Spivak, *Doctors and Family Planning* (New York: National Committee on Maternal Health, 1963). See 66–67, 31, 40, 19.
11. William D. Mosher and Charles F. Westoff, *Trends in Contraceptive Practice:*

United States, 1965–1976 (Hyattsville, MD: National Center for Health Statistics, 1982), 15.

12. Watkins, *On the Pill*, 61–62; Leslie Aldredge Westoff and Charles F. Westoff, *From Now to Zero: Fertility, Contraception and Abortion in America* (Boston: Little, Brown, 1968), 64; Christine A. Bachrach, "Contraceptive Practice among American Women, 1973–1982," *Family Planning Perspectives* 16 (December 1984): 253–259.

13. *The Gallup Poll, Public Opinion 1935–1971* (New York: Random House, 1972), 1785–1786, 1916, 1957.

14. *Griswold v. Connecticut*, 381 US 479 (1965), http://caselaw.lp.findlaw.com/ scripts/getcase.pl?court=US&vol=381&invol=479.

15. *Eisenstadt v. Baird*, 405 US 438 (1972), http://caselaw.lp.findlaw.com/ scripts/getcase.pl?court=US&vol=405&invol=438.

16. "Statements by President Johnson concerning Population." Population Council Records, accession 1, box 115, folder 2109.

17. Raymond A. Lamontagne to John D. Rockefeller III (22 September 1965). Population Council Records, accession 1, box 116, folder 2131.

18. "News of the Pill," *Time* 89 (21 April 1967): 66.

19. http://www.oxforddictionaries.com/us/definition/american_english/ revolution?q=revolution.

20. http://www.oxforddictionaries.com/us/definition/american_english/ therapeutic.

21. Ashley Montagu, "The Pill, the Sexual Revolution, and the Schools," *Phi Delta Kappan* 49 (May 1968): 480.

22. Quoted in Barbara Seaman, *The Doctors' Case against the Pill* ((New York: Peter H. Wyden, 1969), 15.

23. J. D. Ratcliff, "An End to Woman's 'Bad Days'?" *Reader's Digest* 81 (December 1962): 76.

24. Gregory Pincus, "Tell Me, Doctor," *Ladies' Home Journal* 80 (June 1963): 134.

25. "More About Birth-Control Pills," *Good Housekeeping* 155 (November 1962): 155.

26. "Freedom from Fear," *Time*, 7 April 1967.

27. http://www.who.int/topics/essential_medicines/en/. Accessed 19 September 2011.

28. Watkins, *On the Pill*, chapter 3.

29. Watkins, *On the Pill*, chapters 4 and 5.

30. Christine A. Bachrach, "Contraceptive Practice among American Women, 1973–1982," *Family Planning Perspectives* 16 (November/December 1984): 256–257. William D. Mosher, "Contraceptive Practice in the United States, 1982–1988," *Family Planning Perspectives* 22 (September/October 1990): 200, 201.

31. Susan C. M. Scrimshaw, "Women and the Pill: From Panacea to Catalyst,"

Family Planning Perspectives 13 (November/December 1981): 254–262, quote on 261.

32. In 1988 the pill was used by 20 percent of currently married women, 25 percent of formerly married women, and 59 percent of never-married women using contraception. Twenty-five years later (2013), the pill was used by 28 percent of all women using contraception. William D. Mosher, "Contraceptive Practice in the United States, 1982–1988," *Family Planning Perspectives* 22 (September/October 1990),:200, 201. Guttmacher Institute, "Fact Sheet: Contraceptive Use in the United States," August 2013. http://www.guttmacher.org/pubs/fb_contr_use.html. Accessed 17 February 2014.

33. Deborah Anne Dawson, "Trends in Use of Oral Contraceptives: Data from the 1987 National Health Interview," *Family Planning Perspectives* 22 (July/August 1990): 169.

34. Richard Lincoln and Lisa Kaeser, "Whatever Happened to the Contraceptive Revolution?" *Family Planning Perspectives* 20 (1988): 20–24.

35. Luigi Mastroianni, Jr., Peter J. Donaldson, and Thomas T. Kane, eds., *Developing New Contraceptives: Obstacles and Opportunities* (Washington: National Academy Press, 1990).

36. Michael Klitsch, "Still Waiting for the Contraceptive Revolution," *Family Planning Perspectives* 27 (1995): 246–253.

37. American College of Obstetricians and Gynecologists, "ACOG Committee Opinion no. 375: Brand versus Generic Oral Contraceptives," *Obstetrics & Gynecology* 110 (August 2007): 447–448.

38. See Daniel Carpenter and Dominique A. Tobbell, "Bioequivalence: The Regulatory Career of a Pharmaceutical Concept," *Bulletin of the History of Medicine* 85 (Spring 2011): 93–131.

39. Laura Mamo and Jennifer Ruth Fosket, "Scripting the Body: Pharmaceuticals and the (Re)Making of Menstruation," *Signs: Journal of Women in Culture and Society* 34 (2009): 925–949.

40. Watkins, *On the Pill*, 37.

41. "Top 200 Drugs for 2009 by Units Sold," http://www.drugs.com/top200_units.html; "Top 200 Drugs for 2009 by Sales," http://www.drugs.com/top200.html. Accessed 18 May 2011.

42. See, for example, Suzanne White Junod and Lara Marks, "Women's Trials: The Approval of the First Oral Contraceptive Pill in the United States and Great Britain," *Journal of the History of Medicine* 57 (2002): 117.

43. Mamo and Fosket, "Scripting the Body," 937.

44. Natasha Singer, "A Birth Control Pill That Promised Too Much," *New York Times*, 11 February 2009.

45. For a fuller discussion of this topic, see Elizabeth Siegel Watkins, "How the Pill Became a Lifestyle Drug: The Pharmaceutical Industry and Birth Control in the United States Since 1960," *American Journal of Public Health* 102 (August 2012): 1462–1472.

46. Scrimshaw, "Women and the Pill: From Panacea to Catalyst," 262.

47. Elaine Tyler May, "Promises the Pill Could Never Keep," *New York Times*, 25 April 2010.

48. K. Daniels, W. D. Mosher, and J. Jones, "Contraceptive Methods Women Have Ever Used: United States, 1982–2010," *National Health Statistics Reports*, 2013, no. 62, http://www.cdc.gov/nchs/data/nhsr/nhsr062.pdf, accessed 7 March 2014.

49. United Nations Department of Economic and Social Affairs, Population Division, "World Contraceptive Use 2011." Sterilization and the IUD were even more widely used worldwide than was the pill. Nineteen percent of women ages 15–49 who are married or have a heterosexual partner have been sterilized, for a total of almost 223 million women. This permanent form of contraception is most popular in Latin America, China, India, and Thailand. The IUD is worn by more than 168 million women, with greatest use in eastern, central, and western Asia; northern and eastern Europe; and parts of South America. http://www.un.org/en/development/desa/population/publications/family/contraceptive-wallchart-2011.shtml (accessed 10 November 2013).

50. Kate Grindlay, Bridgit Burns, and Daniel Grossman, "Prescription Requirement and Over-the-Counter Access to Oral Contraceptives: A Global Review," *Contraception* 88 (2013): 91–96, esp. 93–94.

THREE

Magic Bullet in the Head? Psychiatric Revolutions and Their Aftermath

NICOLAS HENCKES

About eight years ago, a new type of drug [i.e. neuroleptics] was joined to the clinician's therapeutic armamentarium, and within a short time this resulted in three major consequences:

1. Pharmacology was faced with the task of determining the mechanisms which were responsible for the new and surprising therapeutic effects of these drugs.

2. Psychiatry found itself in possession of a new pharmacological approach which resulted in a veritable revolution in the treatment of psychotic conditions.

3. Business was presented with a boom in "tranquilizers."[1]

If a person who wanted to reform society through revolutionary social change were to be stricken with schizophrenia or depression, he would be much more likely to overthrow the government if he took chlorpromazine or an antidepressant than if he did not.[2]

These two quotations illustrate two widely divergent yet inseparable narratives of revolution and change in the field of psychotropic drug development in the postwar era. The first one, written in 1960 by the German-born Canadian psychiatrist and pioneering psychopharmacologist Heinz Lehmann, is an early and perceptive reflection on the contribution of neuroleptics[3] to the dramatic transformations occurring at the time in the laboratory, the clinic, and industry, which would give birth to what is now often termed the biomedical complex. The second quote comes from a popular account written in 1983 by Marvin Lickey and Barbara Gordon, two promoters of the neuroleptic revolution, at a time of crisis in confidence in psychotropic drugs. It engages critically the belief, widely held at the time, that madness is political and that mad people should not be considered as sick people but rather as the forerunners of revolutions yet to come. The revolution, in this case, referred to a wider social movement similar to the one that had affected Western societies starting in 1968. In this regard, Lickey and Gordon's message was clear: the creation of psychotropic drugs was not only a revolutionary breakthrough for psychiatry; it also had a potentially much wider significance.[4]

This chapter addresses the evolving, divergent, and at times competing narratives of revolution and counterrevolution in the field of North American and European psychopharmacology and psychiatry at large from the 1950s to the 1980s.[5] Focusing on discursive constructions of change and progress, it locates revolutionary claims about psychotropic drugs within the dynamics of pharmacological innovation and industrial marketing, as well as within larger visions of transforming mental health care and changing societies.

Most mental health professionals acknowledged the revolutionary nature of neuroleptics almost immediately after their introduction to psychiatry in the early 1950s. But stabilizing a consensual interpretation of their contribution to the field soon proved to be much harder. If standardizing psychiatric practices and knowledge seemed to many a solution to this challenge, it also created immense problems in an increasingly differentiated field. These challenges were magnified by the expectations surrounding a discipline that claimed for itself a role in guiding societies through processes of modernization. All this was reflected in the diverse visions of the neuroleptic revolution that became popular from the beginning of the 1960s. By the 1970s, as fears of widespread social control through the means of psychiatric technologies became increasingly expressed, neuroleptics had become the target of divisive conflicts regarding both their effects on patients and their wider uses in the management of vulnerable populations. In the end, the turbulent trajectory of neuroleptics reflected, in many ways, the deep involvement of psychiatry with contemporary social movements.

A key parameter in this analysis is the increasing differentiation of the world of mental health professionals in the postwar era. The emergence of a series of new professions including psychologists, psychoanalysts, psychotherapists, occupational therapists, social workers, and psychiatric nurses turned mental health into a much disputed jurisdiction. Even within the discipline of psychiatry, different subfields began to claim strikingly divergent visions of what their profession was about, how it should be practiced, and how it should evolve. Patient and consumer movements, the emergence of a full contingent of civil rights activists with an interest in psychiatry, and the involvement of feminists and sexual minorities in psychiatric matters soon turned mental health into an overcrowded battlefield. Fomenting revolutions, in this context, seemed a reasonable strategy to gain both an audience and a clientele.

Turning Chlorpromazine into a Revolution

In recent years, historians of psychiatry have begun to question the scope of the neuroleptic revolution. The psychiatrist and historian David Healy has produced a comprehensive account of the development of psychopharmacology as a field from the early 1950s to the 1990s.[6] While he does not contest the revolutionary status of neuroleptics and other psychotropic drugs, he shows that many of the changes they brought about in psychiatry relied on commercial interests and heavy marketing rather than science or an interest for the well-being of psychiatric populations. Taking an even more critical stance, the psychiatrist Joanna Moncrieff argues that the pharmaceutical industry and the psychiatric profession have overhyped the revolutionary basis of psychopharmacological innovation, and suggests that understanding their contribution in more modest terms should lead to more democratic treatment practices.[7]

Other scholars have focused less on the shortcomings of earlier accounts of the neuroleptic revolution and more on the continuities in psychiatric therapeutic practices throughout the twentieth century. An important stream of research thus advocates a longer history of drug use in psychiatry. Sedatives such as chloral hydrates, bromides, and barbiturates were at the origin of a first series of psychopharmacological hypes during the last third of the nineteenth century, and remained in widespread use right towards the end of the twentieth century.[8] As the historian Nicolas Rasmussen has demonstrated, amphetamines were marketed as a specific treatment of depression well before the advent of tricyclic antidepressants.[9] Moreover, several historical studies have shown that older therapies were often complemented rather than replaced by new drugs. Chemotherapy was not easily implemented in many institutions plagued by overcrowding, shortage of staff, and limited funding. Benoît Majerus's thorough examination of patient records at the Institut de psychiatrie in Brussels shows that neuroleptics were not homogeneously disseminated in Belgium, and that various shock techniques continued to be used well into the 1950s and 1960s.[10] In his magisterial history of psychosurgery, Jack Pressman argues that the reason shock therapy was abandoned was not because it was less effective and regarded as ethically more questionable than drugs, but rather because it no longer compared well to them in the new understanding of therapy that had emerged over time.[11]

However compelling these arguments, it remains important to take into account the widespread sentiment, already expressed within months of the initial description of the psychiatric effects of chlorpromazine, that this drug would be of tremendous importance for psychiatry. The processes that led to the discovery of chlorpromazine as a psychiatric drug are well known.[12] A derivative of the chemical compound phenothiazine, chlorpromazine had been synthesized in 1950 by the French drug company Rhône-Poulenc and introduced into psychiatry by the French military surgeon Henri Laborit. It was probably the Parisian professor of psychiatry Jean Delay, one of the most respected international authorities in the field, who with his assistant Pierre Deniker contributed most to launching the career of chlorpromazine in psychiatry. In the second half of 1952, Delay and Deniker began to report systematically on the drug's effects on psychiatric patients in a series of publications in French journals. The reason why chlorpromazine was remarkable was that, unlike earlier sedatives used in psychiatric hospitals, it had an effect on delusions, hallucination, and mental confusion without inducing sleep. Moreover, its action on an impressively wide array of symptoms made it a choice treatment for a variety of psychiatric conditions, from schizophrenia to chronic delusions to mania.

Within months, chlorpromazine was made available to French neuropsychiatrists and trials were organized in other countries, including the United States in 1953, while clinicians and industry scientists began systematic testing of other compounds with similar chemical properties in the hope of enlarging the armamentarium. In 1955, the first major conference on neuroleptics was organized by Delay and Deniker in Paris, gathering more than four hundred participants from twenty-two countries and demonstrating the worldwide enthusiasm surrounding the discovery. For better or for worse, the narrative of neuroleptic revolution was well on its way by the end of the decade.

Before proceeding further with this story, let us reflect on the constellation of governmental, industrial, and clinical interests that enabled the spread of neuroleptics and their revolutionary status. Pharmaceutical companies clearly played a central role in shaping the perceptions of both psychiatrists and the general public of the neuroleptic revolution from the early days of the commercialization of chlorpromazine. Available evidence suggests, however, that this shaping occurred in diverse ways in different countries. The marketing of chlorpromazine first targeted hospital psychiatry.[13] Since psychiatric hospitals in most countries were funded by the state, this meant that marketers needed to convince both the physicians who prescribed the drug in institutions and the hospital administrators who paid for it. The strategy chosen in various countries thus reflected the balance of power between the two groups and also rested on the relationship between them and the pharmaceutical company. In the United States, chlorpromazine was marketed by Smith, Kline and French (SK&F) as a "major tranquilizer" for treating agitation in institutionalized patients. Sales representatives set out to convince all state governments that they ought to increase funding for therapy in psychiatric hospitals; they also worked with clinicians to improve their work conditions. Historian Judith Swazey quotes former officials of the company describing these efforts as "not lobbying per se," but rather "a true educative effort." But this account probably underplays other, more commercial strategies used by SK&F, including communications in medical journals and mainstream magazines.[14]

On the western side of the European continent, Rhône-Poulenc and its international branch, Specia, do not seem to have expended the same amount of effort. In France, Rhône-Poulenc did not organize trials with clinicians. It distributed free samples to psychiatrists in the hope that they would adopt the drug. Then it distributed doses on demand to hospitals.[15] The company's sales division also produced a leaflet, distributed by sales representatives, describing the wide spectrum of the drug's effects and its interest for several medical specialties. Advertising presence in medical journals was modest, at least during the 1950s. Perhaps the reason why Rhône-Poulenc did not sustain greater promotional efforts in France relates to the small, homogeneous, and centralized milieu of hospital psychiatrists in that country. Rhône-Poulenc also worked closely with state laboratories and clinicians, and may have sought to preserve its standing as a scientific enterprise.

Specia seems to have had a similar strategy in the Netherlands and Belgium. In these two countries, however, as noted by Toine Pieters and Benoît Majerus, the introduction of neuroleptics was delayed in several important institutions, and discrepancies in their use developed over time.[16] Specia did not try to homogenize local practices, but rather embraced those differences by providing personalized dosages. As a team of German medical historians led by Volker Hess has shown, marketing played out differently in the centralized system of production and distribution of pharmaceuticals in the German Democratic Republic (East Germany).[17] These examples suggest how marketing strategies may have, from the outset, engendered quite different local understandings of the chlorpromazine revolution.

However, what soon proved to be common to these various local stories was a dramatic shift in the understanding of how neuroleptics worked during the first decade after their discovery.[18] In the early 1950s, most pioneering psychopharmacologists shared a holistic vision of neuroleptics. Building on a style of reasoning that had been developed in the interwar period and put to work for shock treatment, they thought that chlorpromazine and other drugs with similar properties worked by modifying the regulatory system of the organism overall. After other tentative labels, the term *"neuroleptique"* was chosen by Jean Delay and Pierre Deniker to designate chlorpromazine in 1955 to reference the ways in which the drug was supposed to "grasp" the nervous system.[19]

The psychological effects of the drug, which were characterized without reference to any specific condition, derived not only from the wider impact of these biological phenomena, but also from the very act of administering the "neuroleptic cure." So did a series of sociological effects. What was revolutionary in neuroleptics was not only their stunning effects on patients, but also the ways in which they helped transform the perception of the psychiatric hospital as a truly therapeutic place. Chlorpromazine gave mental health professionals a new role and generated new kinds of relationships, both among professionals and between professionals and patients. A key role in shaping this understanding was played by sociologists and social scientists who had devoted considerable efforts to analyzing hospitals as small communities during the 1950s. Most notably, this account held little room for the idea that some neurological effects of the drugs might in fact be "side effects." Indeed, most early promoters of neuroleptic chemotherapy, including Delay and Deniker, seemed to believe that the neurological effects of the compound were necessary for the drug to exert its psychological effects, as was a controlled milieu.

By the mid-1960s, holistic approaches to neuroleptics had receded and to a large extent had given way to more specific materialist accounts of how they worked. The hypothesis that neuroleptics acted at a molecular level on a brain mechanism underlying a specific disorder, namely schizophrenia, began to gain ground and eventually replaced earlier concepts. A turning point in this process was the multicentered study conducted by the National Institute of Mental Health in the early 1960s that used, for the first time, a battery of standardized diagnostic scales to assess the effectiveness of three neuroleptics.[20] These drugs appeared to have such a dramatic impact on core schizophrenic symptoms that investigators concluded, "Almost all symptoms and manifestations characteristic of schizophrenic psychoses improved with drug therapy, suggesting that the phenothiazines should be regarded as 'antischizophrenic' in the broad sense."[21] In the following years, the idea that neuroleptics were a specific medicine for schizophrenia was strengthened by the hypothesis that they acted on the brain by modifying the balance of a specific neurotransmitter, namely dopamine, and that the onset of schizophrenia might be related to this phenomenon.[22] For the next two decades, the "dopamine hypothesis" would be the leading neuroanatomical model for explaining the cause of schizophrenia. From the statistical perspective of psychiatric epidemiology, the accelerated discharge of hospitalized patients as part of the process of "deinstitutionalization" seemed to make the efficacy of these medications self-evident. Eventually, the transmutation of neuroleptics was made complete by a change in nomenclature. Beginning in the 1970s, the term "antipsychotics" began to be used as a substitute for "neuroleptics" in the United States, and by the 1990s it had largely replaced the original characterization in most countries.

Early appraisals of the revolutionary nature of neuroleptics focused to a large extent on the change they generated within hospitals. By the 1960s, however, narratives of the neuroleptic revolution underscored the wider change in perspective brought about within the psychiatric profession at large. Neuroleptics and other psychotropic drugs had succeeded in bringing about a completely new way of conceptualizing psychiatry as both a practice and a science. A commentary published in 1964 in the *American Journal of Psychiatry* reflecting on "the current psychiatric revolution" waxed eloquent on the changing status of the discipline and its novel association to medicine.[23] Similar statements were made in France and in Germany. By the end of the 1960s, such perspectives had coalesced into the notion that a new medical model of psychiatry was coming of age.[24] The neuroleptic revolution was a revolution for psychiatry as both a practice and a discipline—a psychiatric revolution indeed.

Revolutionary Standards

The recognition that chlorpromazine had brought about a revolution in psychiatry still left the meaning of this revolution as an open question.[25] For example, there was nothing self-evident in how the emerging standard accounts insisted on both the specific action of the drug on schizophrenia and its role in the deinstitutionalization process. Not only were both phenomena disputable, as generations of critics have claimed, but it can be argued that they only made sense within a framework for evaluating psychiatric practices that was largely created at the same time as neuroleptics themselves. Beginning in the late 1950s, what neuroleptics were good for and what they meant began to be understood within a series of new infrastructures for organizing and evaluating psychiatric practices. These infrastructures included classifications, psychopathological and psychometric scales, databases, and trials and involved all aspects of psychiatric work, from diagnosis to prescription to policy making. Their creation, in turn, was the result of a complex dynamic of innovation and standardization processes occurring in the clinic and in the industry, as well as in the administration of welfare and social services. In the end, the very idea of a neuroleptic revolution would be inseparable from a wider transformation in psychiatry through the standardization of knowledge and practice.

Critical voices have pointed to the role of the pharmaceutical industry in shaping these transformative processes, suggesting that "Big Pharma" was allowed to set the very standards it then used to evaluate its own success. It is true that standardization clearly developed into a key battleground for the interpretation of the neuroleptic revolution. Psychopharmacology as both a scientific field and an industrial venture played an important role in the setting of a wide array of influential psychiatric standards, which in turn also shaped in decisive ways how psychopharmaceuticals should be understood. Nonetheless, accounting for the phenomenon in all its dimensions requires a broader perspective. Beyond psychopharmacology, the impulse for standardization in psychiatry came from complex interactions between scientific and professional interests, the industry and marketing practices, and social movements and politics. These various forces played out differently in different contexts, resulting in distinct local configurations. While the standardization of psychiatric practices was certainly a universal phenomenon, locally it affected the various dimensions of psychiatric work in assorted ways, leading to the paradox of standardization shaping diverse local conceptions of the neuroleptic revolution.

The field of diagnosis illustrated the give and take of these processes. Indeed, beginning in the early 1960s, the accumulation of standards for psychiatric diagnosis could be considered a revolution in itself. At least this is how the most iconic of these, namely the third edition of the Diagnostic and Statistical Manual (DSM-III) of the American Psychiatric Association (APA), was hailed by both its promoters and its critics upon its publication in 1980.[26] However, the sensation over the DSM-III and its influence in American psychiatry and beyond has overshadowed the significance of other less discussed but widely influential instruments also developed in the 1960s and 1970s.

German psychiatrists, for example, had created their own standardized schedule for collecting psychiatric data. The Working Group on Methods and Documentation in Psychiatry (Arbeitsgemeinschaft für Methodik und Dokumentation in der Psychiatrie) released the so-called AMP system in the early 1960s, which was implemented in most German clinics by the end of the decade.[27] Also in the 1960s, the World Health Organization (WHO) devoted considerable efforts, with decisive input from British psychiatrists, to develop a classification schedule that could be used by psychiatrists all over the world within the framework of the International Classification of Disease.[28] In cooperation with psychiatrists from the US National Institute of Mental Health (NIMH), WHO also created a series of new standards for the diagnosis of schizophrenia that would contribute to a profound reformulation of the definition of this disease. Many other influential standardized diagnostic scales were developed by individuals or groups of clinicians during the same years, so much so that by the 1980s a plethora of instruments was circulating in the field, at the cost of some confusion when clinicians had to choose from among these different tools for assessing the same conditions.

Psychiatry as a whole did not immediately embrace diagnostic standardization. The initial resistance to DSM-III within the American mental health community has been well described. However, few critiques really disagreed with the ultimate goal of achieving more reliable diagnostic practices.[29] Some national communities developed more idiosyncratic opposition to diagnostic standardization. For decades French psychiatry continued to be characterized by a form of defiance toward standardized instruments in clinical work, to the point that even psychiatrists who otherwise defended a medical and biological vision of their discipline were reluctant to use them.[30] Significantly, although Delay and Deniker would later call for the development of standardized diagnostic instruments and play an important role in their introduction into French psychiatry, they advocated a clinical assessment of neuroleptics in place of randomized clinical trials in their celebrated 1961 handbook of psychopharmacology, at a time when clinical trials were becoming a standard procedure in English-speaking countries.[31]

France was nonetheless a notable exception and, by the mid-1960s, diagnostic standards had become an essential ingredient in psychiatry in general and in the development of psychopharmacology in particular. The regulation of drug marketing and approval in most countries during the 1960s and 1970s made randomization and the use of standardized diagnostic methods a basic requirement of sound trial methodology.[32] Standardization of diagnostic practices also ranked high on the agenda of psychopharmacologists.[33] The credibility of psychopharmacological research required that clinicians working in different settings give similar diagnoses for similar clinical presentations. At a time when almost every clinician might have had his or her own diagnostic idiosyncrasies, achieving reliable diagnoses between raters was no small feat (measures of "inter-rater reliability" would later be quantified as the "kappa score").

Other branches of psychiatric research shared similar concerns with psychopharmacologists regarding diagnostic reliability. In the United States, psychometrics developed into a major research program at the NIMH immediately following its establishment in 1948.[34] Standardizing diagnosis soon became a priority for American psychiatric practice as well, as several influential and widely publicized studies showed high levels of inconsistency in diagnosis practices by the 1970s.[35] Standardizing diagnosis had become both a solution to the fragmentation of the profession and an answer to the widespread critique that psychiatry lacked a scientific foundation.[36] These factors gave American psychiatry a definitive position of leadership in the development of standardized diagnostic methods, followed closely by the British and German-speaking mental health communities.

Accounting for the exact role of diagnostic standards in changing the understanding of neuroleptics is not an easy endeavor, however. A widespread critique of the pharmaceutical industry has been that diagnostic standards have been a major vehicle for its influence over psychiatry. Classifications and diagnostic scales would have been tailored to demonstrate the superior effectiveness of drugs on given mental disorders.[37] There is no doubt that the pharmaceutical industry helped set and disseminate a number of standards. In Germany, the AMP system began as a collaborative project between German university psychiatrists and the Swiss pharmaceutical industry. In the United States, however, the influence of the pharmaceutical industry over experts participating in DSM committees is only—and perhaps can only be—a supposition. But these critiques do not account for the reasons why practitioners used these scales.

As is demonstrated by the history of the Hamilton scale for depression, the fact that a specific scale developed into a standard for both the industry and the profession was due more often than not to a Darwinian-like process of selection in which experts chose from multiple instruments the one that best matched their needs.[38] On a wider spectrum, the fact that sets of standards had become essential both to the assessment of the effectiveness of given drugs and to clinical practices was due to a series of transformations occurring simultaneously within the pharmaceutical industry and psychiatry, eventually aligning research, marketing, and clinical practices. Research by the historians Lucie Gerber and Jean-Paul Gaudillière on the development of antidepressants by the Swiss firm Ciba-Geigy shows how that company's chain of production and marketing was reorganized at the end of the 1960s with the systematic introduction of animal models and standardized psychopathological testing, not only to screen molecules but also to organize markets.[39] The company's success in selling its products was also predicated on the emergence of a new group of prescribers in need of guidelines, namely general practitioners.

While there are still no archivally based studies of the history of neuroleptics in the 1970s and 1980s, it is clear that the story of this class of psychopharmaceuticals departed significantly from that of the antidepressants. Achieving diagnostic specificity in the field of schizophrenia proved to be a daunting challenge. Neuroleptics as antischizophrenics never achieved full acceptance, and psychiatrists continued to prescribe them for a variety of other conditions. A dimension of the problem came from the increasingly influential idea that the label "schizophrenia" might in fact refer to several clinical syndromes that probably did not share any pathophysiological correlates. From the late 1970s on, the selective effects of neuroleptics on certain clinical presentations helped strengthen this approach. For instance, the concept that schizophrenia might be broken down into two syndromes, positive and negative, was shaped in crucial ways by the notion that positive schizophrenia was affected by neuroleptics while negative schizophrenia was not.[40] Nonetheless, the significance of neuroleptics for the diagnosis of psychosis has remained in dispute to the present. In many ways, neuroleptics have not found their standards.

The battle over diagnostic standardization may have been matched in intensity by the one over mental health policy. Mental health reform was certainly a central issue in most, if not all, countries from the early 1900s. During the first half of the century, this took different forms under divergent national psychiatric and political traditions, even though some approaches were circulating across national boundaries. A major development of the 1950s and 1960s was the emergence of deinstitutionalization as perhaps the universal standard for framing mental health policy. Once again, American psychiatry played a leading role in this development.[41] The idea that care for chronic patients could be organized outside mental hospitals and within communities was put forth by the influential 1961 congressional report of the Joint Commission on Mental Illness and Health.[42] Two years later, the launching of the federal Community Mental Health Centers program seemed to offer a plausible alternative to psychiatric hospitalization in delivering care to long-term mental patients. In the next few years, what was then called "deinstitutionalization" turned into a genuine social movement. A significant segment of the psychiatric profession had enthusiastically endorsed community psychiatry, and activists set out to remove patients from institutions by juridical means borrowed from the struggle for civil rights. By the 1970s, following the example of the Reagan administration in California, deinstitutionalization had also become a way to control costs, if not to downsize social services. Deinstitutionalization expanded beyond the field of psychiatry to become a trend in other sectors, such as criminal offense and disability. As Canadian, British, Italian, and a few other European mental health policies followed the American trajectory—and in some cases, such as Italy, took an even more radical approach to closing mental hospitals—and as the WHO also supported the concept, deinstitutionalization seemed to develop into something of a new international standard.[43]

Not all countries adhered to this standard. Again, France was a notable exception to the trend of downsizing psychiatric hospitals.[44] What was seen by many as the French version of community psychiatry, namely the *"politique de secteur"* launched in 1960, did not envision a reduction in psychiatric hospitalization. Rather, it was intended to establish coordination between the numerous institutions in the mental health field to facilitate the transfer of patients from one to another when needed and to avoid the abandonment of patients in understaffed remote hospitals. French mental health policy also included the largest plan to date for construction of psychiatric beds to relieve overcrowding in psychiatric hospitals. A decrease in the population of psychiatric institutions after 1967 was barely anticipated by psychiatrists and health officials, and it was not until the late 1970s that the French Ministry of Health set a reduced number of psychiatric beds as a goal for mental health policy—much to the dismay, at the time, of most psychiatrists, including those who advocated reform of their institutions. Even then, debates over deinstitutionalization did not achieve the same level of popularity as in the United States and Great Britain.[45]

A key indicator in mental health policy debates in many countries was the number of beds in psychiatric hospitals. While there was a long tradition of discussing and comparing hospital statistics, their use as an instrument for policy making was a relatively new development in the postwar period. Although hospital statistics were more performative in a centralized country such as France, where five-year plans set quantified objectives for the construction and renovation of hospitals, they were also widely circulated at every level of the psychiatric systems of other countries. Internationally, standards for the optimal number of psychiatric beds had been set by WHO publications during the 1950s.[46] Bed numbers were a simple and telling measure of the conditions of the delivery of care to psychiatric patients that could be compared across widely divergent contexts. Yet they were a poor yardstick. They did not say much about the way these beds were distributed in the different regions. Nor did they say anything about the amount of care that was actually given to patients in the institutions. They were also silent about patients' conditions outside the institutions. For these reasons, interpretations of these figures tended to be hotly contested.

The contested role of neuroleptics as a cause of deinstitutionalization was central to these discussions. The idea that chlorpromazine was a cause in the reduction of psychiatric hospitalizations was put forth as early as 1957 by the American psychiatrist Henry Brill, who set out to demonstrate the process with statistical precision for the state of New York.[47] By the 1970s this idea had become a central tenet of standard accounts of the neuroleptic revolution. It also had become a highly debated issue in policy circles, as well as a controversial topic for critics of biological psychiatry. Reservations about Brill's conclusions had been expressed ever since his early publications. The British psychiatrist Sir

Aubrey Lewis discussed in 1958 the respective roles of drugs and psychosocial treatments in the decline of hospitalization in Britain. He argued, "Certainly if we had to choose between abandoning the use of all the new psychotropic drugs and abandoning the Industrial Resettlement Units and other social facilities available to us, there would be no hesitation about the choice." Drugs were dispensable, not social psychiatry.[48] Such comments would grow stronger over time.[49] Increasing critiques of psychiatric hospitals as "total institutions" produced the impression that there was a genuine social movement behind the decline of mental institutions. Similarly, changing welfare policies and reimbursement schemes as well as rising neoliberal justifications for rescaling social policies developed into influential explanations of the phenomenon.[50] Finely grained analyses of hospital demography also tended to suggest that the downsizing of psychiatric hospitals owed much to the transfer of certain segments of their population, including older and mentally handicapped patients, in the context of the development of new social policies for those populations.[51] By the end of the 1970s, the search for the causes of deinstitutionalization had become a key battleground for competing visions of psychiatry.

Competing Revolutions

The number of controversies over these issues suggests that more was at stake than merely an appreciation of the true merits of neuroleptics. The debate over neuroleptics only made sense within a broader, though unevenly shared and differently interpreted, understanding that psychiatry was indeed undergoing a revolution. The popularity of the revolutionary imagery was at once a striking and relatively new dimension of postwar psychiatric discourse. It reflected a widespread sense that psychiatry was in the midst of major transformations and that psychiatrists could play a role in guiding these transformations and give them larger significance. This revolutionary rhetoric also encompassed antagonist accounts of progress and change among different professional groups and national communities. These differences reflected commitments to different visions of psychiatry as a practice and a science, as well as different understandings of change as both process and objective.

In this respect, the revolutionary rhetoric was neither universal nor obvious. This was well illustrated by the British case. Skepticism regarding chemotherapy could be expected from a champion of psychosocial treatment such as Lewis. The reluctance of William Sargant, a noted pioneer in biological treatment in Britain, to endorse the enthusiasm of his American colleagues toward chlorpromazine might seem less self-evident.[52] In fact, the stance taken by British psychiatric elites on psychopharmacological innovation reflected a pragmatic attitude toward treatment that had made Great Britain a pioneer in clinical trials and clinical epidemiology. It was also predicated on a commitment to a realist philosophy of history, perhaps best expressed in 1968 by the Birmingham professor of psychiatry William Trethowan in his review of an American textbook on the history of psychiatry: "Despite what some may claim, there has been no really deep penetration at any point, and no major breakthrough, but steady wide pressure towards solving a number of problems. In the same vein, although it is often repeated that we are in the throes of a psychiatric revolution, it is likely that every generation of enthusiasts feels the same way. The word evolution may perhaps be preferred."[53]

In contrast, true believers in the psychiatric revolution were more numerous in France and the United States. In these countries, the psychiatric revolution developed into a political and moral concept. It referred not only to the need for change in psychiatry, but also to perspectives for social change that resonated with other social movements. In this respect, the psychiatric revolution was a project, a worldview, and a calling all at once. It also meant markedly different things in each country, reflecting different political and therapeutic cultures, and thus established a different framework for appreciation of the neuroleptic revolution.

French alienists had certainly held a measured attitude toward change throughout the early decades of the twentieth century, although some ardent reformers had come from their ranks starting in the late 1890s. In the interwar period, the leader of the Association of Asylum Psychiatrists (Association amicale des aliénistes) described the views of his colleague Edouard Toulouse on psychiatric reform as "revolutionary"—a term not intended as a compliment.[54] Just a few years later, in the wake of the liberation of France from German occupation, a new generation of young psychiatrists organized a meeting in 1945 that would be characterized by its organizers as the ferment of a "psychiatric revolution."[55] It gathered asylum psychiatrists from all over the country and resulted in a draft for a new mental health law, as well as a twenty-four-point charter described by an organizer as "a sort of Tennis Court Oath of the psychiatric revolution we are dreaming of."[56] The draft was not discussed outside psychiatric circles, but from then on, this event would be recalled by psychiatrist reformers and their followers as "the psychiatric revolution of 1945."[57] What was even more revolutionary than the legislative draft itself was the attempt by this small band of psychiatrists to shape their destiny and to inspire social change. They sought nothing less than a social movement.

Revolutionary imagery would remain a core dimension of the worldview of a significant part of French psychiatry for the next fifty years. Not only did it underpin psychiatrists' reform projects for both their discipline and society at large, it was also completely integrated into their very concept of therapy. By the late 1940s, the psychiatric revolutionaries of 1945 had developed a new approach to institutional treatment, which they labeled "institutional psychotherapy."[58] Influenced by American and British wartime research on group dynamics and therapy, institutional psychotherapy entailed the introduction into hospitals of occupational and leisure activities for the rehabilitation of patients. More profoundly, though, institutional psychotherapy was thought of as a technique to create momentum within the institutions. It was based on a series of motivational techniques aimed at stimulating hospital personnel to foment what some called an "internal revolution" in the wards.[59] Therapy, in this regard, coincided with a form of social change, albeit restricted to institutions.

French psychiatric revolutionaries were not laying out a grand project for postwar French society. In fact, most of them were wary of a psychiatrization of society that could be co-opted by conservative interests. By the 1960s, however, as a theory of social change, institutional psychotherapy had become a highly influential doctrine. It did so among a wide range of intellectuals, professionals, and activists in fields such as education, political science, sociology, and disability studies. Psychiatrists now found themselves at the vanguard of both the postwar modernization movement and the May 1968 revolution.[60] Indeed, the French psychiatric revolution seemed to resonate with every social movement of postwar French society. All the same, advocates of institutional psychotherapy did not see in neuroleptics an ally for their revolutionary endeavors. Institutional psychotherapists were puzzled by the ubiquity of neuroleptics in the psychiatric system by the end of the 1960s. Contrary to the dramatic ceremony of shock treatment, which had played a major role in early institutional psychotherapy practices, the more banal distribution of pills did little to display psychiatric charisma to patients and nurses. Rather, as one French psychiatrist wrote, "the virtue of therapy had progressively faded," and psychiatry seemed to have lost its therapeutic outlook.[61] In addition, the science behind the drug revolution was being developed far from psychiatric hospitals, in university clinics and labs, largely without the participation of hospital psychiatrists.[62] In the end, the relationship between the promoters of the neuroleptic revolution and those of the psychiatric revolution in France would be built on mutual ignorance.

On the other side of the Atlantic, the American approach to the psychiatric revolution seems at first glance to have had features strikingly similar to those of the French situation.[63] In parallel fashion, a generation of "Young Turks" took advantage of the changing climate of the immediate postwar period to take over leadership in the profession. They formed the Group for the Advancement of Psychiatry (GAP) and developed a comprehensive vision for how psychiatry should evolve, which in many respects served as a blueprint for the Mental Health Act of 1963 and the launching of community psychiatry. However, these psychiatrists thought of their endeavor as a renaissance rather than a revolution. In fact, revolutionary rhetoric appears to have first flourished outside the ranks of the GAP.

For most of the postwar era, American psychiatry's approach to psychiatric revolutions was framed by the humanistic account of the history of psychiatry published in 1941 by the Russian-born psychiatrist and psychoanalyst Gregory Zilboorg.[64] Zilboorg identified two revolutions in the history of psychiatry, which had resulted in a new understanding of man and a deeper integration of madness as an irremovable dimension of humanity. The first had occurred in the sixteenth century under the impetus of the Renaissance protagonists Juan Luis Vives, Paracelsus, Agrippa, Weyer, and Jean Bodin, whose philosophical writings helped eliminate the practice of burning mad people as witches. The second psychiatric revolution coincided with Sigmund Freud's discovery of the unconscious at the turn of the twentieth century. Zilboorg celebrated Freud's revolutionary breaches in therapy, which stood in sharp contrast to the therapeutic nihilism of his time, and called Freud "the first humanist in clinical psychology." Even more so than psychoanalytic psychotherapy, it was "the principle of psychological determinism" that was truly revolutionary in inspiring a more comprehensive science of man.[65]

Writing just two years after Freud's death, Zilboorg implied that American psychiatry was in the midst of its Freudian revolution and that its full consequences had yet to come. But American psychiatry did not have to wait long for the emergence of a new generation of visionaries prophesying the coming of a third psychiatric revolution. In 1952, the Austrian-born educator and group therapist Jakob Moreno did not fear to claim at the first conference on group psychotherapy that the creation of this technique was an event of the same importance as those that had constituted the first two psychiatric revolutions. The idea was further developed by his followers and remained an important discursive theme in this group for decades.[66] In a characteristic statement in 1966, Moreno predicted the glorious advent of a new society as a result of the dissemination of the technique he had helped to invent:

> While the changes brought about by the First Revolution were institutional, and those by the Second psychodynamic, the changes brought about by the Third Revolution are due to the influence of cosmic and social forces. They are further transforming and enlarging the scope of psychiatry. . . . Their ultimate goal is a therapeutic society, a therapeutic world order which I envisioned in the opening sentence of my opus *Who shall Survive?*, [. . .] 'A truly therapeutic procedure cannot have less an objective than the whole of mankind.'[67]

Moreno's grandiloquence was certainly an expression of his somewhat inflated ego, but it reflected a perspective that was increasingly influential in postwar American society. As historians of psychology have shown, the contribution of psychologists and psychiatrists to the war effort, both within intelligence services and in managing the health of combat forces, had earned them the trust of a wide range of government officials, policy makers, and philanthropists and had helped make them one of the most influential professions of the Cold War period.[68] Psychologists and psychiatrists were not only selling their services to an ever-increasing number of individuals in search of mental well-being. Their analyses were also serving to justify decisions on a broad range of geopolitical, family, public administration, and management issues. Their greatest achievement, however, may have been in convincing a wide range of stakeholders that psychiatric expertise might bring about a new concept of citizenship based on democratic participation, promotion of the individual, and the management of antisocial impulses. The psychiatric revolution was to be a radical transformation of American society, a democratic feat indeed. Critics had no way to refute this vision. In his celebrated essay "The Triumph of the Therapeutic," the sociologist Philip Rieff was left to wonder what concept of culture and what kind of institutions were emerging from the hegemony of the therapeutic enterprise—but even he could do little to offer an alternative.[69]

By the 1960s, the narrative of the third psychiatric revolution had become increasingly popular. At the same time, what was meant by a "revolution" had shifted away from the humanistic perspectives promoted by Zilboorg and more toward a positivistic idea of therapeutic progress. Accordingly, the first revolution was now attributed to reformers of the early nineteenth century, including the British philanthropist William Tuke and the French physician Philippe Pinel, who had invented moral treatment and helped to develop asylum psychiatry. In the early 1960s in the United States, the nascent group of community psychiatrists adopted the narrative of the "third psychiatric revolution" as an appropriate way to pitch the innovative ways of practicing psychiatry that were emerging in community mental health centers set up by the federal government.[70] Psychopharmacology supporters also soon embraced the narrative, so that by the end of the 1960s, mental health had become the playing field of an out-and-out competition between revolutions. A psychiatrist writing in the late 1960s probably thought he would put an end to the dispute by suggesting that the third revolution had been underpinned by psychotropic drugs while community psychiatry had simply inspired the fourth.[71]

Unlike in France, the neuroleptic revolution in America was completely integrated into the psychiatric revolution.[72] Beginning in the 1950s, tranquilizers became a crucial element of psychiatrists' and psychologists' therapeutic armamentarium, and a key determinant of their success. Freudianism and the therapeutic ethos were not refuted, but merely retuned by the pharmaceutical industry to promote their drugs. This also meant that consumerism and marketing, rather than citizenship, characterized the psychological culture of the Cold War period. This tension between American psychiatric revolutions would soon catalyze a reversal of opinions.

The Bitter Fruit of Revolutions

In 1977 the psychiatrist Gerald Klerman, then head of the US federal Alcohol, Drug Abuse, and Mental Health Administration, concluded an uncompromising review on deinstitutionalization in the United States and Europe on a rather grim note:

The fear is that drugs and other behavior control technologies, if not controlled and regulated, combined with the anomie and isolation of urban life, will convert our communities into the ultimate total institution, a totalitarian society. Thus, we are faced with the visions—or nightmares—of 1984 and *A Clockwork Orange*. The dilemma is that without new technologies, long-term changes in the mental health system are unlikely, and the creation of new community alternatives will depend upon the availability of new technologies. Thus, the issue of community treatment of the mentally ill is not only scientific and professional, but also social, ethical, and political in the broadest and most humane sense of those terms.[73]

The next chapter of the revolution is too well known. By the mid-1970s, the possibility that psychiatric revolutions might not liberate patients but, on the contrary, give birth to a nightmarish dystopia of social control had become a widespread concern in the mental health professions and Western societies at large. The specific idea that social control was taking new forms in contemporary societies as community treatment and other technologies for controlling deviant people were replacing former practices of institutionalization was theorized on the European continent by scholars inspired by the work of the French philosopher Michel Foucault.[74] The fact that psychiatric and psychological technologies, including drugs, operant conditioning, lobotomy, and electroconvulsive therapy, had become ubiquitous and might serve authoritarian projects became an even more widespread concern. In the United States, the possible misuse of a wide range of "behavior control technologies" thus became a key focus in the emerging field of medical bioethics.[75] During the 1970s, the question was also increasingly debated in many Western countries as the use of psychiatry to repress dissidents in the Soviet Union and experiments in brain washing by intelligence services became known.

In many ways, this emerging scenario of social control was simply an extension of the idea in psychiatric thinking from the postwar period that therapy was political, and that psychiatry in particular and mental health disciplines in general could play a role in creating a "therapeutic state."[76] What had seemed a rather comforting perspective for a society recovering from total war and entering a new era of well-being and consumerism appeared to be far less captivating three decades later. Concerns about psychiatrists' intentions led to a new climate of social critique and scientific skepticism.

By the early 1980s, an even more bitter perspective had come to pass. The technologies behind the psychiatric revolutions might in fact not be able to control much of anything beyond the noisiest manifestations of psychopathologies. Moreover, their shortcomings created new, more intractable forms of distress among the people they were supposed to serve most: psychiatric patients.[77] Unattainable cures and disabling side effects, lack of funding for psychiatric services, enduring stigmatization of patients and former patients, and poor recognition of their suffering were creating homelessness, poverty, and disability rather than empowerment and participation. The very foundations of both deinstitutionalization and the neuroleptic revolution itself were thus called into question.

Notwithstanding all their unfulfilled promises, the ideals of the psychiatric revolutions remained the only horizon for most protagonists of this unfolding drama. For the pharmaceutical industry the stakes were particularly high. Its interests had clearly played a major role in building the consensus on drugs. Even so, psychiatrists and patients willing to opt out of drug treatment were left with few therapeutic alternatives. The mental health community faced a new dilemma: acknowledging the harm created by neuroleptics without imagining another path to progress.

Again, these perspectives were not universally shared either internationally or within national borders. In the 1970s, in many countries such as France, Germany, and Great Britain, a number of groups emerged that were critical of mainstream psychiatry and eager to develop alternative ways of treating mental patients. Most of these groups, however, consisted of mental health professionals, often psychiatrists, whose radical solutions to the enduring mistreatment of mental patients only rehashed the earlier revolutionary rhetoric conceived by their forerunners. Much has been written on the "antipsychiatry" treatises of influential thinkers such as the American psychoanalyst and psychiatrist Thomas Szasz, the American sociologist Erving Goffmann, the Scottish psychiatrist R. D. Laing, and the Italian psychiatrist Franco Basaglia. Yet even these arguments would have sounded familiar just two decades earlier.[78] Psychiatrists might well have thought that antipsychiatry had become a genuine social movement, but in most countries this movement did not get much support from outside the mental health world.

Developments were more contentious in the United States. The very possibility that physicians contributed to the climate of denial and understatement surrounding the overuse and toxic effects of many drugs became central to the critique of medical power in the late 1970s. The crisis of minor tranquilizers and the politicization of LSD consumption were instances of a profound reversal of perspective on drugs that were once regarded as being as miraculous as neuroleptics.[79] The neuroleptics did not suffer from this dramatic change of mood. Nonetheless, by the mid-1970s, the long-term toxic effects of antipsychotic drugs had become a major source of concern among the psychiatric profession, health authorities, and pharmaceutical companies. The crisis was triggered by the gradual recognition of the severity and widespread character of disabling long-term effects known as tardive dyskinesia.[80] Several lawsuits were filed against companies starting in 1974, and after a period of denial, the American Psychiatric Association (APA) was eventually forced to issue a letter recommending a thorough assessment of the risk-versus-benefit balance before beginning long-term treatment.[81]

By 1980 a number of commentators had described the attitude of the psychiatric profession to tardive dyskinesia as a form of "panic."[82] Much more disturbing, however, were the cases of treatment refusal successfully brought to the courts by patients and civil rights activists in the second half of the 1970s. The fact that patients made use of their agency against the treatments that were supposed to restore this agency provoked true shock among psychiatrists. Over time, the impetus given by these cases to the nascent movement of "psychiatric survivors" also created unease among the professionals who were the target of this movement. The most momentous of these cases was a suit filed by patients from the Boston State Hospital with the help of a social worker in 1977 and won in 1979. The court recognized the patients' right to refuse treatment in cases other than an "emergency," for which it gave a restrictive definition. Psychiatrists were especially disturbed in that the court's decision referenced the most fundamental constitutional right: freedom of speech. As in other cases of treatment refusal in other medical specialties, the judge mentioned the right to privacy and to make decisions significant for oneself. He also based the decision on the First Amendment, and argued that forced prescription of a psychotropic drug would breach the fundamental right to produce a thought.[83] "Whatever powers the Constitution has granted our government," he wrote, "involuntary mind control is not one of them, absent extraordinary circumstances. The fact that mind control takes place in a mental institution in the form of medically sound treatment of mental disease is not, itself, an extraordinary circumstance warranting an unsanctioned intrusion on the integrity of a human being."[84]

These very terms produced an upsurge of protest in American psychiatry. In subsequent years, several other American judicial decisions recognized that competent patients had the right to refuse treatment, even when they had been involuntarily committed into mental health facilities.[85] Psychiatrists prophesied that they would no longer be able to take a therapeutic stance and would have to care for a growing group of patients refusing medication who would, they claimed, "rot on their feet" in psychiatric institutions.[86] These perspectives may have been overstated, but as the former APA president and medicolegal expert Paul Appelbaum noted, this moral panic among American psychiatrists revealed that they were uncomfortable with their own argument that drugs were as effective as they were claimed to be.[87] Appelbaum's recommendation, a decade after the Boston case, that psychiatry should reassert the therapeutic value of drug treatment and be confident in its healing powers was probably not comforting to many.

The crusade against psychopharmaceuticals was more popular outside the psychiatric profession—among psychologists, social workers, and, above all, the newly organized survivors movement. Psychiatrists who embraced this crusade became rapidly marginalized within the psychiatric establishment, but the crisis was not without consequence to the practice of mainstream psychiatry. Surveys conducted during the 1980s suggested that the prescription of neuroleptics had decreased over the previous decade.[88] There might have been different reasons for this trend, not all related to the side effects ascribed to the drugs. In any event, a new public attitude toward neuroleptics became widespread, mingled with growing concerns over the limits of deinstitutionalization and the fear that a significant part of the psychiatric population was being mistreated by virtue of failed therapeutic and inefficient social policies.

Conclusion: Revolutions Yet to Come

Of all the therapeutic revolutions of the postwar era, the neuroleptic revolution was perhaps the most controversial, if not the most consequential. The dream of finding a cure for one of the most intractable and elusive disorders had set excessively high expectations among psychiatrists and broader communities interested in mental health. More profoundly, however, the wider significance of mental health in Cold War societies, as well as increasing differentiation within the psychiatric world, created a foundation for widespread conflict over any single mental health issue. As this chapter has argued, differentiation and conflict, rather than standardization and consensus, characterized the arena of neuroleptic use from the 1960s on. While there were many reasons to see a revolution in the profound transformations that affected psychiatry from the 1950s, there were also many reasons to contest every statement formulated about a singular revolutionary process.

In many ways, the very idea of a neuroleptic revolution overdetermined any discourse about change and progress.

In spite of these many criticisms, the neuroleptic revolution remains alive and well in contemporary psychiatry. The mental health world has been largely shaped by the outcome of the cycle of reforms and transformations from the 1950s to the 1970s. Even though there have been a number of calls for the reinstitutionalization of patients in the last twenty years, the landscape of mental health care is still characterized by fewer beds and the search for community alternatives. Although psychiatric research has developed and explored a number of other avenues to find cures, the dominant approaches to mental disorders today remain those biological models developed in the wake of the neuroleptic revolution. Furthermore, neuroleptics remain one of the main sources of therapeutic innovation and a major generator of profits in the mental health sector.[89] In many ways, there is no escaping the neuroleptic revolution.

Yet perhaps the most significant legacy of the neuroleptic and psychiatric revolutions of the 1960s and 1970s might be the very idea of the revolution itself. While it might be argued that the cycle of changes and reforms that began in the 1950s has now come full circle, revolutionary rhetoric has perhaps never been as pervasive in psychiatric discourses on progress and change as it is today. Virtually every innovative basic science approach to mental illness—from genomics to "phenomics" to brain imagery to big data analysis—is greeted with the promise that it will revolutionize mental health. Other more psychosocially oriented segments of psychiatry are equally quick to use revolutionary rhetoric to publicize their innovations. The contemporary recovery movement in the field of psychiatric rehabilitation is a good example of this tendency. Such grand promises are clearly explained by the need to attract funding at a time of constricted budgets and intensifying competition between divergent approaches. But this rhetoric also testifies to the living spirit of postwar revolutions.

NOTES

Benoit Majerus was an initial collaborator on this project, and I thank him for generously letting me work with his ideas. I also want to thank the editors both for inviting me to contribute this chapter to the volume and for innumerable comments and suggestions that have helped quite substantially to improve it. Finally, I thank Isabelle Baszanger, Jean-Paul Gaudillière, and Marie

Reinholdt for their comments on an earlier version of the chapter, as well as
Marina Urquidi, who corrected my English.

1. H. E. Lehmann, "Psychoactive Drugs and Their Influence on the Dynam-
    ics of Working Capacity," *Journal of Occupational and Environmental Medi-
    cine* 2 (1960): 523.
2. Marvin E. Lickey and Barbara Gordon, *Drugs for Mental Illness: A Revolution
    in Psychiatry* (New York: W. H. Freeman, 1983), 297.
3. Throughout this chapter, the term "neuroleptics" is used generically to
    refer to the class of drugs that was created with the description of the
    psychiatric effects of chlorpromazine in 1952. As will be reviewed below,
    other terms were in circulation in some countries—most notably "major
    tranquilizers" in the United States—and the term "antipsychotics" gradu-
    ally replaced the term "neuroleptics" from the 1970s on in most if not
    all countries. For the period covered in this chapter, "neuroleptics" was
    the most commonly used label, even more so in scientific publications. It
    remains the term used today in the World Health Organization, *Interna-
    tional Pharmacopoeia*, 5th edition (2015). Accessed 11 February 2016 at
    http://apps.who.int/phint.
4. Another version of the convergence of psychiatric treatment and a wider
    social movement is reflected in figure 0.1 in this volume.
5. For an account of the emergence of psychopharmacology as a "counter-
    revolution" see Andrew Scull, "A Psychiatric Revolution," *Lancet* 375, no.
    9722 (2010).
6. David Healy, *The Antidepressant Era* (Cambridge, MA: Harvard University
    Press, 1997); David Healy, *The Creation of Psychopharmacology* (Cambridge,
    MA: Harvard University Press, 2002); David Healy, *Let Them Eat Prozac:
    The Unhealthy Relationship between the Pharmaceutical Industry and Depres-
    sion* (New York: New York University Press, 2004).
7. Joanna Moncrieff, *The Bitterest Pills: The Troubling Story of Antipsychotic
    Drugs* (Palgrave Macmillan, 2013); Joanna Moncrieff, *The Myth of the
    Chemical Cure: A Critique of Psychiatric Drug Treatment* (Basingstoke, NY:
    Palgrave Macmillan, 2008).
8. Benoît Majerus, *Parmi les fous: Une histoire sociale de la psychiatrie au XXe
    siècle* (Rennes: Presses universitaires de Rennes, 2013); Stephen Snelders,
    Charles Kaplan, and Toine Pieters, "On Cannabis, Chloral Hydrate, and
    Career Cycles of Psychotropic Drugs in Medicine," *Bulletin of the History of
    Medicine* 80 (2006).
9. Nicolas Rasmussen, *On Speed: The Many Lives of Amphetamine* (New York:
    New York University Press, 2008).
10. Majerus, *Parmi les fous*.
11. Jack D. Pressman, *Last Resort: Psychosurgery and the Limits of Medicine* (Cam-
    bridge: Cambridge University Press, 1998). Also see Joel Braslow, *Mental
    Ills and Bodily Cure: Psychiatric Treatment in the First Half of the Twentieth
    Century* (Berkeley: University of California Press, 1997).

12. The best account on the development of chlorpromazine by the French pharmaceutical company Rhône-Poulenc and its distribution from the early 1950s on remains Judith P. Swazey, *Chlorpromazine in Psychiatry: A Study of Therapeutic Innovation* (Cambridge, MA: MIT Press, 1974). The value of Swazey's work lay notably in the fact that she had access to industrial archives that are now lost. In recent years, only Viviane Quirke was able to access new material from Rhône-Poulenc. See Viviane Quirke, *Collaboration in the Pharmaceutical Industry. Changing Relationships in Britain and France, 1935–1965* (London: Routledge, 2008), 197–204. David Healy's interviews with psychopharmacologists are also invaluable sources of data on the various groups that have contributed to shaping psychopharmacology over the years: David Healy, ed., *The Psychopharmacologists*, 3 vols. (London: Arnold, 1996–2000). The introduction and standardization of chlorpromazine in markets other than France and the United States was the focus of scholars gathered in the European network DRUGS. See Toine Pieters and Stephen Snelders, "Special Section: Standardizing Psychotropic Drugs and Drug Practices in the Twentieth Century," *Studies in History and Philosophy of Science Part C: Studies in History and Philosophy of Biological and Biomedical Sciences* 42, no. 4 (2011). For a history of phenothiazines before the psychiatric revolution, see Séverine Massat-Bourrat, "Des phénothiazines à la chlorpromazine: Les destinées multiples d'un colorant sans couleur" (Thèse de doctorat en Sciences, Technologies et Sociétés, Université Louis Pasteur (Strasbourg), 2004).

13. Why companies focused on mental hospitals is not clear. The argument most often found in the literature—that office-based psychiatry was either nonexistent or not receptive to drugs for ideological reasons—does not seem to be particularly compelling. Chlorpromazine was also initially marketed for a series of nonpsychiatric purposes, including nausea and vomiting, as well as anesthesia. While these nonpsychiatric uses—and others that emerged later on, including in palliative care—were by no means negligible, they were clearly not a central market for companies. On these issues, see Swazey, *Chlorpromazine in Psychiatry*.

14. Ibid., 203.

15. Ibid., 138–141 and Quirke, *Collaboration in the Pharmaceutical Industry*, 203.

16. Toine Pieters and Benoît Majerus, "The Introduction of Chlorpromazine in Belgium and the Netherlands (1951–1968): Tango between Old and New Treatment Features," *Studies in History and Philosophy of Science Part C: Studies in History and Philosophy of Biological and Biomedical Sciences* 42, no. 4 (2011).

17. Volker Hess, "Psychochemicals Crossing the Wall: Die Einführung der Psychopharmaka in der DDR aus der Perspektive der neueren Arzneimittelgeschichte," *Medizinhistorisches Journal* 42 (2007). Also see Ulrike Klöppel and Viola Balz, "Psychotropic Drugs in Socialism? Drug Regulation in

the German Democratic Republic in the 1960s," *Berichte zur Wissenschafts-geschichte* 33, no. 4 (2010).

18. A straightforward, albeit partial, account of this shift can be found in: Moncrieff, *The Bitterest Pills.*

19. Pierre Deniker, "Qui a inventé les neuroleptiques?" *Confrontations psychi-atriques,* no. 13 (1975). Delay and Deniker were prominent partisans of the diencephalic hypothesis. See Emilie Bovet, "Biographie du diencéphale: Revisiter l'histoire de la psychiatrie à travers le parcours d'une zone céré-brale" (Thèse de doctorat ès sciences de la vie, Université de Lausanne, 2012). On Delay's holism, see George Weisz, "A Moment of Synthesis: Medical Holism in France between the Wars," in *Greater Than the Parts: Holism in Biomedicine, 1920–1950,* ed. Christopher Lawrence and George Weisz (New York: Oxford University Press, 1998).

20. Swazey, *Chlorpromazine in Psychiatry.*

21. The National Institute of Mental Health Psychopharmacology Service Center Collaborative Study Group, "Phenothiazine Treatment in Acute Schizophrenia," *Archives of General Psychiatry* 10, no. 3 (1964): 257.

22. B. K. Madras, "History of the Discovery of the Antipsychotic Dopamine D2 Receptor: A Basis for the Dopamine Hypothesis of Schizophrenia," *Journal of the History of the Neurosciences* 22, no. 1 (2013); Alan A. Baumeis-ter and Jennifer L. Francis, "Historical Development of the Dopamine Hypothesis of Schizophrenia," *Journal of the History of the Neurosciences* 11, no. 3 (2002).

23. "The Current Psychiatric Revolution," *American Journal of Psychiatry* 121, no. 5 (1964).

24. E.g., S. S. Kety, "From Rationalization to Reason," *American Journal of Psy-chiatry* 131, no. 9 (1974).

25. This question was formulated by several prominent figures. See, for instance, the comments made in 1954 by an official of SK&F on the significance of chlorpromazine: Swazey, *Chlorpromazine in Psychiatry,* 190. For France, see comments made by the prominent psychiatrist Henri Ey: Henri Ey, "Neuroleptiques et services psychiatriques hospitaliers," *Con-frontations psychiatriques,* no. 13 (1975).

26. E.g., R. L. Spitzer and J. C. Wakefield, "DSM-IV Diagnostic Criterion for Clinical Significance: Does it Help Solve the False Positives Problem?" *American Journal of Psychiatry* 156, no. 12 (1999); W. M. Compton and S. B. Guze, "The Neo-Kraepelinian Revolution in Psychiatric Diagno-sis," *Eur Arch Psychiatry Clin Neurosci* 245, nos. 4–5 (1995); R. Mayes and A. V. Horwitz, "DSM-III and the Revolution in the Classification of Men-tal Illness," *Journal of the history of the behavioral sciences* 41, no. 3 (2005).

27. Viola Balz, *Zwischen Wirkung und Erfahrung—eine Geschichte der Psychophar-maka. Neuroleptika in der Bundesrepublik Deutschland. 1950–1980* (Bielefeld: Transcript Verlag, 2010). AMP was the acronym for Arbeitsgemeinschaft für Methodik und Dokumentation in der Psychiatrie (Working Group on

Metholodolgy and Documentation in Psychiatry), after the name of the group that had created the system.

28. The history of the chapter on mental disorders in the ICD remains understudied. For an introduction to the issues raised at the time of the first edition of this classification, see Robert E. Kendell, *The Role of Diagnosis in Psychiatry* (Oxford: Blackwell Scientific Publications, 1975). Also for a very brief historical overview, see Michael Shepherd, "ICD, Mental Disorder and British Nosologists: An Assessment of the Uniquely British Contribution to Psychiatric Classification," *British Journal of Psychiatry* 165, no. 1 (1994).

29. The critique of the DSM-III that proved to be the most influential probably came from psychology and bore on the very means by which psychiatrists sought to achieve reliability. See Stuart A. Kirk and Herb Kutchins, *The Selling of DSM: The Rhetoric of Science in Psychiatry* (New York: A. de Gruyter, 1992). .

30. See Nicolas Henckes, "Mistrust of Numbers: The Difficult Development of Psychiatric Epidemiology in France, 1940–1980," *International Journal of Epidemiology* 43, no. suppl. 1 (2014).

31. Jean Delay and Pierre Deniker, *Méthodes chimiothérapiques en psychiatrie* (Paris: Masson, 1961). . The first multicentric randomized trial was organized in France in the 1970s. See Henri Loo and Edouard Zarifian, "Limite d'efficacité des chimiothérapies psychotropes," in *Comptes rendus du congrès de psychiatrie et de neurologie de langue française* (Paris: Masson, 1977). According to a recent survey by the WHO and the World Psychiatric Association (WPA), French psychiatrists today still lag far behind their colleagues in other countries in the use of classifications schemes. See: G. M. Reed et al., "The WPA-WHO Global Survey of Psychiatrists' Attitudes towards Mental Disorders Classification," *World Psychiatry* 10, no. 2 (2011).

32. Dominique A. Tobbell, *Pills, Power, and Policy: The Struggle for Drug Reform in Cold War America and its Consequences* (Berkeley: University of California Press, 2012); Jean-Paul Gaudillière and Volker Hess, *Ways of Regulating Drugs in the 19th and 20th Centuries* (Basingstoke, UK: Palgrave Macmillan, 2013).

33. T. A. Ban, "A History of the Collegium Internationale Neuro-Psychopharmacologicum (1957–2004)," *Progress in Neuro-Psychopharmacology and Biological Psychiatry* 30, no. 4 (2006). See also the comments by Delay at the Fourth World Congress of Psychiatry: Jean Delay, introduction to *Proceedings, Fourth World Congress of Psychiatry: Madrid 5–11 September 1966*, ed. J. J. Lopez-Ibor, et al. (Amsterdam: Excerpta Medical Foundation, 1967).

34. Ingrid G. Farreras, Caroline Hannaway, and Victoria Angela Harden, *Mind, Brain, Body, and Behavior: Foundations of Neuroscience and Behavioral Research at the National Institutes of Health* (Amsterdam ,Washington: IOS, 2004).

35. Hannah S. Decker, *The Making of DSM-III: A Diagnostic Manual's Conquest of American Psychiatry* (New York: Oxford University Press, 2013); Steeves Demazeux, *Qu'est-ce que le DSM ? Genèse et Transformations de la Bible Américaine de la Psychiatrie* (Paris: Ithaque, 2013).

36. See the autobiographical comments of Melvin Sabshin, the medical director of the American Psychiatric Association at the time of the launching of the DSM-III project: Melvin Sabshin, *Changing American Psychiatry: A Personal Perspective*, 1st ed. (Washington: American Psychiatric Publications, 2008).

37. Healy, *The Antidepressant Era*; Moncrieff, *The Bitterest Pills*.

38. Michael Worboys, "The Hamilton Rating Scale for Depression: The Making of a 'Gold Standard' and the Unmaking of a Chronic Illness, 1960–1980," *Chronic Illness* 9, no. 3 (2013).

39. Lucie Gerber and Jean-Paul Gaudillière, "Marketing Masked Depression: Physicians, Pharmaceutical Firms, and the Redefinition of Mood Disorders in the 1960s-1970s," *Bulletin of the History of Medicine* (forthcoming 2016).

40. Healy, *The Antidepressant Era*.

41. Gerald N. Grob, *From Asylum to Community. Mental Health Policy in Modern America*. (Princeton, NJ: Princeton University Press, 1991).

42. The Joint Commission on Mental Illness and Health had been established in 1955 by the US Congress. Presided over by Kenneth E. Appel, professor of psychiatry at the University of Pennsylvania, it gathered representatives from some twenty organizations working in the mental health field.

43. At the WHO, this was especially the case of the Regional Office for Europe, who devised in 1973 an ambitious program for mental health services calling for the development of community psychiatry. See Hugh L. Freeman, Thomas Fryers, and John H. Henderson, *Mental Health Service in Europe: 10 Years On* (Copenhagen: World Health Organization Regional Office for Europe, 1985).

44. Nicolas Henckes, "Le nouveau monde de la psychiatrie française: Les psychiatres, l'Etat et la réforme des hôpitaux psychiatriques de l'après-guerre aux années 1970" (Thèse de sociologie, Ecole des hautes études en sciences sociales, 2007).

45. Other countries were even more distant from the US pattern, with Japan probably being the latest of the developed nations to downsize its psychiatric hospital system.

46. World Health Organization, *The Community Mental Hospital: Third Report of the Expert Committee on Mental Health*, Technical Report Series no. 73 (Geneva: World Health Organization, 1953).

47. H. Brill and R. E. Patton, "Analysis of 1955–1956 Population Fall in New York State Mental Hospitals in First Year of Large-Scale Use of Tranquilizing Drugs," *American Journal of Psychiatry* 114, no. 6 (1957).

48. Henry Brill et al., "The Impact of Psychotropic Drugs on the Structure, Function and Future of Psychiatric Services in Hospitals," in *Neuro-*

Psychopharmacology: Proceedings of the First International Congress of Neuro-Pharmacology, Rome, September, 1958, ed. P. B. Bradley, Pierre Deniker, and C. Radouco-Thomas (Amsterdam, New York: Elsevier, 1959), 21. Jean Delay would later reply to Lewis that, thankfully, there was no need to choose between sociotherapeutic and chemotherapeutic methods and that they were complementary. Delay, introduction, 286.

49. For a classic review, see Leona L. Bachrach, *Deinstitutionalization: An Analytical Review and Sociological Perspective* (Rockville, MD: National Institute of Mental Health, 1976); Kathleen Jones, "Deinstitutionalization in Context," *Milbank Memorial Fund Quarterly. Health and Society* 57, no. 4 (1979).

50. Most notably, Andrew Scull, *Decarceration: Community Treatment and the Deviant—a Radical View* (Englewood Cliffs, NJ: Prentice-Hall, 1977).

51. Gerald N. Grob, *The Mad among Us: A History of the Care of America's Mentally Ill* (New York: Free Press, 1994).

52. W. Sargant, "Aim and Method in Treatment: Twenty Years of British and American Psychiatry," *Journal of Mental Science* 103, no. 433 (1957).

53. W. H. Trethowan, *British Journal of Psychiatry* 114, no. 510 (1968): 660.

54. Georges Demay, "Services ouverts, quartiers d'observation et placement direct à l'asile," *Annales Médico-psychologiques* I (1928): 193.

55. The leaders of the movement began to develop a narrative describing the events from 1945 as a revolution in the following months, with explicit reference to the French Revolution of 1789. See especially Georges Daumézon and Lucien Bonnafé, "Perspectives de réforme psychiatrique en France depuis la Libération," in *Congrès des médecins aliénistes et neurologistes de langue française. 44e session, Genève et Lausanne, 22–27 juillet 1946* (Paris: Masson, 1946). Georges Daumézon, "Informations syndicales: Rapport du secrétaire général," *Information psychiatrique* 23, no. 10 (1947).

56. Daumézon and Bonnafé, "Perspectives de réforme psychiatrique en France depuis la Libération," 586. The Tennis Court Oath (*Serment du jeu de paume*) was the first formal act of defiance of the king's authority in the early days of the French Revolution.

57. See François Fourquet and Lion Murard, "Histoire de la psychiatrie de secteur ou le secteur impossible?" *Recherches*, no. 17 (1975).

58. On institutional psychotherapy, see Henckes, "Le nouveau monde de la psychiatrie française."

59. The "internal revolution" was mentioned by François Tosquellès in Henri Ey et al., "Symposium sur la psychothérapie collective," *Evolution psychiatrique*, no. 3 (1952): 537.

60. Jean-Pierre Le Goff, *Mai 68, l'héritage impossible* (Paris: La Découverte, 1998). Robert Castel, *La gestion des risques: De l'anti-psychiatrie à l'après-psychanalyse* (Paris: Editions de Minuit, 1981); Sherry Turkle, *Psychoanalytic Politics: Freud's French Revolution* (New York: Basic Books, 1978).

61. Pierre Bailly-Salin et al., "La passivité: Approche anthropologique, relationnelle et psychosomatique," *Journal de Médecine de Lyon*, no. spécial

Journée de thérapeutique psychiatrique (1959): 86. See also Georges Daumézon, "Essai d'approche du savoir infirmier sur les psychotropes," in *Comptes rendus du congrès de psychiatrie et de neurologie de langue française. LXXVe session—Limoges—27 juin 1977*, ed. Pierre Warot (Paris: Masson, 1977).

62. Nicolas Henckes, "Reshaping Chronicity: Neuroleptics and the Changing Meaning of Therapy in French psychiatry, 1950–1975," *Studies in History and Philosophy of Science Part C: Studies in History and Philosophy of Biological and Biomedical Sciences* 42, no. 4 (2011).

63. Grob, *From Asylum to Community*; Pressman, *Last Resort*.

64. Gregory Zilboorg and George W. Henry, *A History of Medical Psychology* (New York: W. W. Norton, 1941).

65. Ibid., 509.

66. See especially Rudolf Dreikurs, "Group Psychotherapy and the Third Revolution in Psychiatry," *International Journal of Social Psychiatry* 1, no. 3 (1955).

67. J. L. Moreno, "The Third Psychiatric Revolution and the Actual Trends of Group Psychotherapy," in *The International Handbook of Group Psychotherapy*, ed. J. L. Moreno, et al. (New York: Philosophical Library, 1966).

68. Ellen Herman, *The Romance of American Psychology: Political Culture in the Age of Experts, 1940–1970* (Berkeley: University of California Press, 1995); Grob, *From Asylum to Community*; Nikolas S. Rose, *Governing the Soul: The Shaping of the Private Self* (London: Routledge, 1990).

69. Philip Rieff, *The Triumph of the Therapeutic: Uses of Faith after Freud* (New York: Harper & Row, 1966).

70. See especially Leopold Bellak, "Community Psychiatry: The Third Psychiatric Revolution," *Handbook of Community Psychiatry and Community Mental Health* (New York: Grune & Stratton, 1964).

71. Louis Linn, "The Fourth Psychiatric Revolution," *American Journal of Psychiatry* 124, no. 8 (1968).

72. Jonathan M. Metzl, *Prozac on the Couch: Prescribing Gender in the Era of Wonder Drugs* (Durham, NC: Duke University Press, 2003); David Herzberg, *Happy Pills in America: From Miltown to Prozac* (Baltimore: Johns Hopkins University Press, 2008); Andrea Tone, *The Age of Anxiety: A History of America's Turbulent Affair with Tranquilizers* (New York: Basic Books, 2009).

73. G. L. Klerman, "Better but Not Well: Social and Ethical Issues in the Deinstitutionalization of the Mentally Ill," *Schizophrenia Bulletin* 3, no. 4 (1977): 630.

74. Françoise Castel, Robert Castel, and Anne Lovell, *The Psychiatric Society* (New York: Columbia University Press, 1982); Castel, *La gestion des risques: De l'anti-psychiatrie à l'après-psychanalyse*; Gilles Deleuze, "Post-scriptum sur les sociétés de contrôle," in *Pourpalers* (Paris: Editions de Minuit, 1990); David Armstrong, "The Rise of Surveillance Medicine," *Sociology of Health and Illness* 17, no. 3 (1995); Rose, *Governing the Soul*.

75. David J. Rothman, *Strangers at the Bedside: A History of How Law and Bioethics Transformed Medical Decision Making* (New York: BasicBooks, 1991); Alexandra Rutherford, "The Social Control of Behavior Control: Behavior Modification, Individual Rights, and Research Ethics in America, 1971–1979," *Journal of the History of the Behavioral Sciences* 42, no. 3 (2006).
76. Michael E. Staub, *Madness Is Civilization: When the Diagnosis Was Social, 1948–1980* (Chicago: University of Chicago Press, 2011).
77. For a similar reversal of attitudes regarding the therapeutic effects of antibiotics in the same years, see chapter 1 in this volume.
78. Staub, *Madness is Civilization*.
79. Tone, *The Age of Anxiety*; Erika Dyck, *Psychedelic Psychiatry: LSD from Clinic to Campus* (Baltimore: Johns Hopkins University Press, 2008).
80. Moncrieff, *The Bitterest Pills*, 76 sq. P. Brown and S. C. Funk, "Tardive Dyskinesia: Barriers to the Professional Recognition of an Iatrogenic Disease," *Journal of Health and Social Behavior* 27, no. 2 (1986).
81. Brown and Funk, "Tardive Dyskinesia."
82. Gardos and Cole, quoted by Moncrieff, *The Bitterest Pills*, 79.
83. G. J. Annas, "Refusing Medication in Mental Hospitals," *Hastings Center Report* 10, no. 1 (1980).
84. Paul S. Appelbaum, *Almost a Revolution: Mental Health Law and the Limits of Change* (Oxford: Oxford University Press, 1994). Both parties appealed the decision, and the case was eventually brought to the federal Supreme Court, which did not pronounce and returned the case to state courts. The last decision in the case, in 1983, confirmed the initial judgment.
85. Involuntary commitment was not the case for all the Boston State patients, one reason why their case was relatively easy.
86. P. S. Appelbaum and T. G. Gutheil, "The Boston State Hospital Case: 'Involuntary Mind Control,' the Constitution, and the Right to Rot,'" *American Journal of Psychiatry* 137, no. 6 (1980).
87. Appelbaum, *Almost a Revolution*.
88. D. K. Wysowski and C. Baum, "Antipsychotic Drug Use in the United States, 1976–1985," *Archives of General Psychiatry* 46, no. 10 (1 October 1989): 929–932.
89. According to figures from the healthcare information company IMSHealth, the single best-selling drug in 2013 in America was an antipsychotic. See IMS Institute for Healthcare Informatics, *Medicine Use and Shifting Costs of Healthcare: A Review of the Use of Medicines in the United States in 2013* (Parsippany, NJ: IMS Institute for Healthcare Informatics, 2014).

FOUR

Revolutionary Markets? Approaching Therapeutic Innovation and Change through the Lens of West German IMS Health Data, 1959–1980

NILS KESSEL AND CHRISTIAN BONAH

In 1997, Louis Lasagna, one of America's most influential clinical pharmacologists, looked back over the twentieth century as a time of "veritable pharmacotherapeutic revolution":

Antibiotics have made it possible to cure previously untreatable infections. Psychiatric illnesses ranging from anxiety states and mania to schizophrenia and psychotic melancholia have yielded to new therapies. Elevated blood pressure can be brought to normal by drugs. Hormonal therapy. . . . Immunosuppressants . . . Vaccines . . . Therapeutic and preventive progress in the twentieth century has, therefore, been dramatic.[1]

Lasagna's narrative of the pharmacotherapeutic revolution mobilizes three key elements. First, the biomedical past between 1930 and 1980[2] led to a cumulative and continuous transformation of therapy built on new therapeutics. Over this six-decade period, the list of innovations was expanded, adding beads of pharmacological discoveries onto the string of pharmaceutical advancement. Second, this past was glorious and beneficial to society overall: "Patients and physicians have reason both to be grateful for the fruits of the research and development efforts of the past and frustrated by the many unmet needs of the present," Lasagna wrote. Third, the progress achieved during the period of the pharmacotherapeutic revolution could be regained returning to a methodology of science-based therapies and to forms of less-regulated research. "Problems remaining unsolved," Lasagna continued, "such as AIDS and . . . cancers . . . , should find solutions by reflecting on the lessons of the past revolution."[3]

From the historian's perspective, this narrative of the pharmacotherapeutic revolution is intriguing. In contrast to the notion of the "scientific revolution" of the early modern era, the mid-twentieth-century pharmacotherapeutic revolution may be described not in terms of a fundamental change in scientific thinking and methodology (which in medicine had taken place during the nineteenth century) but rather as the result of technological progress through the *therapeutic use* of pharmaceuticals.[4] In other words, it is less an intellectual history of conceptual change than a material history of things and their uses. Between 1930 and 1980, medical practice was revolutionized by new pharmaceuticals that may be considered a form of technology.[5] The concept of the therapeutic revolution thus implies that people in the mid-twentieth century were facing a material revolution—that is, a revolution of new technologies and their uses.[6]

Lasagna's three-pronged analysis (accumulation, health benefit, and science-based rationale) of the improvement of public health from 1930 to 1980 is representative of numerous Whiggish accounts of the history of pharmaceuticals. Many physicians and historians likewise describe a list of major therapeutic innovations developed and commercialized between the 1930s and 1980s as testament to the revolutionary nature of therapeutics in that era.[7] In so doing, they promulgate a cumulative history of pharmaceutical invention from the perspective of scientists and physicians.[8] According to this history from above, the 1930–1980 therapeutic revolution added more and more powerful treatments which ultimately replaced less effective drugs. Implicit is the assumption that science-based technological change—that is, new pharmaceutical therapy—was easily, rationally, and self-evidently adopted by physicians and patients both in and outside of hospitals.[9]

In contrast, historians of technology have argued that technological diffusion never works in such a simple, straightforward way, but depends on complex alliances and negotiations among inventors, producers, regulators, and consumers.[10] How, then, can we reconcile Lasagna's version of a pharmacotherapeutic revolution with a more nuanced perspective on how new technologies diffuse through society?

Our analysis borrows from the historian of technology David Edgerton's *The Shock of the Old*, which argues for taking into account the importance of older technologies in the modern innovation-driven world. Edgerton demonstrates that novelty-driven accounts of technological modernity lose sight of persistent older forms of technology. Forward-looking observers then experience the disturbing surprise of the continuing use of former technologies alongside the innovation emphasized in progress narratives. Given this perspective, one might describe our received notions of therapeutic revolutions as suffering from "Columbus syndrome." When Christopher Columbus and his contemporaries discovered what to them was a new world, they conceived of these territories as virgin land, ignoring the long history of Native American inhabitants. Similar "pioneer perspectives" can be found in the narratives that scientists and physicians produce to explain the scientization and modernization of medical therapy. Actors' accounts of therapeutic change tend to discount what had existed before and what may still remain. Working against this pioneer perspective, our reading suggests that the twentieth-century therapeutic revolution was not a straightforward uniform process of radical change with a clear-cut boundary between an ancient regime and a new present. Instead, we delineate the multiple forms of therapeutic change that took place in different fields of therapy.

This chapter sheds light on selected drug markets in a twofold manner. First, we investigate to what degree "revolutionary" medicines were present in pharmacy purchases in the 1960s and 1970s. Second, we analyze markets for drugs that do not appear in the classical therapeutic revolution narrative. Antibiotics and cardiovascular drugs serve as examples of the former, and analgesics, hypnotics, and sedatives serve as examples of the latter. Approaching the therapeutic revolution from the less investigated perspective of sales and use, our narrative challenges, along with Edgerton, the convention that "when we are told about technology from on high we are made to think about novelty and the future."[11] We suggest that the therapeutic revolution narrative is highly normative. Like other progress narratives, it mobilizes a glorious past to legitimize action in the present and future. As in many progress narratives, the mobilized arguments and episodes are highly selective in their perception of the past.

Based on access to two decades of pharmaceutical market data amassed in West Germany from the late 1950s to the 1970s by the market research company IMS Health, our contribution approaches the material nature of the era of the therapeutic revolution from the "demand side."[12] Our narrative counters the dominant rhetoric of the pharmacotherapeutic revolution as a tale told largely by pharmaceutical marketers and pharmacologists themselves. It reveals that such rhetoric is hopeful, idealized, and should be understood as a positioned narrative of self-promotion.

IMS and Pharmaceutical Market Data

Using IMS surveys from the last two decades of the alleged revolutionary period, we analyze what pharmacies bought and how drug markets changed in West Germany during this period. Market data may be considered as an alternative view, both to the medical expert view from above and to patient histories from "below," although our quantitative approach can be seen as complementary to qualitative history from below.[13] IMS data provide clues to what was actually on the pharmacy counter, in contrast to what pharmacologists and physicians thought was new, pathbreaking, and revolutionary. These data provide ample information about which drugs were purchased and in what quantities by pharmacies.

Founded in 1959, the Institute for Medical Statistics (Institut für medizinische Statistik, IMS) was the West German branch of an internationally active American company named Intercontinental Marketing Services. The latter had been founded in New York five years earlier, in 1954, and it expanded rapidly in the Western world.[14] Although non-sector-specific market research companies, such as the Society for Consumption Research (Gesellschaft für Konsumforschung), had existed in Germany since 1934, IMS's specific focus on the pharmaceutical market soon gave the company a predominant position in this precious niche. Starting in 1959, IMS produced an annual statistical review of drug sales, called *The Pharmaceutical Market* (DPM: Der pharmazeutische Markt).

The data gathered by IMS was unique for the period in Germany, since neither health insurance companies nor public agencies collected similar information until the late 1970s.[15] Produced for the pharmaceutical industry, these reports were proprietary: contractual obligations on the front page of all IMS reports clearly specified that any distri-

| 06 ANTIRHEUMATICA GICHTMITTEL | | | | B EXTERNE ANTIRHEUMATICA | | | | | DEZEMBER 1968 | | SEITE | 76 | 06B |

HERSTELLER - PRAEPARAT - FORM	EINF. JAHR	PREIS EINH.	MONATLICH					KUMULIERT					±%	INDEX
			EINH.	%	DM	%	%D	EINH.	%	DM	%	%D		
FAPACK														
GR. 0 8X25 CM 01 WD	.10	2S	215	0.7	520	0.8	24	2.872	0.8	7.059	1.1	20	2-	92
GR. 1 22X25 CM 01 WD	1.80	1.70	11	5.1	19	3.7	47	106	3.7	182	2.6	17	100	189
GR. 2 15X40 CM 01 WD	2.20	2.08	32	14.9	67	12.9	25	280	9.7	577	8.2	11	1	96
GR. 2 15X40 CM 01 WD	2.50	2.36	99	46.0	232	44.6	31	1.406	49.0	3.319	47.0	18	3	98
GR. 3 25X40 CM 01 WD	2.90	2.75	73	34.0	202	38.8	15	1.080	37.6	2.981	42.2	23	11-	84
....HEYDEN..................		2S	97	0.3	313	0.5	0	2.180	0.6	6.988	1.0	0	13-	82
INFILTHINA	7.67	2S	91	0.3	307	0.5	0	2.081	0.6	6.886	1.0	0	13-	82
LIQUID 60 ML 01R WB	4.00	3.80	69	75.8	262	85.3	0	1.469	70.6	5.584	81.1	0	14-	81
LIQUID 10X60 ML 01R WB	33.63	31.85	0	0.0	0	0.0	0	1	0.0	32	0.5	0	53-	45
S 25 ML 01R WB	2.19	2.08	22	24.2	45	14.7	0	611	29.4	1.270	18.4	0	6-	89
SALIT	.09	0S	6	0.0	6	0.0	0	99	0.0	102	0.0	0	38-	59
OEL 55 ML 01 WB	1.14	1.10	6		6		0	99		102		0	38-	59
....KREWEL..................		8S	1.484	4.5	2.855	4.6	0	16.710	4.6	30.388	4.5	1	5	99
ANALGIT	.23	0S	61	0.2	78	0.1	0	689	0.2	892	0.1	0	26-	70
I 50 GR. 07 WB	1.12	1.18	17	27.9	20	25.6	0	146	21.2	165	18.5	0	27	120
I 100 GR. 07 WB	1.92	2.04	2	3.3	4	5.1	0	23	3.3	46	5.2	0	21	115
II 50 GR. 07 WB	1.12	1.18	28	45.9	33	42.3	0	264	38.3	303	34.0	0	30-	67
II 100 GR. 07 WB	1.92	2.04	6	9.8	12	15.4	0	114	16.5	227	25.4	0	44-	53
SALBE 30 GR. 07 WA	1.00	1.10	7	11.5	7	9.0	0	121	17.6	122	13.7	0	22-	74
SALBE 50 GR. 07 WA	1.39	1.51	1	1.6	2	2.6	0	21	3.0	29	3.3	0	33-	64
HYPERAEMOL	.36	0S	97	0.3	124	0.2	0	1.282	0.4	1.711	0.3	0	4-	91
50 GR 07 WB	1.10	1.18	56	57.7	66	53.2	0	511	39.9	581	34.0	0	4-	91
100 GR 07 WB	1.89	2.04	15	15.5	30	24.2	0	324	25.3	632	36.9	0	1-	94
SALBE 30 GR. 01 WA	1.01	0.97	20	20.6	20	16.1	0	302	23.6	301	17.6	0	4-	91
SALBE 50 GR. 01 WA	1.42	1.36	6	6.2	8	6.5	0	144	11.2	190	11.1	0	5-	90
JOD SAPEN	.25	0S	2	0.0	2	0.0	0	20	0.0	20	0.0	0	33	126
5VH. 30 GR. 01 WB	0.92	0.88	2		2		0	19	95.0	19	95.0	0	27	120
5VH 50 GR 01 WB	1.47	1.39	0	0.0	0	0.0	0	1	5.0	1	5.0	0	0	
MEDIMENT	.34	2S	508	1.5	748	1.2	0	5.891	1.6	8.148	1.2	0	11-	84
50 GR. 07 WB	1.07	1.15	311	61.2	357	47.7	0	3.964	67.3	4.391	53.9	0	3-	92
100 GR. 07 WB	1.85	1.98	197	38.8	391	52.3	0	1.914	32.5	3.674	45.1	0	20-	79

FIGURE 4.1 DPM exemplary content of pharmaceutical sales data compiled in the form of statistical listings organized by identification code, producer-product-conditioning, price per unit, units sold per month with annual cumulative figures, and value of sales per month with annual cumulated figures (in German marks). Reproduced from DPM 1968. Archives IMS Health Germany.

bution of information was prohibited.[16] Fortunately, the company recently granted us access to its historical archives, allowing us to analyze this rich source of data from the historian's perspective. In order to appraise critically the data that undergird this chapter, it is useful to present the IMS methodology in the context of the German drug distribution system of the 1960s and 1970s.

As shown in figure 4.2, more than 80 percent of all medicinal drugs sold in West Germany were distributed to privately owned pharmacies. Because of this dominant position of pharmacies in the West German drug distribution system, IMS had based its calculation on selected pharmacies' "sell-in."[17] "Sell-in" included all drugs officially ordered by pharmacies from either wholesalers or manufacturers.[18] The data were collected monthly, listing sales quantities by value and by number of product units (packages). By differentiating between monetary value and product units, IMS enabled its manufacturer clients to more fully analyze markets, submarkets, and products.[19] IMS provided the first data source for national pharmaceutical production and sales to be collected and aggregated in a systematic and periodically sustained way.[20]

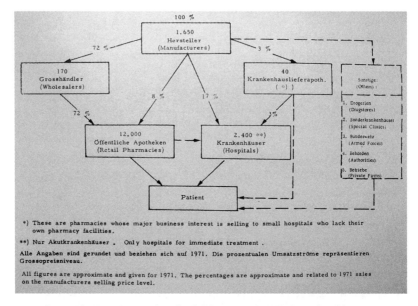

*) These are pharmacies whose major business interest is selling to small hospitals who lack their own pharmacy facilities.

**) Nur Akutkrankenhäuser . Only hospitals for immediate treatment .

Alle Angaben sind gerundet und beziehen sich auf 1971. Die prozentualen Umsatzströme repräsentieren Grossopreisniveau.

All figures are approximate and given for 1971. The percentages are approximate and related to 1971 sales on the manufacturers selling price level.

FIGURE 4.2 The West German drug distribution system in 1971. Reproduced from DPM 1974, p. 12. Archives IMS Health Germany.

These sales figures of package units and prices cannot tell us exactly which medicines sufferers consumed for a given illness, or when they consumed them. Neither can they shed light on the motives of physicians, pharmacists, or patients in prescribing, dispensing, or consuming medications. Given these limitations, it is best to avoid the term "consumption" in the specific context of medicinal drug use, as the question of what actually happens with drugs in the households remains unsolved. Indeed, there is much evidence to suggest that many drugs are never actually taken but spend many years in medicine cabinets before being thrown away. Nevertheless, these sales figures do indicate what price consumers—or their insurance plans—paid for products including both prescription drugs and over-the-counter medications.[21]

In the Revolution's Core: Antibiotics and Cardiovascular Drugs

In rereading the therapeutic revolution through the lens of IMS data, our first step was to examine therapeutic agents central to Lasagna's account of the therapeutic revolution: antibiotics and cardiovascular drugs. As Podolsky and Lie discuss elsewhere in this volume, the development of antibiotics has become a key symbol for medical[22] and popular conceptions of modern biomedical therapeutics.[23] But does the central place occupied by antibiotics in the therapeutic revolution narrative correspond to their relevance on the 1960s and 1970s West German market? One indication of their outsized influence on pharmaceutical markets was their unusual designation as an independent indication group among the fifty indication groups that structured the IMS data in the DPM reports.[24] This category of "antibiotics" was composed of six subgroups: (1) broad- and medium-spectrum antibiotic; (2) penicillin and derivatives; (3) penicillin-streptomycin combinations; (4) antibiotic-sulfonamide combinations; (5) streptomycin, dihydrostreptomycin and their combinations; and (6) other antibiotics.

The novelty and relevance of specific antibiotics can be investigated by combining information about the age of a product and its importance in pharmacy purchases. We use the year of market introduction as the variable to determine the age of the product, but not necessarily that of the active principle. Of the sixty-seven products in the antibiotic indication group on the West German market in 1967, one had been introduced before 1948. New market introductions increased until 1954, with twelve new products that year, and eventually stabilized in the early 1960s with an average of four to six new products each year (see figure 4.3). Older bacteriostatic sulfonamides, such as Gerhard Domagk's Prontosil, do not appear in the antibiotic indication group or elsewhere, thus suggesting that they were replaced by the newer antibiotics based on penicillin, streptomycin, and their derivatives. Sulfonamides remained on pharmacy counters only when used as combination antibiotics or in dermatology.[25] Up to this point, IMS data for antibiotics corroborate the classical revolutionary account in the sense that new products based on new substances—those less than twenty years old—appeared and replaced earlier remedies intended to treat infectious diseases in this therapeutic field.

Beyond product names and numbers, IMS data allow further analysis of how sales of individual products, as well as entire indication groups and subgroups, changed over time in terms of purchased package units and total revenues. For example, the five-year cumulative data for antibiotics' purchase values between 1963 and 1967 indicate an increase from 26.4 million German marks in 1963 to 46.7 million German marks in 1967 for the subcategory of broad and medium antibiotic, a total increase of 77 percent and annual growth rates between 6 and 28 percent.[26] The nine-year report for 1966–1974 shows a rise in

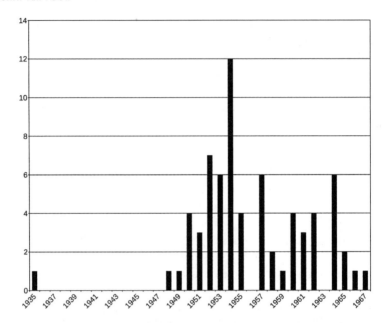

FIGURE 4.3 Year of market introduction (*X* axis) of all antibiotics listed in DPM for 1967. Annual number of antibiotics introduced to the market (*Y* axis) between 1935 (first product) and 1967 (year of the census). Data extracted from DPM Vierjahresvergleich 1963–1967. Archives IMS Health Germany.

the number of purchased package units from 14.9 million in 1966 to 25.8 million units in 1974 for the subgroup "systemic antibiotics," an increase of 73 percent and annual growth rates of 0 to 13.7 percent.[27] Even with respect to biases induced by modifications in classification, the changes were indeed dramatic, and these data on antibiotic purchases implied that new substances were replacing older substances with steadily increasing units and purchase values.

However, a closer look at the IMS data enables us to see this antibiotic therapeutic revolution in broader perspective. The picture becomes more complex when antibiotics are considered in relation to the total pharmaceutical market. In comparison to all medicines purchased by pharmacies in West Germany in the mid-1960s, antibiotics accounted for 1.7 percent of all purchases (see table 4.1).[28] Even when viewed from the perspective of the purchases' value rather than number of units sold, antibiotics still accounted for only 4.5 percent of market share in 1961. They remained at 4.5 percent in 1966, and even decreased to 4 percent in 1971 (see table 4.2).

There is no doubt that antibiotics dramatically changed the treatment of infectious diseases. Physicians, pharmacists, and patients have rightly valued these drugs advertised by the industry as innovations. Our market-demand perspective does not question the impact of antibiotics for individual patients and for infectious disease epidemiology. Rather, it suggests that the narrative is based on the bundling of specific forms of therapeutic change into one all-encompassing therapeutic revolution. Indeed, extrapolation from the case of antibiotics to a general therapeutic revolution seems to confound antibiotics' revolutionary efficacy and iconic status with their relative distribution on the overall drug market. If the history of therapy is broadened beyond experts' visions to include patients' and users' points of view, then it is clear that antibiotics are far too specific to represent overall therapeutic practices in the mid-twentieth century per se (see table 4.1).

We now turn our attention to drugs for cardiovascular diseases, a second group that has been mobilized to illustrate the history of the therapeutic revolution.[29] The identification of risk factors for "silent" diseases affecting the heart and the vascular system, and their effective treatment with new antihypertensive medications that normalize blood pressure, have transformed cardiac disease morbidity and have been considered a major therapeutic achievement.

IMS data corroborate that significant changes in cardiac drug sales

Table 4.1 Market-share percentages of drugs listed by therapeutic group, in units (1966). Calculations are ours, based on IMS DPM 1966. Archives IMS Health Germany.

	% in 1966
Cardiovascular system	12.2
Analgesics	13.4
Vitamins	3.3
Hormones	1.8
Sedatives and hypnotics	5.4
"Cough and cold" drugs	12.1
Antibiotics	1.7
Antirhumatic drugs	5.3
Dermatological drugs	6.7
Antidiabetics	0.8
Laxatives	3.5
Psychotropic drugs	3.1
Antacids	2.5
Antiseptics	0.7
Antiadipositas	0.8
Antihaemorrhea and antivaricosa	3.9
Gynecological drugs (incl. contraception)	0.7
Others	22.1
Total	100

Table 4.2 Market-share percentages of drugs, listed by therapeutic group, in value (1961, 1966, 1971). Calculations are ours, based on IMS DPM 1961, 1966, 1971. Archives IMS Health Germany.

Therapeutic group / year	% in 1961	% in 1966	% in 1971
Cardiovascular system	15.2	14.9	17.1
Analgesics	9.4	6.9	5.2
Vitamins	8.5	5.1	4
Hormones	7.7	3.9	3.3
Sedatives and hypnotics	6.1	3.9	2.8
"Cough and cold" drugs	5.6	6.6	7
Antibiotics	4.5	4.5	4
Antirhumatic drugs	3.7	5.3	5.8
Dermatological drugs	3.6	6.1	5.7
Antidiabetics	3	3.6	3.8
Laxatives	2.6	2.2	1.8
Psychotropic drugs	2.3	4.5	5
Antacids	2.1	1.9	1.7
Antiseptics	1.9	0.9	1.3
Antiadipositas	1.4	1.1	0.9
Antihaemorrhea and antivaricosa	0.7	4.6	4.7
Gynecological drugs (incl. contraception)	0.3	1.1	3.1
Others	21.4	22.9	22.8
Total	100	100	100

occurred in the 1960s, but again these were more complicated than standard narratives of the therapeutic revolution may suggest. In contrast to infections and antibiotics, cardiovascular diseases were manifold and treated with many different preparations, often in combination.[30] Cardiovascular medicines in the 1963–1967 report were listed under the indication category "heart and circulatory therapeutics," in nine subgroups: (1) rauwolfia preparations, (2) rauwolfia combinations, (3) other antihypertonica, (4) coronarospasmolytica, (5) peripheral vasodilatators, (6) heart glycosides, (7) antiarrhythmics, (8) sympathicomimics and analeptica, and (9) other heart and circulation preparations. Clearly, this group encompasses much more than the innovative antihypertensive drugs that came onto the market in the 1950s and 1960s, which makes it difficult to speak of one common drug class in respect to the therapeutic revolution.

For the "heart and circulation" indication group on the West German market in 1967, IMS listed 168 products within the nine subgroups. Contrary to the relatively recent arrival of the antibiotics group, dates of market introduction for cardiacs spread over an extremely long time span, from 1900 to 1967 (see figure 4.4). Twenty-nine cardiac drugs had been on the market for more than twenty years. The

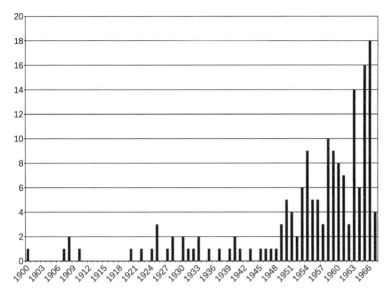

FIGURE 4.4 Year of market introduction (X axis) of all cardiovascular drugs listed in DPM for 1967. Annual number of cardiovascular drugs introduced to the market (Y axis) between 1900 (first product) and 1967 (year of the census). Data extracted from DPM Vierjahresvergleich 1963–1967. Archives IMS Health Germany.

17 percent of cardiac drugs introduced prior to 1947 and still available can hardly be included in the group of revolutionary cardiac drugs as classically portrayed (see table 4.4). If the lens is narrowed to view only those subgroups introduced after World War II, then glycosides and peripheral vasodilators far outnumber innovative antihypertension specialties. It is difficult to tease out the precise impact of the revolutionary antihypertension drugs among the 139 cardiovascular preparations introduced between 1947 and 1967 from the 1963–1967 IMS data, but viewed from this early 1960s census, the "heart and circulation" indication group is certainly a much more heterogeneous category than exclusive references to antihypertension medications suggest. Indeed, cardiac drugs are representative of "the shock of the old," because treatment innovations including glycosides and diuretics progressively transformed therapy over the course of more than half a century. If some single cardiovascular drugs might be understood as revolutionary, their bundling to an all-encompassing therapeutic revolution appears highly simplistic (see table 4.3).

In comparing cardiovascular drugs with other drug classes on the overall drug market in West Germany, we find that cardiacs were the

Table 4.3 Pharmacy purchases of cardiovascular drugs in thousands of packages, 1966–1974. Calculations are ours, based on IMS DPM Neunjahresvergleich 1966–1974. Archives IMS Health Germany.

Cardiac therapeutic subgroups	1966	1967	1968	1969	1970	1971	1972	1973	1974
Cardiacs	73,811	76,532	81,307	87,855	92,821	94,984	98,123	98,583	99,256
Cardiac glycosides and combinations	34,413	35,617	38,445	42,044	45,411	46,657	48,414	48,567	48,332
Other cardiacs	39,398	40,915	42,862	45,811	47,410	48,327	49,709	50,016	50,924
Antihypertension drugs	18,486	19,081	19,276	20,374	20,873	21,555	22,549	23,915	25,872
Diuretics	6,000	6,221	6,048	6,213	5,990	6,172	6,125	6,867	7,664
Peripheric vasodilatators	12,070	13,161	14,166	16,018	17,640	19,731	21,807	23,076	24,583
Vasoprotectors	39,745	41,064	42,452	44,751	48,802	48,656	47,476	44,704	43,921
Other cardiovascular drugs	23,888	22,617	21,763	21,161	19,090	18,644	18,126	16,988	16,560

second most popular sector of preparations sought by patients or pre-scribed by physicians. In terms of market share, they accounted for 12.2 percent of all purchased package units in the mid-1960s (see ta-ble 4.1). In terms of purchase value, cardiovascular drugs were the most significant group, accounting for 15.2 and 14.9 percent of purchase value in 1961 and 1966 and increasing to 17.1 percent in 1971 (see ta-ble 4.2). In contrast to antibiotics, cardiovascular drugs mattered both in terms of units and purchase value. They were deployed to treat dis-eases and conditions with high rates of incidence; moreover, the treat-ments often continued for years or even decades. However, we remain cautious in extrapolating from these data, because the category of car-diovascular drugs is too large to accurately represent medical practice and even market organization. From the pharmaceutical industry's point of view, there was no single market for cardiovascular drugs, but rather several different markets to be analyzed separately.

Thanks to a reorganization of the indication classification system, a much clearer portrait emerges from the 1966–1974 IMS DPM report (see tables 4.3 and 4.4). By then, cardiac medicines had been classified into seven subgroups: (1) cardiac glycosides and combinations, (2) other car-diac drugs, (3) antihypertension drugs, (4) diuretics, (5) peripheric va-sodilatators, (6) vasoprotectors, and (7) other cardiovascular drugs. The reorganization recognized antihypertension drugs as a distinct indica-tion subgroup; this classification allows for a detailed analysis of pack-age units sold (see table 4.3) to reveal the diversity of drugs purchased. In other words, we can test Lasagna's notion that new antihyperten-sive drugs dramatically revolutionized therapeutic practice. It turns out that fully one-fifth of the cardiac drugs on the market were neither new nor innovative. Moreover, the much older cardiac glycosides were sold twice as much as newer antihypertensive drugs until the mid-1970s. Even more surprisingly, the West German market for cardiovascular medicines from the mid-1960s to the mid-1970s was not dominated by diuretics or antihypertensive drugs. Antihypertensive drugs accounted for just over 10 percent of the market share of cardiacs. Observed from the user side, older and perhaps therapeutically questionable products played an important role in a field that was considered by advocates of the therapeutic revolution to be one of the most innovative. Instead of the newer diuretics and antihypertensive drugs, the two most sig-nificant subgroups of cardiac medicines purchased, in terms of units between 1966 and 1974, were glycosides and vasoprotectors, in spite of the controversy over the efficacy of the latter class (see table 4.3).

Why did leading pharmacologists and physicians like Louis Lasagna

Table 4.4 Pharmacy purchases of cardiovascular drugs in thousands of German marks, 1966–1974. Calculations are ours, based on IMS DPM Neunjahresvergleich 1966–1974. Archives IMS Health Germany.

Cardiac therapeutic subgroups	1966	1967	1968	1969	1970	1971	1972	1973	1974
Cardiaca	181,835	211,588	248,469	292,421	355,573	404,721	461,363	528,904	592,227
Cardiac glycosides and combinations	74,571	86,551	100,360	117,825	145,229	165,444	190,871	217,763	235,494
Other cardiacs	107,264	125,037	148,109	174,596	210,344	239,277	270,492	311,141	356,733
Antihypertension drugs	49,694	55,758	60,430	68,841	77,862	89,456	106,385	130,151	159,151
Diuretics	19,254	21,913	24,535	29,955	36,006	44,249	55,438	79,632	102,742
Peripheric vasodilatators	39,715	49,334	59,541	74,765	94,261	125,062	157,248	193,927	248,084
Vasoprotectors	91,989	106,729	117,073	135,033	164,792	184,407	195,668	211,317	235,471
Other cardiovascular drugs	42,318	44,032	44,232	45,037	45,412	48,719	50,262	52,194	55,699

Table 4.5 Purchase figures in package units for cardiovascular drugs in general, and for beta-blockers in particular, in millions, 1966–1974. DPM Neunjahresvergleich 1966–1974. Archives IMS Health Germany.

Indication class	1966	1967	1968	1969	1970	1971	1972	1973	1974
All cardiovascular drugs	174.0	178.7	185.0	196.4	205.2	209.7	214.2	214.1	217.9
Beta-blockers	0.134	0.209	0.338	0.512	0.783	1.592	2.183	2.824	3.610

develop such a different interpretation? Viewing the IMS data at an even more detailed level (antihypertensive drug purchase figures for the period of 1966 to 1974), the revolutionary account becomes visible and plausible. During this period, the subgroup of beta-blockers sky-rocketed from 134,000 packages sold in 1966 to more than 3.6 million eight years later, an increase of 2,700 percent (see table 4.5). Within this specific class, change was rapid and dramatic, as revolutionary accounts rightly suggest.

Finally, the comparison of purchases in terms of units shows that the sale of packages of antihypertensive drugs increased steadily by about 40 percent from 1966 to 1974 (see table 4.3). This clearly indicates that their therapeutic use expanded quickly in a considerable manner.[31] Furthermore their purchase value tripled over the same time span (see table 4.5). As an innovative drug class, beta-blockers were able to sell for high prices with impressive sales figures; e.g., their sales units between 1966 and 1974 multiplied by twenty-seven times (see table 4.5). They are similar to antibiotics in their alignment with the therapeutic revolution narrative: they were new and innovative, and brought a radical transformation into treatment.

Nevertheless, these iconic "revolutionary" drugs represent only a fraction of the overall landscape of cardiac drugs, let alone pharmaco-therapeutics in general. The bulk of cardiovascular drugs by sales volumes were older glycosides and vasoprotectors, not newer antihypertensive drugs and beta-blockers. This broader perspective reveals the narrowness of these stories of therapeutic revolution without consideration of the broad range of therapeutics actually in use.

Traditional Outsiders

Finally, we turn from iconic innovative drugs to the counterexample of medicines usually ignored or obscured in these narratives of therapeutic revolution. Let us reexamine the 1966 pharmaceutical landscape (see table 4.1) and take a broader view of the medicines market that aggregated all of the therapeutic-class purchases for that year.[32] Percentages represent the relative share of purchased packages for each therapeutic class. Table 4.1 shows that analgesic drugs, including millions of headache pills, were the most frequently purchased indication group in pharmacies, followed by cough and cold medicines (equivalent to cardiovascular drugs in terms of units purchased). Several of the drug classes typically mentioned as instrumental in the therapeutic revolution do not even appear on this diagram. Antibiotics account for less than 2 percent of purchased medicines: they disappear within the catch-all category "others," lumped together with anorectic drugs, antinausea drugs, antacid drugs, liver protection drugs, anticoagulant drugs, antianemic drugs, anti-infective drugs, antiparasitic treatments, hormones, and vitamins.

This distribution, while perhaps surprising at first glance, can be explained by reasons both methodological and historical. First, vitamin preparations could be sold outside pharmacies, so their numbers appear unusually low in this account. Second, West German women used sex hormones, such as the pill, in far smaller quantities during the 1960s and the early 1970s than their American counterparts did.[33] Third, drugs for minor or common health problems were overrepresented in DPM since hospital pharmacies were not included in the statistical counts. Drugs used almost exclusively in hospital treatment were correspondingly underrepresented. Still, IMS data suggest that the history of medicine and of the use of drugs has paid much more attention to complex and specialized treatments for serious diseases than to the majority of treatments prescribed by physicians or requested by patients. So historical accounts contribute to the selective vision of a therapeutic revolution held by pharmacologists, physicians, and pharmaceutical industries, thus confirming social historians' critique, formulated more than thirty years ago, that classical accounts "from above" have described medicine as practiced solely by science-based university elites, rather than showing the diversity "from below" of medical practices and patient perspectives.

From the perspective of patient demand, we turn to two indication groups that account for large market shares but are considered to be less relevant in the therapeutic revolution narrative: that of sedatives and hypnotics, and that of analgesics.[34] Respectively, the two groups account for 5 and 13 percent of all drugs purchased in 1966 in West Germany. Following the methodology used for antibiotics (see figure 4.3) and cardiovascular drugs (see figure 4.4), the diagram of hypnotics and

sedatives according to their year of introduction indicates the significant presence of older or traditional specialties on the market in 1967,
with thirty-eight out of ninety-one preparations introduced before
1945 (see figure 4.5).[35]

This observation becomes even more acute by homing in on the
chemical substances contained in the individual products. The class
of hypnotics and sedatives included a number of different chemical
groups. The majority of the substances in the hypnotics indication
group of the 1960s and 1970s had been in use since the beginning of
the twentieth century.[36] Barbituric acids had been first commercialized
as hypnotics at the beginning of the twentieth century, with Bayer's
Veronal (barbital, 1903) and Luminal (phenobarbital, 1912) gaining
significant reputations. Products free of barbituric acid relied on either
the older bromides, which were considered rather unsafe, or the more
recent piperidinedione derivatives, a chemical parent of the barbituric

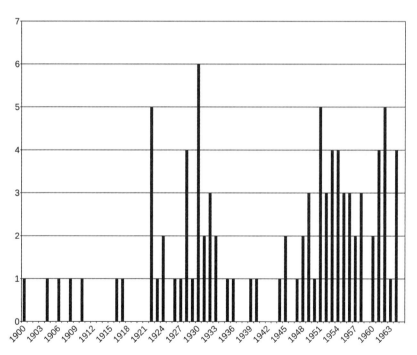

FIGURE 4.5 Year of market introduction for all sedative-hypnotic pharmaceutical preparations listed in DPM for 1967. Annual number of sedatives and hypnotics introduced to the
market (Y axis) between 1900 (first product) and 1967 (year of the census). Data extracted
from DPM Vierjahresvergleich 1963–1967. Archives IMS Health Germany.

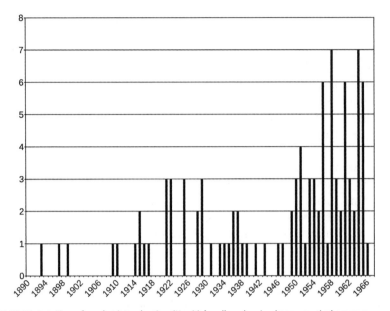

FIGURE 4.6 Year of market introduction (*X* axis) for all analgesic pharmaceutical prepara-tions listed for 1967 in DPM. Annual number of analgesics introduced to the market (*Y* axis) between 1889 (first product) and 1967 (year of the census). Data extracted from DPM Vierjah-resvergleich 1963–1967. Archives IMS Health Germany.

acids, which included gluthethimide (trade name Doriden), and tha-lidomide (trade names Contergan, Distaval, and Softenon).[37]

After thalidomide was removed from the market in late 1961, older bromide hypnotics grew in importance. Beyond the question of the products' therapeutic value, it was precisely their long-standing exis-tence—and their relatively low prices—that gave them value as practi-cally "tested" medicines.[38] Especially in times of uncertainty and di-saster, old well-known drugs have been considered safer alternatives to innovative drugs.

The last indication group we investigate is that of analgesics. As the most important group in terms of market share, it contained a wide diversity of drug products, including the over-the-counter "soft" an-algesics, known pejoratively as "headache pills." Nonprescription an-algesics, together with nonprescription influenza treatments, were by far the most important subgroups in terms of number of packages purchased.[39] Compared to the cardiovascular drugs and the hypnot-ics, the analgesics do not seem to be exclusively "old" drugs. However, an overwhelming majority of the drugs in this class were combination products of two or more active principles, mainly one or two analgesic agents combined with caffeine and either vitamin C (influenza treatment) or sedatives (headache pills); in this way, their composition resembled that of traditional "panacea"-like remedies. And like the hypnotics, new drugs in the category of analgesics (i.e., those introduced after 1950) were almost exclusively based on much older therapeutic substances. In spite of the importance of these therapies—in terms of both users' images of drugs and quantities purchased, prescribed, sold, and probably consumed—none of the nuance or complexity revealed here has been addressed in the narrative of the therapeutic revolution.

When Does Change Become a Revolution?

To think about therapeutic revolution in practice requires a willingness to engage with new forms of data on drug use identified through IMS market surveys. Instead of confirming or invalidating the *concept* of a therapeutic revolution, this chapter suggests a more complex reading of twentieth-century therapeutic change in practice. IMS data indicate that there were clearly radical and dramatic changes in pharmaceutical sales at the pharmacy in the 1960s and 1970s, but they are visible mainly for a few isolated drug groups. The question of whether this change is best described as a revolution remains open, and cannot be decided through quantitative data analysis because it is a qualitative judgment.

For some individual drug categories, historical IMS data bear out Lasagna's theories of therapeutic revolution. Antibiotics clearly reframed fears of infection.[40] Syphilis and tuberculosis, among other diseases, could now be controlled through antibiotics, although condoms and other preventive strategies also contributed. Treatment with beta-blockers, to take another example, fundamentally improved survival rates in heart disease. Yet the symbolic importance of these developments may have left us with an exaggerated sense of their importance in overall pharmaceutical consumption. According to accounts like Lasagna's, neither a single innovation nor a series of specific "miracle drugs," but rather a panoply of new available therapies, shaped the impression of a "therapeutic revolution" during the period under scrutiny.[41]

IMS data suggest that the common narrative of therapeutic revolutions is elitist, normative, and historically problematic. It is elitist since its coherence as a narrative depends on the vision of medical therapy as an expert domain of university physicians and scientists working on and with the latest therapies. Continuity of traditional therapies and tradition-oriented consumer preferences can only be integrated in the narrative as relics of a doomed, less advanced past.

Second, it is normative in the sense that it defines relevance. Lasagna mentions a number of important therapeutic innovations: psychopharmacology, antibiotics, hormones, cardiovascular drugs, vaccines, and, later, immunosuppressants and treatment for arthritis. The drug groups identified by Lasagna and others appear as beads making up one overall chain of therapeutic revolution. Yet, as IMS data have shown, they were in fact poorly represented in terms of purchases, with a couple of notable exceptions (cardiovascular drugs and selected psychoactive drugs). It may not be surprising that highly complex hospital medications for rather rare diseases such as acute leukemia did not appear as significant in terms of overall sales, but even such landmark drug groups as antibiotics and psychotropic drugs can hardly be considered as representative for the whole of therapeutic landscapes from the IMS data perspective.

Third, it is historically problematic to describe a therapeutic regime defined by novel, radical, and quick change as being applicable to entire nations and to humanity in general.[42] As discussed above, the therapeutic revolution narrative suffers from a "Columbus syndrome" of selective perception—fascination with discovery of the new leads to disregard for the already existing therapeutic "other." IMS data suggest the value of looking beyond pharmaceutical marketing to reinvestigate more precisely the nature of the relationship between therapeutic innovation and drug use in the past, present, and future.[43]

The drugs of the so-called psychopharmacological revolution, one last example and highly visible theme in the broader narrative of the pharmacotherapeutic revolution, serve here to sum up our analysis, although a more exhaustive analysis remains beyond the scope of this chapter.[44] According to the narrative of radical change in psychiatric practice, institutional psychiatry and outpatient care were transformed by the new psychoactive drugs made available since the 1950s. The principle argument is similar to that of the therapeutic revolution as a whole. Valium and related benzodiazepines would have produced a series of psycho-pharmacotherapeutic revolutions, and would have been a commercial success as well. At least IMS data supports the latter argument: Purchases of tranquilizers in West Germany increased from roughly 15 million units in 1966 to over 25 million in 1974. Moreover, Valium ranked as the second most profitable product on the West

German drug market, with a purchase amount of 50 million German marks spent on this single product in 1974.[45]

Yet, in terms of units, the revolutionary character of modern psychopharmacology appears less evident. Between 1966 and 1974, annual pharmacy purchases increased slowly for antidepressants, and even showed decreasing sales for neuroleptics.[46] The very idea of a psychopharmaceutical revolution becomes even less convincing when one looks specifically at pharmacy purchases for several hypnotic and sedative subclasses. The drugs that mattered, particularly after the thalidomide scandal, were not just the "modern" benzodiazepines, but the old barbiturate drugs in new combinations and even the bromides from the early twentieth century. These observations raise the question of whether this psychopharmaceutical revolution ever took place, not only in its genesis but also in histories about it, since these latter have been written largely from the point of view of the conquering substances like Valium, as Nicholas Henckes explores in chapter 3 of this volume.

Finally, when we take a closer look at what was actually purchased in great quantities but not yet mentioned, we see that many panacea tonics, cough and cold medicines, and laxatives, in addition to the analgesics, accounted for more unit purchases than many profitable innovations celebrated in both contemporary media and later historical accounts. Many nineteenth-century therapeutic classes typically viewed through the lens of unregulated advertising (and with some suspicion of quackery), such as analgesics, tonics, cough and cold medicines, laxatives, and dermatology and metabolism drugs, were nonetheless a constant presence in pharmacies. The same 1966 list of leading pharmaceutical specialties that ranked Valium as number two for value and number eight for units registered also listed the panacea Klosterfrau Melissengeist[47] at position 19 for units, and Oligoplexe (produced by the herbal alternative medicine manufacturer Madaus) at position 23.

These examples indicate that older drugs often had a much longer existence than narratives of the therapeutic revolution suggest. As discussed above, this phenomenon can be understood in terms of what historian of technology David Edgerton has called the "shock of the old."[48] But these two examples from the user perspective also indicate that what is more generally at stake are the questions of who defines medicines as regular or charlatanesque, licit or illicit, and who decides the perspective from which the history of drugs should be told. Purchase data indicate that patients trusted both the innovative *and* the traditional, while historians might have been swayed to some degree by their actors' discursive framework about the prevalence of novelty.[49] Our use of historical quantitative marketing data, coupled with a close reading of historiography, suggests that most of the existing narratives about the therapeutic revolution have been declarative and normative in nature. Mobilizing the term "therapeutic revolution" functioned not only to analyze the past, but also to prescribe how therapy should be practiced in the present and in the future.

In consequence, pharmaceutical innovations need to be studied within their preexisting markets for medicines in order to allow for a more nuanced understanding of therapeutic change during the so-called revolutionary period of the mid-twentieth century. The interpretations of actors like Lasagna have dominated representations of therapeutic change as the mere diffusion of novelty. Instead, our interpretation of IMS data suggests that many, if not the majority, of medicines in West German pharmacies in the 1960s and 1970s were older products that had been on the market for decades. This finding hints at a second set of attributes, beyond novelty, that have made therapeutic agents trustworthy to physicians and patients: long-standing usage and experience as indices of safety and value.

Of course, we must also acknowledge the limitations of the IMS data and their analysis that inform our historical market perspective. Our analysis is based on aggregating all fields of therapy, thereby lumping together what other scholars might try to split. We do not intend to argue that the specific therapeutic agents on Lasagna's list did not have a significant impact on a specific segment of society, or that they are of minor interest because they represent only a minor and limited proportion of overall sales across a larger population. All the medicines listed by Lasagna mattered very much to individual patients, whose lives were improved or even saved by these therapeutically efficient products. Yet we suggest that the term "pharmacotherapeutic revolution" implies a form of generalization about changes in medical therapeutics in which individual beads of the pharmaceutical progress string are made to stand for the whole. It is this synecdoche that we call into question.

Conclusion

Based on a unique archive of drug purchases in West Germany, this chapter has employed quantitative historical methods to provide a synthetic view of the drugs available and in use in the 1960s and 1970s, and to suggest a different reading of the therapeutic changes that occurred in the middle decades of the twentieth century. It supports the argument made by historians of technology that innovations are neither self-evident nor implemented without resistance. On the contrary, in most cases innovations have to be implemented *against* the resistances of those who are used to or treasure the old and empirically proven.

It may seem unlikely or unwise to apply this claim to the medico-pharmaceutical sector, where the therapeutic and epidemiological benefits witnessed during the twentieth century have been largely uncontested. But questioning these narratives of therapeutic revolution does not necessitate the questioning of the lived experiences of physicians or patients. Cardiovascular therapy actually saved many patients' lives. Insulin and oral antidiabetics improved the daily existence of millions of diabetic patients. Antibiotics enabled the cure of previously untreatable infections.[50] The therapeutic value of all these medicines for individual therapy is not at stake here, nor is their role in the epidemiological transition.[51]

Rather, this chapter showcases the difference between the cultural and historical visibility of the new miracle drugs of mid-century on the one hand and the continued material existence, popularity, economic value, and therapeutic significance of older medicines on the other. University-based, technologically-sophisticated scientific medicine coexisted and continues to coexist with general practice that combines new approaches, older traditions of scientific medicine, and even traditional healing practices and patients' self-medication. All of these therapeutic practices and products must be integrated into a less selective historical analysis of what physicians prescribed and what patients bought in the search of relief and treatment in the twentieth century and beyond.

NOTES

1. Louis Lasagna, "Recent Trends in Drug Development," In *The Inside Story of Medicines: A Symposium,* ed. Gregory J. Higby and Elaine C. Stroud (Madison, WI: American Institute of the History of Pharmacy 1997), 217–222, quote on 217. The authors would like to express their gratitude to Monika Diefenbach and Dr. Gisela Maag of IMS Health in Germany who made possible research in the company's archives. Nils Kessel's research has been supported by a GIS Mondes Germaniques research grant.

2. Lasagna extends his revolutionary period from 1930 to 1990. Other authors refer to the "therapeutic revolution" as the period of 1928 to 1968, or 1930 to 1970. See for example, Pierre Theil, *Le médicament: Mission humaine et fonction sociale* (Paris: A.M.P.S. 1969), 27. For the rest of this chapter we will refer to the "therapeutic revolution" as defined in the context of this book as being essentially situated between 1930 and 1980.

3. Lasagna, "Recent Trends," 217. The complete formulation is, "Cancers not removable by surgery or radiotherapy are only amenable to chemotherapy in a minority of cases." Lasagna, like many of his contemporaries, was concerned about the "drug lag" that resulted from what he saw as overly restrictive regulation of biomedical and pharmaceutical research since the 1970s. Daniel Carpenter, *Reputation and Power: Organizational Image and Pharmaceutical Regulation at the FDA* (Princeton, NJ: Princeton University Press, 2010).

4. See Charles E. Rosenberg's 1977 seminal paper and his contribution in this volume revisiting the original paper. Charles E. Rosenberg, "The Therapeutic Revolution: Medicine, Meaning and Social Change in Nineteenth Century America," *Perspectives in Biology and Medicine*, 20 (1977): 485–506. See also this volume's introduction for a panoramic view of the history of the therapeutic revolution concept. On the notion of scientific revolution, see Thomas Kuhn, *The Structure of Scientific Revolutions* (Chicago: Chicago University Press, 2012). Steven Shapin criticizes the idea of a single process called scientific revolution in his book *The Scientific Revolution* (Chicago and London: University of Chicago Press, 1996).

5. David E. H. Edgerton deals with contraceptives as technologies in his influential book *The Shock of the Old: Technology and Global History since 1900* (Oxford: Oxford University Press, 2007). See Madeleine Akrich, "Le médicament comme objet technique," *Revue internationale de Psychopathologie* 21 (1996): 135–158. See also Jonathan Simon, *Serum as a Technological Object*, volume 2, Habilitation a diriger des recherché Manuscript (Lyon: University of Lyon, 2013).

6. Ideas also have a material nature—paper technologies, ways of knowing and ways of writing, for example. See John V. Pickstone, *Ways of Knowing. A New History of Science, Technology and Medicine* (Chicago: University of Chicago Press, 2001); Volker Hess and Andrew Mendelsohn, "Case and Series: Medical Knowledge and Paper Technologies, 1600–1900," *History of Science* 48 (2010): 287–314.

7. See, for example, Ralph Landau, Basil Achilladelis and Alexander Scriabine: *Pharmaceutical Innovation, Revolutionizing Health* (Philadelphia: Chemical Heritage Press, 1999); Miles Weatherall, *In Search of a Cure: A History of Pharmaceutical Discovery* (Oxford: Oxford University Press, 1990).

8. Historical timelines usually include antibiotic therapy (sulfonamides and penicillin), cardiovascular diseases (in particular hypertension), and

asthma. See, for example, "A Brief History of Pharmacology" Canadian Society of Pharmacology and Therapeutics, CSPT Pharmacology History, http://pharmacologycanada.org/history-of-pharmacology-therapeutics. Accessed 19 March 2014.

9. On the larger question in the history of technology of who "pushes" technologies, see the two volumes of the "Making Europe" series: Martin Kohlrausch and Helmuth Trischler, *Building Europe on Expertise: Innovators, Organizers, Networkers* (Basingstoke, UK: Palgrave Macmillan, 2014) and Ruth Oldenziel and Mikael Hård, *Consumers, Tinkerers, Rebels: The People who Shaped Europe* (Basingstoke, UK: Palgrave Macmillan, 2013). Oldenziel and Hård clearly show the major impact users have on technology use and development.

10. On diffusion of technologies, see Edgerton, *Shock of the Old.*

11. Edgerton, *Shock of the Old*, p. x. For an application of the notion of "shock of the old" to the history of drug marketing, see Nils Kessel, "Beyond Innovation: The Marketing of 'Old Drugs,'" in *The Development of Scientific Marketing in the Twentieth Century,* ed. Jean-Paul Gaudillière and Ulrike Thoms (London, Pickering et Chatto, 2015), 15–27.

12. Sporadic information on individual drugs or classes from IMS sources has been mobilized in George Weisz, "Diagnosing and Treating Premenstrual Syndrome in Five Western Nations," *Social Science and Medicine* 68 (2009): 1498–1505.

13. Patient history is a very productive field of medical history, particularly since Roy Porter's ground breaking studies on eighteenth-century Britain's medical markets. Roy Porter, ed., *Patients and Practitioners: Lay Perceptions of Medicine in Pre-Industrial Society* (Cambridge: Cambridge University Press, 1985). For Germany, see Martin Dinges, ed., *Patients in the History of Homoeopathy* (Sheffield, UK: European Association for the History of Medicine and Health Publications, 2002); Robert Jütte, *Ärzte, Heiler und Patienten: Medizinischer Alltag in der frühen Neuzeit* (München/Zürich: Artemis & Winkler 1991); Martin Dinges and Robert Jütte, eds., *The Transmission of Health Practices, c. 1500 to 2000* (Stuttgart: Steiner 2011).

14. Today the company has evolved towards consultancy in the health sector and is known as IMS Health, deemphasizing the importance of statistics per se and instead identifying itself as an information and strategic advice service company.

15. Similar information does not seem to be available for other countries for a period as early as the late 1950s, nor accessible in a systematic form like the IMS data we have been able to obtain for West Germany.

16. The political relevance of the data made it repeatedly an issue of controversy as companies knew much more about drug markets than did health insurances or government agencies. Dietrich Nord, *Arzneimittelkonsum in der Bundesrepublik Deutschland: Eine Verhaltensanalyse von Pharma-Industrie, Arzt u. Verbraucher* (Stuttgart: Ferdinand Enke Verlag 1976), 73.

17. In order to keep the amount of data manageable, IMS collected raw data from just 300 West German pharmacies in the late 1950s; soon that number grew to 360. The company aimed for representative coverage of the West German nation by taking into account both population distributions by geographical region and by rural or urban location. In the 1960s more than four thousand data sets were collected annually from 360 panel pharmacies identified in proportion to geographic and rural-urban population distribution. Extrapolating from this double-layered demographically adjusted system allowed IMS to claim nationwide representativeness for its census, and at the same time to be able to offer some selective regional data. Data did not include hospital pharmacies.

18. IMS health data indicate manufacturer prices and allow calculation of wholesaler and pharmacy purchases price levels. As pharmacy surcharges were fixed, it is possible to calculate consumer prices. A sales tax was introduced in 1969. A detailed discussion of drug consumption and IMS data methodology can be found in Nils Kessel's forthcoming PhD thesis.

19. Yet client companies were critical of the accuracy of the data produced during IMS's first years of existence, revealing the limitations of such a project in its early stage. In the very early years, Ciba officials in France complained about margins of error. See minutes of the coordination committee meeting of Ciba Laboratories, Paris, 15 June 1962, p. 9, in Novartis Company Archives, CIBA, KGK 2: Fr 41, Konzerngesellschaften Frankreich, 56 Comité de coordination. This criticism does not question the IMS data's validity in general. It is mainly focused on specific products with a minor market share. As most drug markets are dominated by less than a dozen products with data strong enough to support accurate statistical calculations, projections with data of rarely sold drugs have been less accurate in the beginning because only a few packages have been sold in the panel pharmacies.

20. In contrast to consumer research surveys mostly based on questionnaires and interviews.

21. West German health insurance coverage provided easy access to all kinds of drugs either via self-medication or by prescription. Since the 1940s, the German health insurance system had progressively extended coverage to new groups such as elderly people and family members. West Germany's health insurance changed considerably during this period.

22. For the history of antibiotics, see Eric Lax, *The Mold in Dr. Florey's Coat: The Story of the Miracle of Penicillin* (New York: Holt, 2004); Robert Bud, *Penicillin: Triumph and Tragedy* (Oxford, UK: Oxford University Press, 2007); J. T. Macfarlane and M. Worboys, "The Changing Management of Acute Bronchitis in Britain, 1940–1970: The Impact of Antibiotics" *Medical History* 52 (2007): 47–72.

23. Most recently the BBC television series *Pain, Pus and Poison*. BBC4: *Pain, Pus & Poison: The Search for Modern Medicines*, broadcast 3, 10, and 17

October 2013, accessed 17 July 2014. http://www.bbc.co.uk/programmes/
p01f51s5.

24. Examples for indication groups are 01 amoebia remedies, 02 analgesics,
03 anaesthetics, 04 antiacids and stomach therapeutics, 06 antirheumatics
and gout remedies, etc.

25. They were then listed among the drugs of the indication group "dermatology," which included (1) antibiotics and sulfonamide including their
combinations, and (2) corticoids in combination with antibiotics and
sulfonamides. Although they continued to be used in different therapeutic fields such as dermatology, the sulfonamides' importance had strongly
declined in favor of the new antibiotic therapies.

26. Inflation and other end consumer relevant price changes do not influence
sales growth in IMS data as changes in drug pricing can be reconstructed
within the data. In consequence, annual growth rates of 28 percent in
value do not indicate companies' benefits.

27. In this report another classification of antibiotics had been introduced.
The latter included antibiotic treatments that could be found previously
in groups such as dermatology.

28. Despite the relevance of certain periodical infectious diseases for antibiotic
prescription, the picture does not vary considerably during the decade.

29. For the history of cardiac therapeutic agents see, for example, Carsten
Timmermann, "A Matter of Degree. The Normalisation of Hypertension,
c. 1940–2000," in *Histories of the Normal and the Abnormal: Social and
Cultural Histories of Norms and Normativity*, ed. Ernst Waltraud (London:
Routledge 2007), 245–261; Jeremy A. Greene, *Prescribing by Numbers: Drugs
and the Definition of Disease* (Baltimore: Johns Hopkins University Press,
2007); Christian Bonah, "'We Need for Digitalis Preparations What the
State Has Established for Serumtherapy . . .': From Collecting Plants to
International Standardization: The Case of Strophantin, 1900–1938," in
Evaluating and Standardizing Therapeutic Agents, 1890–1950, ed. Christophe
Gradmann and Jonathan Simon (Basingstoke, UJ: Palgrave MacMillan,
2010), 202–228; Cay-Rüdiger Prüll, Andreas-Holger Maehle, and Robert
Francis Halliwell, *A Short History of the Drug Receptor Concept* (Basingstoke,
UK: Palgrave Macmillan, 2009).

30. Few therapeutic indication groups in the IMS statistics were as homog-
enous as the antibiotics.

31. Neither dosage changes determining the daily numbers of pills/packages
taken nor therapeutic change can explain this variation.

32. The 1966 figures do not vary considerably from those of other years in
that decade.

33. See chapter 2 in this volume. In her multigenerational study of contraceptive pill use, historian Eva-Maria Silies states that the first generation of
West German women using the pill did not consider it a tool for female
emancipation. See Eva-Maria Silies, *Liebe, Lust und Last: Die Pille als*

weibliche Generationserfahrung in der Bundesrepublik 1960–1980 (Göttingen: Wallstein, 2010).

34. From a therapeutic point of view, it is irrelevant to split "sedatives" and "hypnotics" into separate indications, as those drugs had both effects in the period we observe.

35. The hypnotic-sedative group of the DPM 1963–67 includes neither minor tranquilizers such as Meprobamate (Miltown) nor benzodiazepines (e.g., Librium or Valium). Even if this drug group included minor tranquilizers, global results would not change significantly.

36. As late as 1974, fifty-nine out of seventy-two drugs of this class were made of active principle dating back before 1950. See Kessel, "Beyond Innovation."

37. The latter was recalled after causing a worldwide drug disaster. Thalidomide caused more than ten thousand malformations in fetuses. The best documented history of Thalidomide in Germany is still Beate Kirk, *Der Contergan-Fall: Eine unvermeidbare Arzneimittelkatastrophe? Zur Geschichte des Arzneistoffes Thalidomid* (Stuttgart: Wissenschaftliche Verlagsgesellschaft, 1999). Carpenter, *Reputation and Power,* shows the FDA's reaction to thalidomide in chapter 4. See also Arthur Daemmrich, "A Tale of Two Experts:. Thalidomide and Political Engagement in the United States and West Germany," *Social History of Medicine* 15, no.1 (2002): 137–158. The entangled history of the competing products Doriden and Contergan (Thalidomide) is studied in Nils Kessel, "Doriden von CIBA: Sleeping Pills, Pharmaceutical Marketing and Adverse Drug Reactions," *History and Technology,* special issue, 29, no. 2 (2013): 153–168.

38. Relying on Louis Lasagna's statement that older medications can be more useful because of the knowledge about them, Canadian historian of psychiatry Edward Shorter has argued that the barbituric acids could still provide useful medication for psychiatric disorders and could even be a better alternative than certain newer psychotropic drugs. See Edward Shorter, *How Everyone Became Depressed: The Rise and Fall of the Nervous Breakdown* (Oxford: Oxford University Press, 2013), 149.

39. IMS DPM 1966–74, vol 1.

40. See chapter 1 in this volume.

41. Physicians at this time were deeply impressed by the number of new treatments available. The chairman of the British Committee for the Safety of Drugs, Sir Derrick Dunlop, insisted on the important victory over infectious diseases: "Diphtheria, from which as late as 1940 there were 2,500 fatal cases in England and Wales alone, has disappeared; typhoid, typhus, tetanus, cholera, plague, yellow fever, rabies, smallpox, measles, whooping-cough, and poliomyelitis can be prevented; many tropical diseases such as malaria have been controlled; and the lives of patients suffering from diabetes and pernicious anaemia can be preserved and considerable relief given to sufferers from hypertension, arthritis, asthma,

and many nervous and mental disorders. The list is far from comprehensive . . . ." Derrick M. Dunlop, "Use and Abuse of Drugs," *British Medical Journal* 5459:2 (1965): 437–441.

42. Studies on national and regional specificities of therapeutic change are therefore of greatest importance. Chapters 8 and 9 in this volume illustrate the many different ways of appropriating therapies.

43. A special issue of *History and Technology* has recently shed light on pharmaceutical marketing. See Jean-Paul Gaudillière and Ulrike Thoms, "Pharmaceutical firms and the construction of drug markets: From branding to scientific marketing," *History and Technology*, special issue (2013) 29, no. 2: 105–115.

44. See chapter 3 in this volume.

45. Between 1966 and 1974 Valium's purchase value increased annually by 77 percent.

46. Once again, the validity of those figures is limited by the fact that only outpatient psychiatry is taken into account while institutional psychiatry (as part of hospital markets) is excluded.

47. A tonic called "Carmelites' Balm."

48. Edgerton, *Shock of the Old.*

49. Ibid. Nils Kessel has argued elsewhere that current scholarship is largely focused on innovation as the driving force of pharmaceutical markets. Studying pharmaceutical development, regulation, and marketing, recent scholarship in the history of drugs has undoubtedly added much to a better understanding of how therapeutic agents reach their markets. Nevertheless those accounts remain largely captured in an exemplary-drug-story-accounting-for-the-whole approach that has been cast originally by proponents of the "revolution" and of "progress." This is equally true for historiographical accounts such as the "biographies of remedies" or the "drug trajectories." Studies on therapeutic agents such as sex hormones, insulin and oral antidiabetics, new antidepressant drugs, cardiovascular drugs, and cancer medication have been particularly inclined to overestimate the importance of those medications outside the hospital.

50. Lasagna, "Recent Trends," 217.

51. Diverging interpretations of epidemiological transition are discussed by Paul Farmer, Matthew Basilico, and Luke Messac in chapter 7 of this volume.

FIVE

Recurring Revolutions?
Tuberculosis Treatments
in the Era of Antibiotics

JANINA KEHR AND FLURIN CONDRAU

There are no more revolutions in store to impel a continued forward flight.
—BRUNO LATOUR, *WE HAVE NEVER BEEN MODERN*, 1993

On 12 May 2007 Andrew Speaker, son of a Centers for Disease Control epidemiologist, boarded an Air France flight from Atlanta to Paris with his fiancée. After two days, the couple continued their journey again on Air France to Athens, and then by ferry and Olympic Air flights to the island of Santorini in the Aegean Sea, where they got married. Their return journey led the Speakers via Mykonos back to Athens, and then by Czech Airlines first to Rome and then to Montreal, from where they crossed back into the United States in a rental car.

Before he had first left Atlanta, however, Speaker was already the subject of a diagnostic investigation, based on the assumption that he might be carrying a strain of XDR tuberculosis, the extensively drug-resistant form of the disease, which had recently become the focus of renewed public health interest in tuberculosis.[1] Speaker would ultimately be isolated (against his will) in New York under the Public Health Service Act.[2] His case became a full-blown public health scare and scandal, involving a congressional hearing about the exact nature of his disease and the roles played by the various agencies he encountered. National news outlets jumped on the story, lawsuits were filed by passengers on the same flights, and thinly veiled references to Typhoid Mary were made, as were suggestions that the travels of this (healthy?) carrier of TB had something to do with bioterrorism.[3]

The Speaker story went viral across the Western world, from the *New York Times* to *Le monde* to online news outlets. The TB Alliance recognized a few weeks later that it had "ignited a media firestorm and a minor health panic."[4] However, despite being an eventful case with extensive media coverage producing public statements oscillating between angst and reassurance, the Speaker case is far from unique. Similar cases—in which air travel by tuberculosis patients has triggered highly visible public fears and a less visible plethora of public health interventions of isolation, contact tracing, and preventive treatment—happened before and afterwards.

Such was the case in France in 2006, when a Chechen refugee arrived at the Paris international hub Charles de Gaulle Airport with severe respiratory distress. After having fled Chechnya via Beirut with his wife and his two children, he got off the plane coughing blood and was taken care of by the French emergency rescue services who drove him to the nearest hospital. Despite surgery to remove those parts of his lung most destroyed by mycobacteria, the patient died from severe bleeding a few days later in a hospital. He left behind not only his wife and two children, but also the question of how to handle this "new tuberculosis"[5] in its mobile, multiresistant form, in the clinic as much as in public health. Drug sensitivity testing had indeed shown that the patient was infected with the mycobacterium tuberculosis strain, which was resistant to isoniazid, rifampicin, streptomycin, kanamycin, amikacin, capreomycin, fluoroquinolones, ethambutol, and thiacetazone—a result that "met the WHO case definition criteria of XDR TB."[6] Given the antibiotic resistance of the strain, the pharmaceutical treatment options available to prevent the disease among the refugee's family remained rather limited. The preventive treatment initiated for the mother and her children had to rely on alternative antibiotics with heavy side effects. The public health doctor responsible for the follow-up characterized the drugs as "extremely strong and with a high level of toxicity for the women's body." Treating the refugee's wife and his children thus necessitated "a very tight clinical surveillance of the side effects during the course of the treatment, a treatment that lasts for 2 years."[7]

These stories confirm that public health and clinical medicine faced serious challenges at the beginning of the twenty-first century to treat and control tuberculosis. It is a key concern of this chapter to demonstrate that the contemporary crisis has deep historical roots. By intertwining a historical with an anthropological analysis, we hope to be able to reconnect the contemporary tuberculosis problem to its long history, whilst also making visible the limitations of a historical analysis ending prematurely.

A half-century earlier, in 1952, Selman Waksman received the Nobel Prize in medicine for the discovery of streptomycin as the first antibiotic useful in the treatment of the disease. Waksman was by no means alone when he proclaimed that "medical science and clinical practice have been revolutionized"[8] through antibiotics. He emphasized that much more progress was to come in the near future: "One may look forward to further discoveries of agents that will combat diseases not now subject to therapy, to more active and less toxic agents than those now available, and to combined therapy of several antibiotics or of antibiotics and synthetic compounds which will prove to be more effective than the use of single substances."[9]

Waksman's revolutionary language and heroic optimism proved realistic in the short term, as new antibiotics against tuberculosis did indeed become available in the following years. Short-course therapy against tuberculosis, combining four antibiotics, was implemented in the 1970s, and for a new generation of chest physicians and public health officers, TB lost not only its dangerousness but also, according to historian Anne Hardy, its "news value" and "medical interest."[10] Historians of medicine became interested in tuberculosis within the paradigm of a declining disease quite literally of the past.[11] Pierre Guillaume, the French historian of tuberculosis, entitled his book *From Distress to Salvation*, and argued that "the happy ending of the history of tuberculosis in the developed countries leaves the role of tuberculosis as a bugbear, which it held for almost two centuries, to other diseases."[12] National epidemiological statistics in the West showed a regular decline in disease incidence, and Abdel Omran, a professor of public health and head of a WHO reference center in epidemiology, supported such notions of decline through his framework of an epidemiologic transition from infectious to chronic diseases.[13] Tuberculosis was, in Omran's view, the quintessential disease of transition: chronic yet contagious, significant as much because of social circumstances as because of bacteriology. Interestingly, this new framework of an epidemiological transition hinged on Omran's work among the Navajo in Arizona as he looked for a field in which to study an old disease that had effectively been relegated to a Third-World disease in the West.[14] All in all, the end of TB was thought to be near, at least in the West. This assumption was shared by contemporary tuberculosis control experts in the 1950s and 1960s, and by historians of medicine who became interested in the disease in the 1980s and 1990s.

Unfortunately, as Paul Farmer has pointed out, the historical narrative of decline and disappearance of tuberculosis has turned out to be somewhat Eurocentric.[15] It seems to us that notions that tuberculosis has come back are also missing the point.[16] In fact, our interdisciplinary perspective allows us to grasp that TB has never really gone away, and that the rhetoric of a return has masked the persistent presence of the disease on a worldwide scale, albeit with strong geographic and social inequalities in its epidemiology.[17] But news media in all formats have accepted the notion of a return, and stories of particularly threatening forms of TB and of patients spreading ultraresistant superbugs are frequent. Have multi-resistances subverted the dream of TB control through pharmaceuticals? Have they perhaps even halted the faith in drug-based solutions or revolutions, in "magic molecules"[18] and "miracle drugs"? The story we aim to tell in in this chapter is not a story about the return of tuberculosis perhaps after a failed revolution, diagnosed in hindsight. Nor do we aim to investigate ultraresistant TB as a novel phenomenon. Rather, our chapter investigates the tensions between treatability and untreatability, between acute emergencies and chronic states of disease and care, and between the threatening (and thus exciting) episodes and the declining (and therefore uninteresting) periods that have been characteristic of the fight against TB since the nineteenth century. In our view, these tensions are central to comprehending the practices and problematizations of modern biomedicine regarding diseases and patients. Within an increasingly technologized and pharmaceuticalized field, tuberculosis exists as an awkward and backward spatio-temporal object—an object whose age as well as low status has made it marginal and uninteresting for much of the contemporary clinical avant-garde in the global North. Yet imaginaries of traveling and untreatable superbugs in a world without borders render this old disease interesting for biomedical research and practice once again—an interest whose duration is nevertheless as uncertain as the future of tuberculosis.

This chapter combines historical and ethnographic methods. From an interdisciplinary approach we hope to gain historical depth for the contemporary enquiry, as well as analytical sharpness for the historical enquiry. We analyze the tensions around the treatable and untreatable, the emergent and the chronic, in two distinct time frames, conceptualized as historical and ethnographic case studies. The first case, "pharmaceuticals and the end of tuberculosis's future," is historical, and the second case, "tuberculosis's second modernity," is ethnographic. The first period concerns the decline and disappearance of TB from Western countries in the first half of the twentieth century, whereas the second period touches on the reappearance of TB in the wake of HIV/AIDS, multiresistance, and globalization at the end of the twentieth and the beginning of the twenty-first century. Throughout both sections, we interrogate the narratives of revolution and stagnation and of exciting acuteness and boring routine, which we regard as significant for the relation between tuberculosis and its treatments, as well as for the ways in which a disease and its biomedical treatments and public health approaches are defined, problematized, and understood in medicine and society, past and present.

Pharmaceuticals and the End of Tuberculosis's Future

The tension between treatability and untreatability has been a characteristic of the history of tuberculosis for a long time. The clear effectiveness of antibiotics has contributed to the labeling of previous treatment regimes as "failures," both in real time and in hindsight.[19] That notion is a classic example of the dangers of retrospective analysis, where the contemporary serves as the measuring stick for the past. We aim, however, to look to the past for examples of debates between the treatable and the untreatable, with the hope of uncovering patterns that are of importance not just in the historical realm, but also for the contemporary viewpoint. A first major example for this general theme in the history of tuberculosis is Robert Koch, the German bacteriologist credited with the discovery of mycobacterium tuberculosis in 1882.[20] His work was part of a broader change in the medical understanding of diseases that began to resolve the old debate between contagionist and anticontagionist explanations of infection. The older medical hygiene approach, championed in Germany by Max von Pettenkofer, one of Koch's oldest and fiercest rivals, favored local environmental explanations and thus triggered an unprecedented investment in local sanitary infrastructure during the second half of the nineteenth century. It failed, however, to link the proposed disease etiology to a coherent therapeutic regime.

Koch believed that, with the new etiology of infectious diseases, therapeutic success must be near. In August 1890, a mere eight years after the discovery of mycobacterium tuberculosis, he announced tuberculin as an effective treatment against tuberculosis at the Tenth International Congress of Medicine in Berlin.[21] Koch's tuberculin therapy created immediate headlines in Germany and around the world. A veritable euphoria broke out in Germany, which the historian Barbara Elkeles called "tuberculin rapture" (Tuberkulinrausch); it celebrated how, almost overnight, tuberculosis had been beaten by the genius of Koch.[22] Alfred Grotjahn, a leading German social hygienist and a member of the Reichstag for the Social Democratic Party (SPD), described the arrival of tuberculin in his autobiography:

Finally the great day also arrived for Greifswald on which the Clinic for Internal Medicine was to carry out the first inoculations with tuberculin. It was celebrated like the laying of a foundation stone or the unveiling of a monument. Doctors, nurses and patients dressed in snowy white and the director garbed in a black frock coat stood out against a background of laurel trees: ceremonial address by the internist, execution of the vaccination on selected patients, a thunderous cheer for Robert Koch![23]

Yet clinical introduction preceded careful testing, and it soon turned out that tuberculin did not improve the survival rates of tuberculosis patients. Urgent testing in German and international hospitals soon confirmed that tuberculin was not the medication Koch had hoped it to be. In 1891 the pathologist Rudolf Virchow performed a careful study of pathologic samples of treated patients and revealed not only that tuberculin did not influence the disease in the way Koch had described, but also that tuberculin may well have accelerated the disease process.[24] Almost as quickly as it had appeared, tuberculin disappeared as a mainstream medical treatment, while remaining a viable diagnostic tool. This episode had a profound influence on the development of clinical testing of medical effectiveness, and Koch was rewarded with his own research institute as well as control over hospital beds at the Charité Hospital in Berlin to continue his research on infectious diseases.[25] The tuberculin episode also triggered renewed interest in the social conditions of disease. Without a treatment solution, the whole logic of bacteriology as the sole explanation for disease came into question, which allowed the aforementioned social hygienist Grotjahn and others to claim the authority of social hygiene in the control of tuberculosis.[26] The tuberculin disaster thus is not only of relevance to a history of therapeutic revolutions, but also in relation to therapeutic versus preventive concepts of infection control.

Another consequence of this failed therapeutic innovation was an increase in popularity for alternative treatment regimes. Beginning in the 1890s, the sanatorium became the flagship institution in campaigns against tuberculosis. The idea of the sanatorium was of course older, and linked tuberculosis treatment with existing notions of convalescent homes and physical treatment regimes. The German social insurance scheme (1889) and the English national insurance (1911) brought sanatorium treatment into the center of social policy debates and funded huge networks of these institutions. The kind and quality of the treatment provided remained unclear. An experienced German sanatorium pioneer, Peter Dettweiler, argued that, in the absence of a specific remedy, the sanatorium regime was the only treatment option remaining: "Unfortunately it seems to me personally that the expectation, that bacterial tuberculosis, the most complex of diseases, may be treatable with a specific remedy, is highly unlikely."[27] Exposure to open air, a rich and plentiful diet, and rest or gradual exercise was supposed to strengthen the body in its fight against tuberculosis. Marcus Paterson, the superintendent of the Brompton Hospital Sanatorium at Frimley, outside of London, labeled the treatment regime "auto-inoculation against tuberculosis."[28] In its explicit reference to vaccination, this label contributed to the blurring between effective treatment and individual prevention in the sanatorium. R. C. Wingfield, Paterson's successor at Frimley, commented in 1924: "Sanatorium treatment is not a drug to be handed out in known doses. It is not a cure for pulmonary tuberculosis. It is a form of treatment intended to refit the consumptive for the ordinary conditions of life, and to educate him how to live it."[29]

The sanatorium system required substantial investment in infrastructure outside of the existing hospital systems, and this triggered unprecedented questions about the validity of the treatment. The initial claims were made on the basis of individual cases as success stories. Peter Dettweiler, for example, published a report in 1886 of more than seventy successfully cured patients, as justification for further investment in the sector.[30] But suspicions about the economic self-interest of doctors began to be raised as well. German sanatorium doctors were occasionally called "business-minded hoteliers" by medical colleagues and the lay public.[31] With an unclear and often changing treatment philosophy, and with concerns about the obvious self-interest of sanatorium doctors, questions surfaced about the efficacy of such treatment. From around 1900, an increasing number of studies began to address post-sanatorium survival rates. In particular, the German social insurance organizations, the *Landesversicherungsanstalten*, were keen to know whether there was any economic benefit to sending patients to a sanatorium.[32] It soon became clear that within five years of discharge, around 50 percent of all patients had either left employment or died. Similarly, British studies focusing on length of survival found that about half of the sanatorium patients had died within five years of discharge. Grotjahn commented dryly: "How anyone could ever have any joy with such numbers is beyond me."[33]

With treatment success not forthcoming, the sanatorium system reinvented itself as a place for social policy and education. Ernst von Leyden, an influential clinician and spokesperson for the sanatorium movement, argued: "We have never claimed to cause miracles with our sanatoria, but we have promised to give the same care to the most needy and the poorest, that is readily available to the affluent and the rich."[34] The justifications changed considerably under the influence of adverse statistical data. Nonetheless, a lack of alternative treatment options allowed these institutions to remain important until the early 1950s despite heavy criticism by bacteriologists. George Cornet, a disciple of Robert Koch, argued in his influential textbook on tuberculosis: "The campaign against tuberculosis in the sanatorium is worth no more than trying to fight a famine with caviar and oysters rather than bread and bacon."[35] Fundamentally, it seems that even to this eminent bacteriologist the tensions among medical science, specific treatment, and social determinants of disease remained unresolved.

What changed after World War II? The news of a potentially effective antibiotic treatment against tuberculosis coincided with the war's end.[36] In a striking parallel to the sensationalist terms that we have encountered earlier in the tuberculin disaster, streptomycin developers were "besieged by panic requests for the drug."[37] Selman Waksman, who won the Nobel Prize in medicine for his discovery of streptomycin, recounted the story of tuberculosis treatment in his Nobel Lecture of 1952. In his narrative, Waksman contrasted the centuries-long burden of disease against his observation that medicine had been "revolutionized."[38] While he was certain that more breakthroughs were to come, he linked future developments to a veritable "antibiotic gold rush," suggesting, quite rightly, that there was a lot of money to be made with the discovery and clinical introduction of antibiotics.

The development of sulfonamides, largely credited to Gerhard Domak of IG Farben (later to become Bayer AG), had prepared the ground.

In particular, in the control of streptococcal childhood and maternal infections, sulfonamides proved to be quite effective, attracting a first wave of celebrity patients such as Franklin D. Roosevelt, Jr., son of the US president, whose illness was successfully treated with Prontosil in 1936.[39]

The subsequent "antibiotic goldrush" predicted by Waksman did take some time to take hold in the field of tuberculosis control.[40] Issues of scarcity, emerging Cold War rivalries, and international trade limitations certainly played a role. In addition, tuberculosis physicians who had already witnessed so many promising therapies come and go remained somewhat skeptical about the long-term benefits of streptomycin. In one of the first assessments of the clinical efficacy of the new drug, published in 1946, Corwin Hinshaw, a clinical researcher at the Mayo Clinic, argued that the new drug was "a potentially useful adjunct to approved and timetested therapeutic procedures in tuberculosis, but by no means a substitute for them."[41]

Also, the first wave of the clinical introduction of antibiotics led not necessarily to permanent cure, but to relapses over time. In fact, some tuberculosis sanatoria continued their operations into the 1960s, mainly aiming to safeguard the long-term recovery of patients after hospital treatment with antibiotics.[42] Eminent researchers and bacteriologists were not surprised. They recognized that the problem of tuberculosis was very closely related to resistance. René and Jean Dubos wrote, "Unfortunately, streptomycin and PAS (4-aminosalicylic acid) rarely bring about a permanent and complete cure of pulmonary tuberculosis. . . . the bacilli often become resistant to the drugs, in particular to streptomycin, after a few weeks to a few months of treatment."[43] But together with isoniazid, streptomycin and PAS began to be seen as the answer to tuberculosis without running much risk of resistance.[44]

But while clinical practice adjusted to the newly available antibiotic therapy in line with the striking growth of antibiotic production during the early 1950s, questions remained about the validity of the treatment outside of institutions. Here, two stories intersect: one about antibiotic development and subsequent clinical introduction, and the other about the emergence of the clinical trial regime as a standardized way to measure the efficacy of medications. In the United States, tests were conducted as large-scale treatment trials, which showed great results. In Britain, a trial under the auspices of the Medical Research Council (MRC) was conducted with a control group receiving sanatorium treatment but no streptomycin. Comparable tests had been done before, but Austin Bradford Hill, the lead statistician, randomized the selection of patients into each of the two groups. John Crofton, a researcher on the trial team, recalled later, "These results were better than anyone had achieved anywhere in the world, indeed far better than we ourselves had expected, and for a number of years our figures were not believed. Perhaps because of this . . . we received large numbers of visitors from abroad [who] were more interested in learning about the treatment methods we had based on the results of our studies."[45] On this basis, the concept of statistical measurements of treatment outcomes received the boost it needed to lay the foundation for a new framework to evaluate medicines.[46] Indeed, Bradford Hill later developed stringent criteria of causality based on his experience with the streptomycin trials and his later work on smoking and lung cancer.[47]

The streptomycin trials resolved the question of whether or not antibiotics were effective against tuberculosis. In the late 1950s, tuberculosis studies undertaken in Madras and Bangalore by the Indian Council for Medical Research in conjunction with the MRC and the World Health Organization (WHO) engaged wider questions of tuberculosis control. Despite clear signs of a continuation of antibiotic resistance, which had earlier dogged the single-antibiotic treatment regime, these studies were publicized as proof of the universal effectiveness of antibiotics regardless of social situation: a global cure for tuberculosis.[48]

Yet antibiotics researchers themselves remained aware that tuberculosis continued to be a problem even if it had moved out of the Western limelight. Crofton, who had participated in the streptomycin trials and later isoniazid research at the MRC, argued that "it is clear that tuberculosis is very far from being defeated, even in economically developed countries. Indeed, in underdeveloped countries it is often said that little can be done until the standard of living is raised. One can hardly accept that."[49] The WHO tuberculosis specialist and future director-general, Halfdan Mahler, summarized this new position in tuberculosis control in 1968: "The technology for controlling tuberculosis [has been] standardized and simplified to such an extent [that the solution lay] in setting up an effective . . . sales organization with standardized consumer goods."[50] The WHO Expert Committee on tuberculosis stated in their Ninth Report of 1974 that the disease remained "a problem in many technically advanced countries," where it "often causes more deaths than all other notifiable infectious diseases combined."[51] This historical perspective helps to explain how antibiotics were established as the favored intervention of tuberculosis control. The therapeutic innovation, the statistical modeling of success, and the international collaboration in effectiveness research marginalized alternative control strategies focused on poverty or on the continuous need for institutional treatment for tuberculosis to prevent resistance.

Tuberculosis's Second Modernity

Tuberculosis had thus disappeared as a major public health concern on a worldwide scale by the end of the 1960s, as had the search for new ways to prevent and treat this infectious disease in the medical field. But a "new tuberculosis"[52] appeared at the end of the 1980s, in a "deadly liaison"[53] with AIDS and in multi- or ultraresistant forms. These novel forms of disease posed a significant challenge to the contemporary clinic and public health. Beginning around the turn of the twenty-first century, media outlets as well as public health officials concerned themselves with issues such as multi- if not ultraresistance, often arguing against the perceived dormancy of this disease in the scientific, medical, and political realms.[54] In Western clinical practice, though, tuberculosis did not lose its image as a disease of the past. The historical narrative of decline and disappearance was coupled with a routinized standard treatment regimen, making it uninteresting, even boring, from the medical practitioner's point of view. As we will show, the tediousness of TB in the clinic in the global North contrasts dramatically with the contemporaneous emergency scenario and call for novel revolutions in the field of public health in the global South.

"Tuberculosis is simple—it's nothing difficult."[55] This expression from an intern in a German hospital mirrors a widely held perception in the fields of both infectious diseases and lung health in French and German hospitals at the turn of the twenty-first century. The extensive distribution of this perception was observed and validated through interviews with doctors, nurses, and social workers at different career stages in a multi-sited ethnography between 2006 and 2010. In this view, TB is nothing difficult because once diagnosed, it is easily curable through a standardized treatment regimen consisting of six months of combination therapy with isoniazid, rifampicine, pyrazinamide, and ethambutol. This "one-size-fits-all"[56] pharmaceutical treatment approach is indeed applied in French as well as German hospital settings on a standard basis.[57] The approach has been promoted globally by the WHO since 1995 under the name of DOTS (directly observed treatment, short course), just two years after the WHO declared tuberculosis a "global emergency" in 1993. As the policy analysts Jessica Ogden, Gill Walt, and Louisiana Lush have shown, this emergency scenario—in the wake of TB/HIV coinfection as well as during the intermittent rise of resistant tuberculosis in metropolitan centers in the North, like New York City and London—was accompanied by the global "branding" of DOTS, thereby bringing TB successfully "back to the attention of policy makers" and making it "a problem in need of redress."[58]

Despite such renewed attention to tuberculosis on an international policy level, in the field that came to be known as "global health" during this time, the view of TB as tedious prevailed in European and North American biomedical settings, particularly outside the realm of the designated TB control institutions. Not much was at stake, either scientifically or clinically, for doctors treating a tuberculosis patient in clinics in the North, as this German resident physician illustrates:

Once this is clear, then there is this algorithm; then we go to the algorithm that the patient gets educated. He will have to sign that he was educated also in the side effects of the drugs. Then he receives the drugs, adjusted to his weight, first in standard combination of four different drugs. I don't know the actual dosage off the top of my head; I have to read it up always. Then I program consultations with the otolaryngologist and the ophthalmologist. Yes, that's it, really. . . . The positive thing for me as the ward physician is that, well, they cause a lot of work, really, only for the first two days; but then you don't really have much to do with them anymore (laughs). Then you can just go in, a short consultation, then there is not much diagnostics that you will have to do; you'll have to monitor the taking of the drugs and eventually . . . you can just discharge them.[59]

As this explanation shows, treating a patient for TB is merely an algorithm, a standardized procedure to follow, for which one does not need to be particularly specialized or engaged. This technical take on TB control through the administration of pharmaceuticals allowed for the mainstreaming of DOTS on a global level. The WHO medical officer Demot Maher argued in a widely read *Lancet* article in 1994 that DOTS relied "on the implementation of old, tried and tested technologies."[60] This technical take also perpetuated disinterest in this old disease from much of the medical community, at least in the North. As the anthropologist and health care worker Paul Draus confirmed in his ethnography of TB control in New York City, "In the modern medical world, tuberculosis is not a difficult condition to treat . . . and therefore [is] boring from a medical point of view."[61]

Tuberculosis, with its old diagnostic gold standard of sputum culture[62] and its relative absence of novel clinical trials or pharmaceutical innovation, had indeed become an uninteresting disease for much of the clinical avant-garde in the global North since the 1970s. As sociologist Adele Clarke and her colleagues have argued, this medical elite increasingly defined itself as "techno-scientific."[63] However, interviews conducted with French and German lung specialists reveal doubts about the technoscientific value of tuberculosis treatment and its control. The head of a German lung health department argued, "In the tuberculosis clinic, we are as far from research as birds from ornithology,"[64] referring to the disarticulation of clinical research and routine treatment in his hospital. Meanwhile, the head of a French infectious diseases department concluded, "There is mostly empiricism in the treatment of tuberculosis,"[65] in a discussion of the relative absence of novel evidence-based treatment guidelines beyond the standard combination therapy, as well as of the practical difficulty of accompanying socially and culturally diverse patients until the completion of their treatment. On the potential of prestige and remuneration, a German lung specialist noted that "you can't make money with infectious diseases like tuberculosis,"[66] pointing out that TB had been effectively relegated to the margins of interest in his department during the previous two decades by an increasing attention to chronic pulmonary diseases like asthma or chronic obstructive pulmonary disease (COPD).

These declarations sit uneasily with Clarke and her collaborators, who argue that since the 1980s, healthcare has found itself in a process of biomedicalization, defined as "the increasingly complex, multi-sited, multidirectional processes of medicalization that today are being both extended and reconstituted through the emergent social forms and practices of a highly and increasingly technoscientific biomedicine."[67] Without doubting the validity of these claims, we conclude that the case of tuberculosis nevertheless clearly demonstrates that some diseases are more suitable for processes of biomedicalization than others, and differently so around the world and across time. Suitable objects in the global North are lucrative chronic diseases, for which patients need continuous surveillance and are encouraged to take "drugs for life,"[68] or highly research-intensive fields such as cancer.[69] Yet tuberculosis, with its straightforward treatment regime, has ceased to be a suitable object for contemporary biomedical attention in the North, having already been biomedicalized decades earlier. Biomedical research and the clinic as the prime site of practice had indeed focused on other diseases after the 1970s. Having found a pharmaceutical treatment "solution" in the

1950s, biomedical science turned away from old diseases like tuberculosis, and toward newer, more interesting, and more profitable challenges. The fact that almost no historical studies of TB research and treatment between the 1970s and the 1990s exist makes it difficult to actually describe how this gradual disinterest in the disease, still palpable in Northern clinics today, was orchestrated.[70]

In the early twentieth century, tuberculosis control took place within strong bureaucratic institutions. But after the advent of antibiotic combination therapy, and with declining disease rates in the 1970s, public health infrastructures, like dispensaries and mobile screening units, decreased in number and scope, along with their staff. The status of tuberculosis control changed in Western Europe and in North America. A national priority of many Western nation-states until the 1960s,[71] TB control and related institutions steadily moved toward silent functioning, loss of professional reconnaissance, and relative invisibility in the United States,[72] the United Kingdom,[73] France,[74] and Germany. The flagships of early twentieth-century public health, the local tuberculosis dispensaries, have lost their status. Instead, these almost forgotten, often overlooked health bureaucracies have become characterized by the administrative and low-tech functioning typical of state apparatuses and national programs. They portray the image of backwardness, stagnation, and low prestige so often attached to tuberculosis itself. In sum, TB started a shadow existence in medical departments of the North in the 1970s—a shadow existence that continued into the twenty-first century.

Yet despite this shadow existence within clinical biomedicine and public health in the global North, the growing attention to tuberculosis on a worldwide scale since the mid-1990s, materialized in the 1993 WHO declaration of tuberculosis as a "global health emergency," cannot be ignored. During the decades of its shadow existence, a nonnegligible percentage of tuberculosis cases had become challenging or even untreatable, albeit very unequally so around the world.[75] Untreatable forms of tuberculosis, like XDR-TB, have contributed to slowly reversing the historical trend of treatment stagnation and Northern disinterest in the disease.[76] The strategic focus on untreatable forms of TB on a global scale and the establishment of an emergency scenario of antibiotic resistance led to calls for shorter, safer, more effective TB drug therapies. This brings us back to Waksman. He, too, had demanded and hoped for future therapeutic progress in spite of streptomycin's overwhelming success.

In March 2014, with World TB Day approaching for the thirty-

second consecutive year, Doctors without Borders (Médecins sans fron-
tières, or MSF) issued a comprehensive "crisis alert" on drug resistant
tuberculosis. MSF's crisis document subtitle read, "The New Face of an
Old Disease: Urgent Action Needed to Tackle Global Drug-Resistant TB
Threat." The MSF medical director held that "it doesn't matter where
you live; until new short and more effective treatment combinations
are found, the odds of surviving this disease today are dismal."[77] Charts
of duplicating lines of pill-clones filling almost a page and graphic depictions of the amount of pills that MDR-TB patients need to ingest
presented a case against the current MDR-TB treatment regimen. According to the TB Alliance, a major actor in the field of tuberculosis
drug development, the current MDR treatment "takes too long to cure,
is too complicated to administer, and can be toxic."[78] The labeling of
TB as an old disease paradoxically parallels its representation as a truly
new and unprecedented phenomenon in its drug-resistant form.

A depiction of drug-resistant TB as a global threat goes hand-in-hand
with calls for immediate, urgent action in the realm of pharmaceutical
development and political and economic mobilization. Such calls for a
novel wave of "mass mobilization" in the field of tuberculosis control
and pharmaceutical research appeal to imaginaries of an overdue therapeutic revolution—imaginaries that are as common among physicians
working in public TB control centers as in international policy briefs.
Through ethnographic field research we met Anne, a pneumologist and
director of a center for TB control in France, who regularly announced
in conversations with her colleagues, nurses, and physicians:

Nobody is interested in tuberculosis. For AIDS, there are lots of protocols, there is
progress, there are studies. For tuberculosis, there are no studies, there is nothing.
You know the article from 1963 you gave me about contact tracing, we are still doing this. This is unbelievable isn't it? In 50 years, there will still be no studies. It is
political will that is lacking. There are no mobilizations, neither from scientists nor
from pharmaceutical laboratories.[79]

In her remarks, Anne refers not only to the absence of novel drug developments, but also to a general lack of research in the domain of tuberculosis, a domain with much less scholarly investment than other
infectious diseases such as HIV/AIDS, at least in France and other European countries.

This nurse was not the only one commenting on the relative lack of
research and interest in TB; such observations were also common at the
level of international policy. "Despite the flaws with and growing resis-

tance to current TB treatments, no new TB drugs have been developed in nearly 50 years,"[80] stated one brief by the TB Alliance in 2014, which also claimed that "the current TB therapy is highly inadequate and is growing increasingly resistant to available therapies."[81] The Critical Path to TB Drug Regimens Initiative (CPTR) stated, also in 2014, that "new drug regimens are long overdue."[82] And an article in *Nature Reviews* entitled "Tuberculosis Success," published in 2013, observed that the introduction of the new compound "ended a four-decade-long lull in the hunt for a new tuberculosis . . . therapy."[83] This article referred to the US Food and Drug Administration's 2012 approval of a new compound against MDR-TB, bedaquiline—a long-awaited positive message in a field characterized by inactivity and stagnation. As movement in drug development has recommenced, calls have been made for future revolutions.

An assortment of actors—the TB Alliance, Doctors without Borders, drug developers, and microbiologists—have portrayed TB research as a field of investigation that is awakening in the 2010s after decades of dormancy. They thereby represent "new treatments" as the solution to the tuberculosis problem, and depict them as "key to unlocking the global DR-TB crisis."[84] These calls for action and mass mobilization have been accompanied by testimonies of optimism. The MSF crisis alert noted, "Today, there is reason for hope. The first new TB drugs in 50 years, along with development in diagnostic tests and new approaches to care, have real potential to radically improve patient outcomes."[85] In a similar vein, Mel Spiegelman, president of the Global Alliance for TB Drug Development, commented on the FDA approval of bedaquiline, "This is a major step forward, and the beginning of what should be a dramatic improvement in TB therapy that I hope we're going to see over the next 5–10 years."[86]

Such calls, now very common internationally, go back to the Cape Town Declaration of 2000, signed at a meeting in South Africa convened by the Rockefeller Foundation that brought together a wide range of delegates from the field of global health, representing philanthropic organizations (the Bill and Melinda Gates Foundation), international organizations (the World Bank and the World Health Organization), the humanitarian sector (Doctors without Borders), and the pharmaceutical industry (the Association of the British Pharmaceutical Industry). In the declaration, these actors committed themselves "to accelerate the development of new TB drugs to improve the prevention and treatment of this disease." They viewed issuing a "report on the Pharmaco-economics of TB Drug Development that clarifies the size and characteristics of TB drug markets" to be of paramount importance, clearly framing the problem of TB drug development as one of the pharmaceutical market.[87] The declaration announced a commitment to the creation of a Global Alliance for TB Drug development—known since then as the TB Alliance—involving partners from academia, industry, major agencies, nongovernmental organizations, and donors from around the world. Finally, the declaration calls on all of the major players in global health to invest in the development of new drugs, through novel institutional formations, "new partnerships for drug development" (PDPs).[88] With such global and ambitious goals, the Cape Town Declaration can be seen as a milestone in the international politics of TB control, in which the traditional approach that viewed TB as a disease of poverty began to be replaced with its reframing as a disease of inadequate scientific investment.

These statements from a diverse group of global health institutions—humanitarian organizations, international organizations, philanthropic institutions, and pharmaceutical companies—have in common not only their objective (to generate more funding on TB research by sketching scenarios of threat and hope, fear and redemption) but also the urgent call for scientific investment in the development of new antituberculosis drugs. They have portrayed this investment in TB as an object of research as being long overdue and wrongly overlooked in preceding decades. They have also included hopeful statements on the contemporary progress of TB treatments and brand new developments—"major steps forward,"[89] "breakthroughs,"[90] and "magic molecules"[91]—in other words, promissory proof of a novel therapeutic revolution in the making.

Conclusion

We note a stark contrast between the alleged "boringness" of TB in the clinic and in public health in the North with the corresponding tone of emergency in global policy statements on TB in the South and in novel efforts in research and development. The Stop TB Partnership of the WHO has been revived, and the DOTS strategy has been progressively implemented in the global South.[92] One consequence of this renewed institutional and public health interest in tuberculosis, clearly linked to increasing numbers of antibiotic resistance on a global scale, has been a steady rise of voices calling for a new "revolution" in TB therapeutics and diagnostics: an intensification of demands to invest—

again—in basic as well as clinical science and the search for new drugs, and thus to realize a rebiomedicalization of the disease.

Our history and ethnography of tuberculosis control past and present demonstrates that it is not possible to speak of one single therapeutic revolution in the realm of TB treatment. Instead, we observe recurring revolutions over the last seventy years, in which tropes of emergency and threat, urgency and fear, hope and progress have been repeatedly employed, albeit by shifting actors. We argue that the novel demand for a therapeutic revolution and the renewed investment and interest in tuberculosis in the early twenty-first century amounts to a rebiomedicalization of the disease. The social and political determination of tuberculosis—as represented in the McKeown hypothesis—has once again been relegated to the margins, to be replaced by pharmaceutical economies of hope on a global scale. These novel and techno-scientifically mediated visions of drug-based solutions are increasingly being tested in large-scale clinical trial infrastructures in the South, involving new "drug partnerships"[93] and North-South collaborations. Such trial assemblages put pharmaceutical progress and easy solutions at the forefront of disease control—again. This would suggest that biomedicalization occurs in cycles, as tuberculosis may well be the first biomedicalized disease to go through the process again in a different time-space context. It has also contributed to a revival of tuberculosis as an interesting object of contemporary biomedicine gone global—at least in its multiresistant form.

Over the last few decades the problem of tuberculosis did not just "return," as has often been stated. The anthropological perspective helps us to understand that diseases are not historically constant, but rather that they change over time and in context and geography. The tuberculosis of today is not the tuberculosis of yesterday, and the multi-drug-resistant forms of the disease are not the same as those forms of TB that are sensitive to antibiotics. Especially in its new resistant forms, TB has become *interesting* again as an object for contemporary biomedicine. Global health, it seems, has recombined old and new aspects of the history of tuberculosis to form a new entity. In other words, tuberculosis in its MDR-TB form has acquired an acute global configuration that it may never have had as a historical disease of poverty. As such, it has regained interest as an object of research and investment that bears close resemblance to the object that produced mass mobilization in the first half of the twentieth century, when TB was not yet pharmaceutically curable. This recombination of old imaginaries and old rhetoric of revolution, paralleled by truly novel developments in the area of pharmaceutical science, makes TB a seminal case for investigating the histories and presents of therapeutic revolutions. While the rhetorics of revolution have been highly modernist, analyzing them from both the historiographic and ethnographic perspectives reveals their similarities and differences throughout time, allowing for critical reflection on the discourses and practices around disease transformation and therapeutic change—or, in other words, the historical and anthropological relationships between diseases and their associated drugs.

NOTES

1. Carol Dukes Hamilton et al., "Extensively Drug-Resistant Tuberculosis: Are We Learning from History or Repeating It?" *Clinical Infectious Diseases* 45 (2007): 338–342; Salmaan Keshavjee and Paul Farmer, "Tuberculosis, Drug Resistance, and the History of Modern Medicine," *New England Journal of Medicine* 367 (2012): 931–936.
2. H. Markel, L. O. Gostin, and D. P. Fidler, "Extensively Drug-Resistant Tuberculosis: An Isolation Order, Public Health Powers, and a Global Crisis," *Journal of the American Medical Association* 298 (2007): 83–86; Centers for Disease Control and Prevention, Press Release, 8 June 2007.
3. Alice Park, "The TB Scare. A Broken System?" *Time*, 31 May 2007.
4. http://www.tballiance.org/newscenter/view-brief.php?id=691.
5. Paul Farmer, "Social Scientists and the New Tuberculosis," *Social Science and Medicine* 44, (1997): 347–358.
6. J. Chemardin et al., "Contact Tracing of Passengers Exposed to an Extensively Drug-Resistant Tuberculosis Case during an Air Flight from Beirut to Paris, October 2006" (12 June 2007), http://www.eurosurveillance.org/ViewArticle.aspx?ArticleId=3325.
7. Interview with public health officer, 12 December 2006.
8. Selman A. Waksman, "Streptomycin: Background, Isolation, Properties, and Utilization. Nobel Lecture, December 12, 1952," 1952, 386, http://www.nobelprize.org/nobel_prizes/medicine/laureates/1952/waksman-lecture.html.
9. Ibid.
10. Anne Hardy, "Reframing Disease: Changing Perceptions of Tuberculosis in England and Wales, 1938–70," *Historical Research* 76, no. 194 (November 1, 2003): 554, doi:10.1111/1468-2281.00189.
11. David Barnes, *The Making of a Social Disease, Tuberculosis in Nineteenth-Century France* (Berkeley: University of California Press, 1995); Linda Bryder, *Below the Magic Mountain: A Social History of Tuberculosis in Twentieth-Century Britain* (Oxford, UK: Oxford University Press, 1988); Flurin Condrau, *Lungenheilanstalt und Patientenschicksal: Sozialgeschichte der Tuberkulose in Deutschland und England im späten 19. und frühen 20. Jahrhundert*

(Göttingen: Vandenhoeck & Ruprecht, 2000); Caroline K. Grellet, *Histoire de la tuberculose: Les fièvres de l'âme 1800–1940* (Paris: Ramsay, 1983); Pierre Guillaume, *Du desespoir au salut: Les tuberculeux aux XIXe et XXe siècle* (Paris: Aubier, 1986); Katherine Ott, *Fevered Lives: Tuberculosis in American Culture since 1870* (Cambridge: Cambridge University Press, 1996).

12. Guillaume, *Du desespoir au salut*, 327.

13. Abdel Omran, "Epidemiological Transition in the U.S." *Population Bulletin* 32 (1977): 3–42.

14. George Weisz and Jesse Olszynko-Gryn, "The Theory of Epidemiologic Transition: The Origins of a Citation Classic," *Journal of the History of Medicine and Allied Sciences* 65 (2010): 287–326.

15. Paul Farmer, "Social Inequalities and Emerging Infectious Diseases," *Emerging Infectious Diseases* 2 (1996): 259–69. See also chapter 7 in this volume.

16. Matthew Gandy and Alimuddin Zumla, eds., *The Return of the White Plague: Global Poverty and the "New" Tuberculosis* (London: Verso, 2003), 1.

17. Nicholas B. King, "Immigration, Race, and Geographies of Difference in the Tuberculosis Pandemic," in Matthew Gandy and Alimuddin Zumla, eds., *Return of the White Plague: Global Poverty and the New Tuberculosis* (London: Verso, 2003), 39–54.

18. Stewart T Cole and Giovanna Riccardi, "New Tuberculosis Drugs on the Horizon," *Current Opinion in Microbiology* 14 (2011): 574, doi:10.1016/j .mib.2011.07.022.

19. René Dubos and Jean Dubos, *The White Plague: Tuberculosis, Man, and Society* (New Brunswick, NJ: Rutgers University Press, 1996/1952).

20. Christoph Gradmann, *Laboratory Disease: Robert Koch's Medical Bacteriology* (Baltimore: Johns Hopkins University Press, 2009).

21. Donald S. Burke, "Of Postulates and Peccadilloes: Robert Koch and Vaccine (Tuberculin) Therapy for Tuberculosis" *Vaccine* 11 (1993): 795–804.

22. Barbara Elkeles, "Der 'Tuberkulinrausch' von 1890," *Deutsche Medizinische Wochenschrift* 115 (1990): 1729–1732.

23. Alfred Grotjahn, *Erlebtes und Erstrebtes: Erinnerungen eines sozialistischen Arztes* (Berlin: Herbig, 1932), 51; translated by the authors.

24. Gradmann, *Laboratory*, 136.

25. Christoph Gradmann, "Money and Microbes: Robert Koch, Tuberculin and the Foundation of the Institute for Infectious Diseases in Berlin 1891," *History and Philosophy of the Life Sciences* 22 (2000): 59–79.

26. Alfons Labisch, "Experimentelle Hygiene, Bakteriologie, Soziale Hygiene: Konzeptionen, Interventionen, soziale Träger—eine idealtypische Übersicht," in Jürgen Reulecke and Adelheid Gräfin zu Castell Rüdenhausen, eds., *Stadt und Gesundheit: Zum Wandel von "Volksgesundheit" und kommunaler Gesundheitspolitik im 19. und frühen 20. Jahrhundert* (Stuttgart: Steiner, 1991), 37–47.

27. Peter Dettweiler, "Das Kochsche Verfahren im Verhältnisse zur klima-

tischen Anstaltsbehandlung," in Ernst von Leyden und E. Pfeiffer, eds., *Verhandlungen des 10. Kongresses für Innere Medizin* (Wiesbaden: Bergmann, 1891), offprint.

28. Marcus Paterson, *Auto-Inoculation in Pulmonary Tuberculosis* (London: Nisbet, 1911).

29. Rodolph Charles Wingfield, *Modern Methods in the Diagnosis and Treatment of Pulmonary Tuberculosis* (London: Constable, 1924), 46.

30. Peter Dettweiler, *Bericht über zweiundsiebzig seit drei bis neun Jahren völlig geheilte Fälle von Lungenschwindsucht* (Frankfurt am Main: Johannes Alt, 1886).

31. Franz Wehmer, "Rückblick auf Brehmers Lebensarbeit," *Beiträge zur Klinik der Tuberkulose* 31 (1914): 457–479.

32. Flurin Condrau, "Urban Tuberculosis Patients and Sanatorium Treatment in the Early Twentieth Century," in Anne Borsay and Peter Shapely, eds., *Medicine, Charity and Mutual Aid: The Consumption of Health and Welfare, c.1550–1950* (Aldershot, UK: Ashgate, 2007), 183–206.

33. Alfred Grotjahn, "Die Lungenheilstättenbewegung im Lichte der Sozialen Hygiene," *Zeitschrift für Soziale Medizin* 2 (1907): 196–233, 201; translated by the authors.

34. Ernst von Leyden, *Die Wirksamkeit der Heilstätten für Lungenkranke: Populäre Aufsätze und Vorträge* (Berlin: Neelmeyer, 1907), 66–85.

35. George Cornet, *Die Tuberkulose* (Vienna: Hölder, 1907), 879.

36. Harry M. Marks, *The Progress of Experiment. Science and Therapeutic Reform in the United States, 1900–1990* (New York: Cambridge University Press, 1997).

37. Ibid., 114.

38. Waksman, "Streptomycin," 386.

39. Cynthia Connolly, Janet Golden, and Benjamin Schneider, "'A Startling New Chemotherapeutic Agent': Pediatric Infectious Disease and the Introduction of Sulfonamides at Baltimore's Sydenham Hospital," *Bulletin of the History of Medicine* 86 (2012): 66–93; Irvine Loudon, *Death in Childbirth: An International Study of Maternal Care and Maternal Mortality 1800–1950* (Oxford, UK: Oxford University Press, 1993).

40. John T. Macfarlane and Michael Worboys, "The Changing Management of Acute Bronchitis in Britain, 1940–1970: The Impact of Antibiotics," *Medical History* 52 (2007): 47–72.

41. Marks, *Progress of Experiment*, 114f.; Corwin Hinshaw, "Report of the Committee on Therapy," *American Review of Tuberculosis* 54 (1946): 442.

42. Flurin Condrau, "Who is the Captain of All These Men of Death? The Social Structure of TB Sanatorium Patients in Postwar Germany," *Journal of Interdisciplinary History* 32 (2001): 243–262.

43. Dubos, *White Plague*, 155.

44. Helen Bynum, *Spitting Blood: The History of Tuberculosis* (Oxford, UK: Oxford University Press, 2012), 189.

45. John Crofton, "The MRC Randomized Trial of Streptomycin and Its

Legacy: A View from the Clinical Front Line," *Journal of the Royal Society of Medicine* 99 (2006): 531–534. Available online in James Lind Library Bulletin, 2004: http://www.jameslindlibrary.org/illustrating/articles/the-mrc-randomized-trial-of-streptomycin-and-its-legacy-a-view.

46. Marks, *Progress of Experiment*, 132.

47. Robyn M. Lucas and Anthony J. McMichael, "Association or Causation: Evaluating Links between Environment and Disease," *Bulletin of the World Health Organization* 83 (2005): 792–795.

48. Sunil Amrith, "In Search of a 'Magic Bullet' for Tuberculosis: South India and Beyond, 1955–1965," *Social History of Medicine* 17 (2004): 113–130; Helen Valier, "At Home in the Colonies: The WHO-MRC Trials at the Madras Chemotherapy Centre in the 1950s and 1960s," in Flurin Condrau and Michael Worboys, eds., *Tuberculosis Then and Now: Current Issues in the History of an Infectious Disease* (Montreal: McGill-Queens University Press, 2010), 213–234.

49. Unpublished lecture by Sir John Croften, quoted in Frank Ryan, *Tuberculosis: The Greatest Story Never Told* (Bromsgrove, UK: Swift, 1992), 379.

50. Halfdan Mahler, quoted in Sunil Amrith, "In Search of a 'Magic Bullet' for Tuberculosis: South India and Beyond, 1955–1965," *Social History of Medicine* 17 (2004): 113–130, quote on 129.

51. World Health Organization, *Report of the World Health Organization 1973*, PCO/74.1. (Geneva: World Health Organization, 1974), 17. Accessed online at http://whqlibdoc.who.int/hq/pre-wholis/PCO_74.1.pdf, 17.

52. Farmer, "Social Scientists and the New Tuberculosis," 347.

53. Stefan H. E. Kaufmann and Bruce D. Walker, eds., *AIDS and Tuberculosis: A Deadly Lisaison* (Weinheim, Germany: Wiley-VCH, 2009).

54. On the impossibility of treatment to contrast to incurability in historical and contemporary perspective, see Thomas Gorsboth and Bernd Wagner, "Die Unmöglichkeit der Therapie am Beispiel der Tuberkulose" *Kursbuch* 94 (1988): 123–146.

55. Conversation with an "interne," Department of Infectious and Tropical Diseases, University Hospital, France, 7 December 2006.

56. J. Ogden, G. Walt, and L. Lush, "The Politics of 'Branding' in Policy Transfer: The Case of DOTS for Tuberculosis Control," *Social Science and Medicine* 57, no. 1 (July 2003): 185.

57. U. Greinert et al., "Therapie der Tuberkulose," *Der Pneumologe* 4, no. 3 (May 2007): 175–186, doi:10.1007/s10405–007–0147-y; Société de pneumologie de langue française, "Recommendations de la Société de pneumologie de langue francaise sur la prise en charge de la tuberculose en France," *Revue des Maladies Respiratoires* 21 (2004): 414–420.

58. Ogden, Walt, and Lush, "The Politics of 'Branding,'" 182.

59. Andreas, resident physician, interview transcript from 7 June 2007.

60. Dermot Maher, "Global Challenge of Tuberculosis," *Lancet* 2, no. 8922 (27 August 1994): 610.

61. Paul Draus, *Consumed in the City: Observing Tuberculosis at Century's End* (Philadelphia: Temple University Press, 2004), 59.

62. Tuberculosis Coalition for Technical Assistance, *International Standards for Tuberculosis Care (ISTC)* (The Hague: Tuberculosis Coalition for Technical Assistance, 2006).

63. Adele E. Clarke, Laura Mamo, and Jennifer Ruth Fosket, eds., *Biomedicalization: Technoscience, Health, and Illness in the U.S.* (Durham, NC: Duke University Press, 2010).

64. Interview, head of German lung health department, 7 July 2007.

65. Interview, head of French infectious disease department, 21 December 2006.

66. Interview, senior physician, German lung health department, 25 July 2007.

67. Adele E. Clarke et al., "Biomedicalization: Technoscientific Transformations of Health, Illness, and U.S. Biomedicine," *American Sociological Review* 68, no. 2 (April 1, 2003): 162.

68. J. Dumit, *Drugs for Life: How Pharmaceutical Companies Define Our Health* (Durham, NC: Duke University Press, 2012).

69. Peter Keating and Alberto Cambrosio, *Cancer on Trial: Oncology as a New Style of Practice* (Chicago: University of Chicago Press, 2012).

70. Christian McMillen's book is one example of a new generation of TB histories that starts to investigate the period after the 1970s. Christian W. McMillen, *Discovering Tuberculosis: A Global History, 1900 to the Present*, (New Haven: Yale University Press, 2015).

71. Peter Baldwin, *Contagion and the State in Europe, 1830–1930* (Cambridge: Cambridge University Press, 1999); Guillaume, *Du desespoir au salut*.

72. Richard Coker, "Lessons from New York's Tuberculosis Epidemic," *British Medical Journal* 317, no. 7159 (5 September 1998): 616–20.

73. Alistair Story and Ken Citron, "Private Wealth and Public Squalor: The Resurgence of Tuberculosis in London," in *The Return of the White Plague: Global Poverty and the New Tuberculosis*, ed. Matthew Gandy and Alimuddin Zumla (London: Verso, 2003), 147–162.

74. Annie Thebaud-Mony and France Lert, "Submitting to Disease, Controlling Disease, Industrialization and Medical Technology: The Case of Tuberculosis," *Social Science and Medicine* 21, no. 2 (1985): 129–137.

75. Whereas in France and Germany MDR-TB cases constitute only 2 percent of the cases, in other parts of the world, such as Russia and Eastern Europe, MDR rates rise to 20 percent.

76. For a debate about the right way to respond to the "global challenge of tuberculosis," opposing demands for new and better tools, and demands for the more efficient use of old tools, see Kevin M. De Cock et al., "Global Challenge of Tuberculosis," *Lancet* 2, issue 8922, 344, no. 8922 (27 August 1994): 608–10.

77. MSF Crisis Alert, March 2014.

78. "TB-Alliance: Inadequate Treatment." Accessed 5 November 2014 at http://www.tballiance.org/why/inadequate-treatment.php

79. Anne, head physician, field notes, 12 March 2009.

80. "TB-Alliance. Inadequate Treatment." Accessed 5 November 2014 at http://www.tballiance.org/why/inadequate-treatment.php.

81. Ibid.

82. "Critical Path to TB Drug Regimens," accessed 5 November 2014 at http://www.c-path.org/pdf/CPTR-a-new-paradigm-for-TB-drug-development-10_11.pdf.

83. Dan Jones, "Tuberculosis Success," *Nature Reviews Drug Discovery* 12, no. 3 (March 2013): 175, doi:10.1038/nrd3957.

84. MSF Crisis Alert, 2014.

85. Ibid.

86. Jones, "Tuberculosis Success," 175.

87. For investigations into the relation between global pharmaceuticals and the market, see Stefan Ecks and Ian Harper, *Public-Private Mixes: The Market for Anti-Tuberculosis Drugs in India* (Princeton, NJ: Princeton University Press, 2013); Adriana Petryna, Andrew Lakoff, and Arthur Kleinman, eds., *Global Pharmaceuticals: Ethics, Markets, Practices* (Durham, NC: Duke University Press, 2006); Kaushik Sunder Rajan, ed., *Lively Capital: Biotechnologies, Ethics, and Governance in Global Markets* (Durham, NC: Duke University Press, 2012).

88. Susan Craddock, "Drug Partnerships and Global Practices," *Health & Place* 18, no. 3 (May 2012): 481–489.

89. Jones, "Tuberculosis Success," 175.

90. Ibid.

91. Cole and Riccardi, "New Tuberculosis Drugs on the Horizon," 574.

92. Ian Harper and Melissa Parker, "The Politics and Anti-Politics of Infectious Disease Control," *Medical Anthropology* 33, no. 3 (24 April 2014): 198–205, doi:10.1080/01459740.2014.892484; Ian Harper, "Anthropology, Dots and Understanding Tuberculosis Control in Nepal," *Journal of Biosocial Science* 38, no. 1 (January 2006): 57–67; Ian Harper, "Extreme Condition, Extreme Measures? Compliance, Drug Resistance, and the Control of Tuberculosis," *Anthropology & Medicine* 17, no. 2 (2010): 201–214, doi:10.1080/13648470.2010.493606.

93. Craddock, "Drug Partnerships and Global Practices."

SIX

Pharmaceutical Geographies: Mapping the Boundaries of the Therapeutic Revolution

JEREMY A. GREENE

On a clear and very cold January day in 1979, Walsh Mc-Dermott, editor of *Cecil's Textbook of Medicine,* professor of public affairs in medicine at the Cornell Weill College of Medicine, and architect of the Institute of Medicine (IOM), asked a large audience gathered at the National Academy of Sciences in Washington to consider the vast transformations that twentieth-century therapeutics had wrought in medical science, clinical practice, and society.[1] Recalling an event that had taken place thirty years earlier to the month, and just a few blocks away, he asked his audience to imagine a walk in time and space down the Mall to the front portico of the Capitol, where, on another "clear and very cold January day," Harry S. Truman had presented a list of programs in the first inaugural address to be covered by television. McDermott explained:

Few people can today recall points one, two, and three of his list; but Point Four has had a certain immortality. For the fourth point was the announcement of a program in which the highly valued technology of the United States would be made available for the development of the badly impoverished nations of the world. Biomedical or health technology, if you will, was just coming into flower at that time. Its outstanding attribute was that virtually for the first time medicine was developing the capacity to intervene decisively in the course of a wide range of diseases."[2]

The very power of these new technologies to intervene decisively in health and disease also created new political and moral responsibilities to ensure that they reached *all* people whose lives they could save. A pharmacological revolution may have taken place in the United States, McDermott continued, but the benefits of this change had not diffused evenly between rich countries and poor countries, between different regions of the United States, or even among different ethnic, racial, and socioeconomic strata of his own hometown, New York City.

By the late 1970s it had become a commonplace thing for elder statesmen in academic medicine, whose careers had spanned the development of the sulfa drugs, penicillin, and subsequent generations of antibiotics, to look back across the middle decades of the twentieth century and chart the milestones of a therapeutic revolution. But McDermott challenged his audience to conceptualize this therapeutic revolution as a thing that took place in space as well as time, and one that was woefully incomplete. The hundreds of distinguished guests from academia, government, the pharmaceutical industry, civil society groups, the World Health Organization, the World Bank, and prominent foundations in the fields of development and health were assembled at the IOM to discuss and debate the science and politics of pharmaceuticals for developing countries. The IOM conference brought the academic study of development, or modernization theory, into direct conversation with the engines of biomedical modernity.

"Never before in history have the opportunities for rational scientific and technologic approaches to health problems of developing countries been so great as they are now," a spokesman for the Robert Wood Johnson Foundation reminded the delegates.[3] And yet, at the close of the 1970s, no element of American medicine—not industry, government, profession, or academia—appeared to have any clear idea about how to capitalize on these opportunities. As McDermott concluded, this was all the more tragic because modern therapeutics and new medical technologies offered new opportunities to uncouple the broader relationship between poverty and disease, to control disease *in spite of* economic deprivation.

What I am saying here goes quite against the conventional wisdom, for that wisdom has it that if a disease is characteristically bred, or greatly facilitated, in the conditions of poverty, it is foolish to try to attack it with a technology—one must do something about the conditions in which it is bred. As a general law, that is not a bad one. My point is that it is not always so. Occasionally, the poor get lucky.[4]

Twenty years before the distorted map of global HIV/AIDS mortality came to be understood in terms of disparities in access to lifesaving antiretroviral medications, McDermott articulated a robust critique of what we might call pharmaceutical geography: drawing attention to the uneven distribution of access to modern medicines in different parts of the world. In the developed West, everyday expectations of living and dying had been transformed by virtue of powerful new therapeutics. Living and dying in the contemporaneous developing world, however, bore little difference to the premodern past, in which treatable diseases remained untreatable. McDermott and other attendees of the IOM conference believed that public and private American institutions—universities, pharmaceutical firms, the National Institutes of Health—should play a larger part in bridging that gap.[5]

This chapter will trace how this globalizing discourse of pharmaceutical geography took shape in American domestic health policy, and in American approaches to international development over the course of the 1960s and 1970s. This is not a comprehensive global history of pharmaceutical disparities. It is, rather, an attempt to read the globalizing narrative of American pharmaceutical policy and international relations in close relation to one another in the Cold War era. I focus on a few key episodes in which the voices of several stakeholders present in debates over the role of pharmaceuticals in health and development—physicians, patients, lawmakers, regulators, researchers, and manufacturers—became audible.

Most of these parties agreed that a veritable therapeutic revolution had taken place in the mid-twentieth century. Most believed, furthermore, that the problem of uneven access to new and lifesaving drugs was an urgent and pressing issue for their times. They differed significantly, however, in their understanding of what kind of objects pharmaceuticals were, and how and why they moved or did not move across domestic and global scales. To McDermott—as to Truman, perhaps—the pharmaceutical was a technology in need of more efficient strategies of diffusion. To drug manufacturers, the pharmaceutical was a commodity, which would circulate most efficiently in free markets with strong intellectual property protections. To others in civil society groups and academic medicine, the pharmaceutical was foremost to be understood as a lifesaving therapeutic agent, increasingly essential in the definition of human rights and projects of humanitarian assistance.

Over the past half-century, access to new medicines has become increasingly important as a marker of development and modernity across scales of city, state, nation, and globe. If the map of who benefited from such therapeutics was inconsistent and problematic, so too were the solutions proposed for what to do about it—and the explanations of what made this terrain uneven in the first place.

Who Benefited from the Therapeutic Revolution?

On the eve of Truman's inauguration, analysts estimated that fifty cents of every dollar spent on pharmaceuticals sold in the United States went towards products that had not been invented ten years earlier. By the year 1960 this number had jumped to somewhere between seventy and ninety cents. In the decade of the 1950s alone, more than 4,500 new products were launched on the American market.[6] By the early 1960s, as the tide of new therapeutics began to recede, the question of access to modern pharmaceuticals became a subject of increasing importance in American medicine, public health, and public policy.[7]

Not all people had equal access to these powerful new therapeutic agents, even within the boundaries of the United States. Access to new pharmaceuticals was especially vulnerable at marginal sites of clinical practice, such as rural general practice, an arena in which observers by the 1950s had already noted that the average family doctor increasingly "cannot know everything about therapy nor evaluate the very optimistic claims made for the many medications pressed upon him."[8] As the outcomes researcher Osler Peterson lamented in a study of rural medical practice in North Carolina, "a good conscientious doctor may easily fail to keep up with the vital new knowledge in medicine."[9] Certainly, by the end of the decade, Americans as a population were taking more drugs and paying more for drugs than ever before. But with so many expensive, potentially dangerous, and possibly unnecessary new medications and no nationwide insurance system, how could anyone know that the benefits of the truly wonderful drugs were reaching all those who might most gain from them? Concerns over the limited reach of the so-called therapeutic revolution were publicized by a series of critical essays in the *Saturday Review*, "Taking the Miracle out of Miracle Drugs," and amplified by Senator Estes Kefauver's televised hearings on administered prices in the pharmaceutical industry.[10]

Kefauver began with the corticosteroids, powerful new agents that had rapidly transformed the treatment of a wide range of inflammatory disorders and chronic diseases after their synthesis in the late 1940s, but whose prices varied by orders of magnitude between brand-name and generically named versions. His hearings ended in 1961 with a series of legislative proposals to abolish pharmaceutical trademarks, encourage the use of generic names, and steeply curtail patent protection to ensure that truly innovative new drugs be made available to everyone at reasonable prices.[11] In their repeated trips to Capitol Hill in the 1960s and 1970s to testify in Kefauver's hearings on administered prices and Gaylord Nelson's subsequent Senate hearings on competitive problems in the drug industry, pharmaceutical executives regularly invoked narratives of therapeutic revolutions to explain how antibiotics, antitubercular agents, psychotropics, antidiabetics, and other drugs had transformed the epidemiological, demographic, and economic profile of the United States.[12] "These drugs have—and I use the word advisedly—revolutionized the healing art," Smith, Kline and French's Walter Munns protested in Kefauver's opening hearings.[13]

The rhetoric of therapeutic revolution was used by critics of the industry as well. "There has been an amazing revolution in the drug field in the last two decades that has contributed enormously to combating ill health and has been a major factor in the dramatic advances of medical science," testified Secretary of Health, Education, and Welfare Abraham Ribicoff in front of Kefauver's committee in 1961. In that time, life expectancy had gone up, death rates from infectious diseases had gone down, and many chronic diseases had been made newly manageable. But Ribicoff warned of the drug revolution's "negative side": planned obsolescence, trumped-up claims of "me-too" drugs, extended brand-name monopolies, extravagant advertising and promotional campaigns, and patenting practices that contributed to the high price of drugs.[14] Were all modern drugs simply snake oils in updated packaging, pharmaceutical marketing would be much simpler to regulate. But the problem Ribicoff raised was complicated by the fact that *some* of these new drugs represented highly effective therapeutics with the power to alter the balance of life and death. Current practices of pharmaceutical marketing were a problem precisely because they "have contributed to an unreasonably high cost to the consumer *for many drugs which are essential to the maintenance of health and even life itself.*"[15] The very power of these medicines turned inequity of access into a moral problem. As Ribicoff and other critics suggested, the patent, promotional, and pricing practices of the pharmaceutical industry had to be reconsidered to allow for a basic equity of access to health for the American populace.

The uneven map of therapeutic access could be redrawn by industry apologists, however, to depict a very different landscape of innovation and piracy, in which the territories at stake were not American haves and have-nots but the socialist East and the free-market West. Like other socialist schemes, one industry executive claimed, proposals for patent and pricing reform were "merely another proposal to kill the goose that lays the golden eggs of progress."[16] Austin Smith, head of the Pharmaceutical Manufacturers of America (PMA), explicitly contrasted the geography of American pharmaceutical innovation against that of Soviet pharmaceutical access. As Smith reminded Kefauver's committee, "There has been no missile gap in pharmaceutical research," for the simple fact that the socialist system did not reward therapeutic innovation:

While the U.S. pharmaceutical industry has been leading the world in the development of new medicaments, spurred by the incentives of the free enterprise system, the Soviet Union has all but dropped from the race. No single drug is attributable to Russia in the 42 years that have passed since the October revolution. On the other hand, the U.S.S.R. has freely pirated American developments and is selling identical drugs in world markets at a price advantage, presumably as part of its effort to buy the friendship of uncommitted nations.[17]

In a later exchange, conservative Republican Senator Roman Hruska oriented this East-West map along the poles of future and past. Recalling that leeches had been present at the deathbed of Stalin a few years earlier, Hruska painted a picture of a Soviet medical system hopelessly mired in ancient forms of therapeutics. "They are still resorting to that type of treatment," he declared, "and they also show a backwardness in many other things which we consider apparently in the judgment of some as being a curse and a drag on our development medically and in our health picture."[18] The new objects of the therapeutic revolution were private-sector commodities, products of the free-market system. If American biomedical industries were advancing towards a healthier future, Hruska claimed, state-based socialist medicine was stuck in the premodern past.[19]

The pharmaceutical industry's portrayal of pharmaceutical innovation as a distinctly American endeavor was contested by critics in academic medicine. "The drug business makes many references to the patients benefited by the revolution in drug therapy over the past 25 years," alleged Frederick Myers of the University of California, San Francisco. "The progress is real, but how should we distribute our gratitude?" Myers contended that most important therapeutic innovations between 1930 and 1960 were the result not of American private industry, but of broad transnational academic and state-funded projects across a wide variety of terrains. Publicly funded research gave the world anticoagulants, anterior pituitary hormones, thyroid hormone, and the large-scale manufacture of penicillin. Many other truly innovative classes of drugs emerged from European laboratories: antihistamines, analgesics, local anesthetics, antimalarials, synthetic estrogens, major tranquilizers, oral antidiabetic drugs, and the most potent treatments for high blood pressure. Since the American drug industry played only a minor part in effecting the therapeutic revolution, Myers argued, their patenting, trademarking, and pricing practices should not be allowed to keep the benefits of these innovations out of the reach of the average American.[20]

Myers's geography of innovation was largely shouted down by the Pharmaceutical Manufacturers Association (PMA) and its supporters. As the Republican counsel on Kefauver's subcommittee protested, "the record will show that American discoveries and development of drugs are greater than all other nations combined in the world since 1940."[21] A few years later, in the early stages of Gaylord Nelson's Senate hearings on competitive problems in the drug industry, PMA executive director C. Joseph Stetler would argue that the lion's share of credit for the therapeutic revolution was due to American pharmaceutical manufacturers. Waving a list of 823 single-entity drugs newly available on the American market between 1940 and 1966 in front of Nelson's Senate subcommittee, Stetler proclaimed:

This compilation is significant in your consideration of this great industry for it shows that the United States originated 502 of the 832 new weapons against disease which have been placed in the physician's armamentarium in the last 27 years. And the U.S. shares credit with foreign sources for several others. Of the U.S. discoveries, the laboratories of American manufacturers were responsible for 87 percent. The others came from university, non-profit or government sources.[22]

Though critics would object that many of the items counted in Stetler's therapeutic arms race were "molecular manipulations" or "me-too" drugs only trivially different from existing therapies, by and large the PMA was able to consolidate public opinion that the American drug industry was indeed the goose that laid the golden eggs—as long as burdensome regulation did not kill it first.

Pharmaceutical executives and their critics also disagreed over the forces that helped or hindered the spread of new medicines in practice.

Osler Peterson, in his initial study of therapeutic modernity among rural physicians in North Carolina, found that pharmaceutical advertisements and salesmen had become some of the most important means by which physicians learned of new drugs; this finding was substantiated by a number of other studies funded by the American Medical Association in the 1950s.[23] Arthur Sackler, a psychiatrist and advertising executive at William Douglas McAdams, Inc., the leading medical advertising firm of its day, cited advertising as a key modern tool to close this medical information gap. As an effective and efficient field of communications, Sackler argued, "advertising has made one of the major contributions to the rapid dissemination of new therapeutic information"; he warned of the many lifesaving drugs still "inadequately used and only applied to a small percentage of those patients who require them."[24] This approach to the pharmaceutical as technology saw marketing as the most efficient mechanism for diffusion.

Yet in the wake of Vance Packard's bestsellers *The Hidden Persuaders* (1957) and *The Waste Makers* (1960), critics of pharmaceutical marketing increasingly depicted branding, marketing, and advertising as economic waste that diluted rather than diffused the public benefits of new therapeutic agents. This new critique of the brand—accentuated by twentyfold price differentials between brand and generic versions of chemically equivalent drugs—would become the impetus for Wisconsin Senator Gaylord Nelson's initial investigations into competitive problems in the drug industry in May 1967. On the first day of the hearings, Nelson expressed his concern that the "ordinary, hard-working little consumer having trouble meeting his grocery bill" was unable to access the real benefits of modern medicine.[25] The journalist William Haddad, who had been an aide to Kefauver in the early 1960s, was Nelson's first invited speaker. Haddad represented the New York Citizens' Committee on Metropolitan Affairs, an activist group that had collected spatial data to show that the price of the same prescription drug varied widely from city to city—and within neighborhoods within cities—to create an inequitable geography of therapeutic accessibility (see figure 6.1).[26]

Nationally, Haddad's study showed that key drugs cost as much as forty times more in Atlanta than in New York City; in New York state it documented that residents in Albany paid up to eight times more than those in New York City. Even within the island of Manhattan, as Haddad's colleague John L. S. Holloman testified, residents of Harlem paid more on average for basic drugs than did those living on the Lower East Side (see figure 6.2). As president of the National Medical Association, Holloman was one of the most prominent African American physicians

FIGURE 6.1 Variation in basic drug prices between urban and suburban areas in the state of New York, as documented by the Citizens' Commission on Metropolitan Affairs. *Competitive Problems in the Drug Industry: Hearings Before the Subcommittee on Monopoly of the Select Committee on Small Business, United States Senate, Ninetieth Congress* (Washington: Government Printing Office, 1967), v. 1, p. 7.

in the country. He declared that it should be plain to see that the benefits of the therapeutic revolution were unevenly distributed across geographies of race and class: "As a member of a minority group, and as a man who is involved with the problems of the minority poor, I am concerned that those medications which are safe and effective be made available to all people. I can tell you right now that poor Negroes are not getting the medicines they need."[27]

It is important to note that throughout these well-publicized debates over pharmaceutical policy in the 1960s, politicians on both sides of the aisle generally agreed that the recent "drug revolution" had produced a new set of modern therapeutics qualitatively different in efficacy and utility from the medicines of a generation or two before. All agreed, furthermore, that the benefits were not equally shared among the American people, that not all "wonder drugs" were wonderful, and that even those that were wonderful could have adverse effects on both physiologies and pocketbooks. Significant disagreement erupted, however, over the proper role of the pharmaceutical industry in the inception and translation of this therapeutic revolution, and whether existing patent, pricing, and promotional structures helped or hindered the fullest possible dissemination of its benefits. In the decade that followed, these same questions would be asked on a far broader scale.

Pharmaceutical Diffusion: Solution or Problem?

In March of 1977, Senaka Bibile sent a letter to the Pharmaceuticals Division of the World Health Organization (WHO), relating the domestic debates over pharmaceutical policy in the United States to the broader problems of pharmaceutical access across the Third World. Bibile was the first professor of pharmacology in Sri Lanka (formerly Ceylon) and had become something of an international celebrity for helping to break his island nation's dependence on the importation of brand-name drugs from European and North American manufacturers. As Bibile claimed, his work in Sri Lanka over the past two decades had both informed and been informed by the policy processes of the Kefauver

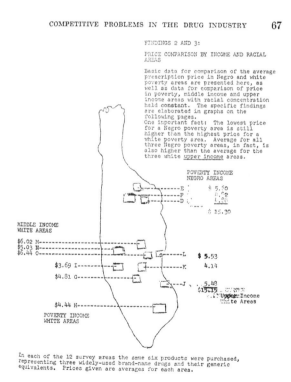

FIGURE 6.2 Variation in basic drug prices between neighborhoods in Manhattan, as documented by the Citizens' Commission on Metropolitan Affairs. *Competitive Problems in the Drug Industry: Hearings Before the Subcommittee on Monopoly of the Select Committee on Small Business, United States Senate, Ninetieth Congress* (Washington: Government Printing Office, 1967), v. 1, p. 67.

and Nelson hearings and their aftermaths.[28] Well beyond the District of Columbia, from Colombo to Colombia, the transcripts of the Kefauver and Nelson hearings circulated as part of a broader exchange of evidence and critiques about the patchiness of pharmaceutical access and excess.[29]

In retrospect, Gaylord Nelson's Senate hearings into domestic pharmaceutical marketing practices (1967–1976) had unfolded during a period of increased attention to the role pharmaceuticals should play in policies of international development. In the late 1960s, modernization theorists at the World Bank and elsewhere began to shift their metrics of development from gross domestic product to more subtle indices of health and well-being. One particularly well-cited model of health and modernity was the epidemiological transition theory of Egyptian-born demographer Abdel Omran.[30] Using the health statistics of the United States and United Kingdom as an epidemiological goal of development, Omran posited a variant of modernization theory that placed the health profile of a given society at the center of an evolutionary narrative from a first "age of pestilence and famine" to a second "age of receding pandemics," to arrive at a third and final "age of degenerative and man-made disease." While Omran praised some countries, like Japan, for passing from stage 1 to stage 3 in an "accelerated transition" through rapid Westernization, he singled out other countries of the developing world—such as Bibile's Ceylon—as being locked in a "delayed model" of underdevelopment. But Omran did not believe such delays had to be permanent: the modernization of the health profiles of the developing world would be "significantly influence[d] by medical technology," especially "imported medical technologies" like modern pharmaceuticals.[31]

To a physician in Ceylon like Bibile, however, the importation of medical technologies represented as much of a problem as it did a solution. Graduating from the Medical College of Colombo in 1945, he had pursued a PhD in pharmacology in Edinburgh, returned to the newly independent Ceylon in 1953, and become the country's first professor of pharmacology in 1958.[32] Bibile attempted to counter Ceylon's dependence on increasingly expensive European and American brand-name drugs with lists that separated essential from inessential medicines. By the late 1950s he had established a short list of inexpensive, generically available drugs that would be stocked in public hospitals, and after 1962 he managed to extend a slightly longer list to the private sector. In October of 1970, Bibile was tasked by the prime minister (along with socialist MP and fellow physician S. A. Wickremansinghe) "to look into and correct the needless loss of foreign exchange in the import of drugs."[33] The resulting Bibile-Wickremansinghe report, published in March 1971, argued that the benefits of modern pharmacotherapy could only be realized in Sri Lanka by a strong central policy prioritizing essential over inessential drugs and generic drugs over brand-name versions, and bringing costs down through competitive bidding, local formulation, and rational use.[34] In the aftermath of this report, the State Pharmaceuticals Corporation of Sri Lanka (SPC) was created in 1972 to implement Bibile's plan.

Within a year the price of the antibiotic ampicillin had dropped by nearly 85 percent in Sri Lanka while rising in all other countries. Bibile's plan was widely discussed at the WHO, the United Nations Conference on Trade and Development (UNCTAD), and the United Nations Centre on Transnational Corporations (UNCTC).[35] Yet after Bibile and Wickremansinghe published an account of the Sri Lankan pharmaceutical policy in the *British Medical Journal*, they were sharply rebuked by George Teeling-Smith, health economist at the British Office of Health Economics, for only assessing "the narrow sense of how to cut down the drug bill" rather than thinking about the role of pharmaceuticals in international development more broadly. The multinational pharmaceutical industry had made substantial economic investments in Ceylon before and after independence, Teeling-Smith warned, and if the Sri Lankan government was not careful, "their recommendations will drive out the few who already do manufacture there."[36]

Further resistance came from Sri Lankan branches of multinational firms and from the organized medical profession. When increased demand for antibiotics during a cholera epidemic exceeded the capacity of local bulk encapsulating plants to package generically purchased tetracycline, the Sri Lankan branch of Pfizer insisted that it would only sell brand-name tetracycline imported from abroad at a sharp markup.[37] A series of criticisms from physicians simmered throughout the program, and erupted again when the governing socialist coalition effectively collapsed in 1975. Though the new government maintained the State Pharmaceutical Corporation, its policies were relaxed to include importation of any drug at any price—or, in the words of UNCTAD's principal development economist, they were "emasculated."

Bibile accepted an invitation to move to Geneva to serve as senior consultant on pharmaceuticals for the Transfer of Technology Division of UNCTAD, and to oversee the development of the Sri Lankan model for pharmaceutical self-sufficiency for thirteen other least-developed countries. In the process of mobilizing this plan at a conference for developing world pharmaceutical technology transfer in Georgetown, Guyana, he was found dead in his hotel room from sudden cardiac arrest. His body was returned to Sri Lanka amid rumors of poisoning and assassination. At Bibile's memorial service, the secretary-general of UNCTAD eulogized that his sudden death was "a major blow to the implementation of the Non-Aligned Heads of State's directives for pharmaceutical policy, of which he was an architect."[38] The impact of Bibile's work contributed substantially to the formation of a collaborative pharmaceutical policy framed at the Fifth Conference of Non-Aligned Countries held in Colombo in August 1976.[39]

But Sri Lanka was only one of many sites for articulating the problems of pharmaceutical access in developing countries; different solutions were proposed in India, Brazil, Tanzania, and other locations.[40] As Bibile's colleague at UNCTAD, the Oxford economist Sanjaya Lall, would later point out, drug policies in the much larger and more industrialized country of India were tied to plans for building an "Indianized" pharmaceutical industry instead of merely finding more favorable terms for importing drugs.[41] The Hathi Committee, India's 1975 regulatory response to therapeutic independence, emphasized that a future of Indian biomedical self-sufficiency could be achieved by focusing public sector expenditures on 117 essential drugs, with a focus on purchasing generic instead of brand-name drugs. As Lall would later report, this focus helped to position India as the first postcolonial nation to, "in a modest way, become an exporter of pharmaceutical technology" that "set up its own 'mini-multinationals,'" which in turn invested in local firms in Sri Lanka, Bangladesh, Cuba, and other locations across Africa, Asia, and Latin America.[42]

On the other side of the Arabian Sea, very different postcolonial critiques of unequal access to modern pharmaceuticals were also developing in east Africa. Meeting in Nairobi in 1973, the East African Medical Research Council devoted its Tenth Annual Scientific Conference to the problem of the use and abuse of drugs and chemicals in tropical Africa. As Research Council President G. L. Monekosso announced, "The theme of this conference is very timely since the full impact of the great pharmaceutical revolution is now being felt in Africa and we should take stock and see what is really happening." Precisely *because* of the increased efficacy of modern medicines, pharmaceutical policies were even more important to public health planning in the postcolonial era than they had been in the days of empire:

Young medical doctors who leave medical school in 1971 and have time at their disposal must ask themselves (if they had the leisure to read medical history) how anyone had the conscience to practice medicine and the effrontery to call himself a doctor before the 1914–18 World War. For apart from a few potent pain-relieving drugs and relatively rudimentary surgery, there was little effective therapy, and our professional grandfathers (should they return) would be amazed to find what we can offer our modern patients (the same patient who does not hesitate to sue his doctor for alleged damages!). I have grown up with this pharmacological revolution, practiced and taught clinical medicine, and even undertaken research, and as a spectator watched this revolution take place.[43]

Monekosso's biographical self-presentation—like those of Senaka Bibile or Walsh McDermott, for that matter—hinged on a twentieth-century therapeutic revolution. Like Bibile, Monekosso left his native country of Cameroon while it was still a colony, for medical training in the United Kingdom, and returned to take up a career in a newly independent nation that would launch him into increasing visibility in national and international public health circles.

Yet in Cameroon, as elsewhere across sub-Saharan Africa, many new therapeutic tools were only available in hospitals in urban centers, and when available were often of dubious quality, or were too expensive for much of the population to afford. Part of the Nairobi conference was devoted to finding "appropriate technologies" for use in the sub-Saharan African context through rapprochement of "traditional" and "modern medicines." As historians Helen Tilley and Abena Osseo-Asare have shown, from Ghana to Madagascar the sciences of botany and pharmacology were invested with an newly Africanized nationalism in the 1960s and 1970s, as new heads of state from Kwame Nkruma to Julius Nyerere founded institutes for the scientific exploration of African traditional medicine.[44] The 1973 Nairobi conference would be followed a few years later by a conference in Brazzaville on African traditional medicine and its role in building health systems.[45] The report from this meeting—the first technical report to be published by the WHO's regional AFRO office—led to a major conference in Geneva and a special issue of *World Health* reevaluating the relationship between "traditional" and "modern" medicines:

The Meeting held that *all medicine is modern* in so far as it is satisfactorily directed towards the common goal of providing health care, despite the setting in time, place and culture. In this light, it was observed that the essential differences among the various systems of medicine arise not from the difference in the goal or effects, but rather from the cultures of the peoples who practice the different systems. It was further stated that traditional medicine is nothing new, since it has always been an integral part of all human cultures. However, as traditional medicine in some developing countries has tended to stagnate through not exploiting the rapid discoveries of science and technology for its own development, it has kept a slow pace of change in comparison with medicine as practiced in the industrialized countries, which keeps abreast of scientific and technological innovations to the extent that it is often exclusively referred to as *modern medicine*.[46]

In this map of therapeutic modernity, all medicines were modern, but some were more modern than others. Similar conflicts over the definition of what constituted modern therapeutics can be found in parallel WHO reports on the syncretic use of traditional Chinese medicine alongside essential Western medicine by Mao's "barefoot doctors," and on the practice of Ayurvedic medicine in South Asia.[47] These paradoxical appeals to the modernity of traditional medicine were packaged as a pragmatic response to the economic and logistical inaccessibility of biomedical pharmacotherapeutics. In that respect, the modernity of traditional medicines represented yet another face of the doctrine of "appropriate technologies" for developing countries when many essential pharmaceuticals were not otherwise available, or were available only in a confusing array of brand names at high prices.[48]

Problems of therapeutic overuse and underuse emerged in different forms across Latin America. In the late 1960s and 1970s, Latin America became an important site for new critiques of modernization as a form of dependency, reinterpreting American and European financial aid and development policies as a set of structures that produced dependent economic relations rather than self-sufficiency.[49] Many *dependistas* pointed to transnational corporations in general (and pharmaceutical firms in particular) as key drivers of the worsening economic and epidemiological asymmetries between global North and South.[50]

For modernization theorists, pharmaceuticals were understood as technologies to be transferred. To dependency theorists, pharmaceuticals were commodities that created increasingly unfair economic relations between North and South. Robert Ledogar's *Hungry for Profits: U.S. Food and Drug Multinationals in Latin America* (1975) is a good example of how in the 1970s dependency theory became an important lens for critics of the American pharmaceutical industry at home and abroad. Whereas Ledogar's work focused on Brazil, "a country whose pharmaceutical market is almost totally controlled by multinational firms," other instances of pharmaceutical dependency were soon documented from Mexico to Argentina.[51] In 1975, Milton Silverman, an investigative journalist whose account of the Kefauver and Nelson hearings, *Pills, Profits, and Politics* (1974), sketched an unflattering portrait of the American pharmaceutical industry, turned his attention to Latin America with *The Drugging of the Americas* (1975). In this work, he compared different promotional claims made by the same company for the same drug in Latin America and the United States and found that promotional language inflating benefits and neglecting risks for top-selling drugs was commonplace in Latin American markets. By the time Silverman's next book, *Prescriptions for Death: The Drugging of the Third World*, was published in 1982, it joined a growing genre documenting the devastating impact of the global pharmaceutical industry on the economic and epidemiological status of the poor in developing countries.[52]

Like Bibile's model formulary in Sri Lanka, and Monekosso's invocation of the modernity of traditional medicines in Kenya, the work of dependency theorists across Latin America drew sustained attention at the United Nations. By the late 1970s, the United Nations Industrial Development Organization (UNIDO)—along with the UNCTAD and UNCTC—commissioned an in-depth analysis of the role of transnational pharmaceutical corporations in helping or hindering international development.[53] Both Ledogar's and Silverman's books were prominently cited as examples of why the international community needed to focus its critique on the American pharmaceutical industry.[54]

American Pharmaceuticals in the Developing World

By the late 1970s, global disparities in pharmaceutical access had become a highly visible problem in international health and international politics across Asia, Africa, and Latin America and a series of UN organizations. As the American pharmaceutical industry became increasingly portrayed as part of the problem rather than the solution, the Carter administration and key members of Congress began to take up these questions as matters of national policy. In the summer of 1977, Democratic Senator Edward Kennedy, of Massachussetts—who had inherited from Estes Kefauver and Gaylord Nelson the role of convening Senate hearings on the marketing practice of the American pharmaceutical industry—reached across the aisle to his Republican colleagues

Jacob Javitz, of New York, and Richard Schweiker, of Pennsylvania, to voice concern about American pharmaceutical interests abroad. As they declared in a joint statement, "Neither component of America's large biomedical research establishment—the pharmaceutical industry's research laboratories and those in academic institutions—was devoting enough attention to the enormous unsolved health problems of the developing countries."[55]

A year earlier, in 1978, Kennedy had been a US delegate to the WHO's Alma-Ata Conference of 1978, which endorsed a philosophy of "essential drugs" as part of a platform for building primary health care capacity. On the domestic front, Kennedy was in the process of building bipartisan support for his sweeping (and ultimately unsuccessful) Drug Reform Act of 1978, which included measures to expand access to inexpensive generic drugs in the United States. For his part, Javitz had put forward two bills to try to lower the barriers to the export of American drugs to developing countries, and with Schweiker he was trying to build momentum for a third attempt. This, then, was the context for the Institute of Medicine, the Department of Health, Education, and Welfare, and the Pharmaceutical Manufacturers of America to organize a three-day stakeholder conference entitled "Pharmaceuticals for Developing Countries" in January 1979.

Kennedy charged the assembled audience to contemplate the widening global disparities in access to effective new drugs. Although the poorest nations spent 50 percent of their health budget on pharmaceuticals, more than 70 percent of their populations had "virtually no access to them at any time."[56] This problem should matter to Americans, Kennedy continued, because "behind these statistics are the faces of millions of people suffering *from symptoms we can alleviate*, dying *from diseases we can treat*, developing *diseases we can prevent entirely*." Kennedy quoted the president of the World Medical Association to paint a picture of a Third World where problems of health care disparity were best understood in terms of the lack of appropriate technologies, like pharmaceuticals: "You only fully understand the vital importance of drugs for health care when you see the long queues of the sick in front of a little dispensary out in the bush which has nothing on its shelves, not a single tablet of an antimalarial or . . . antibiotic. Yet this is the situation in entire regions of the world."[57] It was precisely because these therapeutic agents had such power, Kennedy continued, that America should have moral and pragmatic responsibilities to refocus development efforts around pharmaceutical access.

The Role of Pharmaceuticals in Public Health

As Walsh McDermott noted at the opening of the IOM conference, it was a strange and rather new thing that so much attention should now be focused on the role of pharmaceuticals in international public health—a field that had traditionally been far more concerned with the populational or "public health system" of delivery than with the individual or "personal care system" approaches to health. In the past, even when chemical solutions like DDT or fluoride had been brought to bear on public health problems, they had historically been administered through mass interventions such as public spraying campaigns or supplements in the drinking water supply. But antibiotics, antihypertensives, antidiabetic drugs, and other short- and long-term pharmaceutical supplies could not be administered en masse. Rather, they had to be delivered through one-on-one medical care, through that same system that Kennedy had called "the industrialized model of high-technology, urban-based medical care which exacerbates existing problems in developing nations."[58] This presented a paradox: How to know *which* forms of biomedical technologies were wasteful and which were appropriate in places lacking in resources and infrastructure?

Vittorio Fatturosso, WHO's representative at the conference, explained that the only possible answer was to support the new policy of "essential drugs" articulated by a WHO expert committee the year before. "The concept of the selection of essential drugs to meet the basic health needs of the population is so closely linked with the concept of primary health care," he noted, "and so important, that you must forgive me for repeating some of these obvious statements."[59] If one picked the right drugs, at the right price, to meet the priority health needs of a given population, the limited health budget of a developing country could be leveraged to make a more positive impact.

To that end, the history of therapeutic transformations in the United States could again be mobilized as an inspirational narrative for the developing world to emulate. McDermott charted the decline in overall mortality over the twentieth century by race and sex, revealing a steeper rate of decline in mortality among nonwhite women than among white women, and a sharper drop still—with a clear inflection point after the introduction of sulfonamide in 1937—for nonwhite men. Here, then, was a case in which the introduction of a new drug "applicable to the problems of large numbers of people, young and old"[60] could be demonstrated to achieve significant public health benefits. The "first miracle drug" may have worked its greatest miracles among the population of African American men, a population much less likely to have easy access to modern medical care. Perhaps, McDermott suggested, the improvements would have been even more significant had a real and sustained commitment to therapeutic access been present.[61]

On a more local register, McDermott also argued that the advent of the antitubercular drug streptomycin in 1947 helped to ameliorate health disparities in the city of New York. By 1947, tuberculosis killed fewer than 35 of 100,000 white New Yorkers, while the rate for African-Americans was roughly four times as high. Yet after streptomycin was introduced, not only did mortality fall in both populations, but the disparity was also reduced by half. By 1970 the TB mortality rate among African Americans was only 3 per 100,000, compared to 1.5 per 100,000 in the white population.[62] That similar patterns held true among marginalized indigenous populations in the developing world, he argued, could be demonstrated with reference to a corresponding graph of mortality reductions among indigenous Maori and European-born populations in New Zealand between 1930 and 1960.[63] For McDermott and his collaborators, African Americans in Harlem and Navajo in Arizona each formed a sort of "Third World country within the continental United States." What worked for socially marginalized nonwhite populations at home should also work overseas.[64]

And yet pneumonia and tuberculosis were cosmopolitan diseases; sulfa drugs and antitubercular chemotherapy could help people across all social geographies. What about those diseases of the developing world, like sleeping sickness and onchocerciasis, that had no First-World counterparts? Complicating the challenge of therapeutic rollout in Third World nations was the unique topology of their disease profiles, or what McDermott called their "technologic substrate":

Their disease pattern can be separated into an outer skin and a central core. The core consists of the disease found virtually everywhere in the developing countries. . . . This large core is covered by an outer skin of varied thickness that differs from one locality to another and thus provides distinctive local coloration to the disease pattern of a particular region. The group forming this outer layer consists principally of helminthic or protozoan diseases—such diseases as malaria, Chagas disease, hookworm, and schistosomiasis. This then is the technologic substrate of the developing countries; for some of it we have effective technologies and for some we have not.[65]

McDermott's arguments about the public health significance of innovative pharmaceuticals—what he called the "technological fix"—applied to both "core" and "skin." But for many representatives of academic and pharmaceutical laboratories, the problem of pharmaceutical access in the developing world was now a problem for new approaches to technology transfer. Merck's Lewis H. Sarett equated the lack of pharmaceuticals in the developing world with the lack of "research towards therapeutic agents that are uniquely tropical in their distribution . . . or, at least, endemic in the poor world."[66] Researchers from other firms agreed: the problem of drugs in the Third World simply required more funding of research and development. Two principals from Janssen Pharmaceuticals outlined promising new areas that could exist for pharmaceutical R&D in tropical diseases if a market for purchasing such agents could be guaranteed.[67] When the Nigerian physician Adetokunbo Lucas, head of the new United Nations Development Program / WHO / World Bank program on tropical disease research, called attention to this problem of neglected diseases of the tropics, representatives from the White House and the National Institutes of Health pointed to the need for American scientists to lead in these new research endeavors.[68]

Even within the pharmaceutical industry, however, not all parties agreed that the principal problem of drug supply in the developing world was a problem to be solved by more R&D led by American scientists. As John Urquhart, the chief scientist at the Bay Area drug-delivery firm ALZA, suggested, many existing drugs of proven efficacy weren't getting to people who needed them in the developing world, simply because in "the long journey from factory to receptor," the science of delivery systems had not been properly applied to the context of the developing world:

Between the worlds of pharmacology and of therapeutics lies a series of distribution and delivery systems. Their role is to convey an active substance, with its potency preserved, to appropriate drug receptors in patients throughout the world. With suitable packaging, shipping, transfer through customs, local distribution and transport, and with preservation from temperature extremes and excessive moisture, a pharmaceutical can travel, and remain within specified potency, from a plant in Kalamazoo, Michigan, to a dispensary in El Golaa, Tunisia.[69]

The sciences of drug delivery addressed more than mere physical geography, as seen in the distance between oral contraceptive pills (which Urquhart considered an appropriate technology for developed nations) and Depo-Provera contraceptive shots (which he considered an appropriate technology for developing countries).[70] The development of drugs for the developing world, in Urquhart's terms, was not simply a matter of formulating new chemicals, but of finding delivery devices that would allow for more practical translations between factory and receptor: heat-stable vaccines, single-dose antibiotic regimens, fixed-dose combination drugs. These delivery sciences carried pragmatic assumptions about what was and was not possible in a developing country.

And yet while Urquhart and other entrepreneurs described the problem of drug delivery in terms of the distance between sites of industrial production and sites of cell-surface function, CDC director William H. Foege described it in terms of political will. He reminded the assembled group of dignitaries that "2.5 million children will die this year from four diseases which can be prevented by currently available vaccines," largely because of a lack of sufficient attention to the problem among donor nations.[71] Similarly, the WHO's Vittorio Fatturoso urged American pharmaceuticals firms not just to focus on making new drugs, but to collaborate with the new Action Program on Essential Drugs to also make older drugs available to all who might benefit from them. "Is it real progress," he asked rhetorically, "if most people in the world continue to be excluded from the benefits of modern drugs?"[72]

Patents, Promotion, and Pricing

Why, Fatturosso asked, were the fruits of the therapeutic revolution still not available equally to all people long after the original patents on innovative medications had expired? According to free-market theorists, in a competitive market the invisible hand should balance demand and need. But no invisible hand was balancing demand and supply for essential drugs in the developing world. "The most general conclusion I arrive at," UNCTAD economist Sanjaya Lall concluded, "is that the 'free market' system as it now exists is incapable of resolving the various conflicts that exist between drug multinationals and developing countries without substantial modification."[73]

Lall blamed the spread of American-style product patents for producing this market failure. Yet Hoecht's Max Tiefenbacher, the president of the Geneva-based International Federation of Pharmaceutical Manufacturers (IFPMA), quickly countered that Lall had drawn the arrow of causality backwards. Lax defense of patent laws was discouraging pharmaceutical firm investment in the developing world, and not the other way around. In a survey of pharmaceutical patent protection in forty developing countries, Tiefenbacher found only two nations with "satisfactory" patent protection, nine with no drug patents at all, and twenty-nine with patents that were "legally possible, but worthless."[74] In this account, critiques of product patents by civil society groups and UN organizations had only made things worse. Tiefenbacher alleged that the spread of Bibile's celebrated Sri Lankan policies exacerbated the already questionable quality of pharmaceuticals in developing countries, leaving a sporadic supply of substandard generic drugs in its wake. Lack of attention to product patents and to the utility of pharmaceutical branding and marketing, he concluded, would only create greater disparities across the map of therapeutic access:

Ultimately, the hostile environment must lead one day to withdrawal of multinational companies from the least developed countries. I do not foresee a sudden and dramatic exodus, but a gradual retreat. The companies will fade from the scene by virtue of thousands of daily decisions taken on their operational levels. In fact, I wonder whether multinational corporations have not already begun this process of withdrawal from the developing world?[75]

In the immediate wake of the 1979 IOM meeting on pharmaceuticals for developing countries, the shifting politics of the United Nations seemed to favor Lall and Fattorusso's perspective over that of Tiefenbacher. The WHO moved ahead with the development of its Action Program on Essential Drugs to help developing countries implement the essential drug concept.[76] A few months after the IOM meeting, the UN Centre on Transnational Corporations issued a report, "Transnational Corporations and the Pharmaceutical Industry," that emphasized the problems of dependency that transnational pharmaceutical corporations created for developing countries, and presented strategies for resistance: invalidation of pharmaceutical product patents, reining in of the promotion of brand-name drugs, encouragement of local production by domestic industries, and, when necessary, bulk purchasing of generically-named drugs through competitive bids.[77] Additional pressure was put on multinational firms to adopt an international code of marketing ethics, and on the WHO to host an international conference on the rational use of drugs in Nairobi in 1985.[78]

Yet by the time of the Nairobi conference, a series of broad shifts in the international economics and politics of the early 1980s had begun to move the pendulum back in favor of Tiefenbacher's position. In spite of months of buildup, a United Nations Conference on Science and Technology in Vienna in late 1979 failed to produce any lasting agenda for increasing pharmaceutical access in the developing world.[79] Moreover, as Kristin Peterson mentions in chapter 9 of this volume, by 1985 many multinational companies began to pull out of developing countries like Nigeria, leaving limited production and a logistical morass in their wake. So, too, did Ronald Reagan's second term (under the influence of the American Enterprise Institute) witness a move to shift US funding for development work away from UN projects (such as UNCTAD, the UNCTC, and the WHO's Essential Medicines Program) that were seen as threatening to free-market principles, and toward bilateral and alternate multilateral structures such as USAID, the World Bank, the General Agreement on Tariffs and Trade (GATT), and the first stirrings of what would become the World Trade Organization (WTO). As the WTO grew in strength with each passing round, the relative power and visibility of UNCTAD, UNCTC, and the other nodes of the New International Economic Order of the 1970s began to fade. By the time the WTO accomplished the signing of the Agreement on Trade Related Aspects of Intellectual Property Rights (TRIPS) in the mid-1990s, most of Lall's plans in promoting patent reform through multilateral organizations had been undone by the signing of a multilateral agreement that reinstated American-style pharmaceutical product patents on a global scale.

This broad shift in the late 1980s from Keynes to Hayek, from UNCTAD to WTO, by no means displaced the problem of pharmaceutical access from the politics of international health. Rather, it shifted the terms of the debate from one set of solutions to another. Attention to widespread disparities in pharmaceutical access continued to grow during the 1990s and 2000s, often with no reference to earlier accounts from the 1960s, 1970s, or 1980s. From antiretroviral medications for HIV/AIDS to the call for new drugs for neglected tropical diseases and the expansion of long-term pharmacotherapeutic systems for chronic mental health and noncommunicable diseases, twenty-first-century global health has been concerned with "getting drugs into bodies" or, following Urquhart, getting drugs "to appropriate drug receptors in patients throughout the world."[80]

Conclusion

In a 2002 article entitled "Pharmaceuticals and the Developing World," and in the ensuing 2004 book *Strong Medicine,* the Harvard economist

Michael Kremer concluded that over the second half of the twentieth century, new drugs brought tremendous health benefit to developing countries—even if they were often underused, abused, or oriented towards problems that did not always reflect the disease burdens of target nations. "The role of pharmaceuticals and medical technology in improving health in developing countries," Kremer noted, "stands in contrast to the historical experience of the developed countries." As McDermott had done before him, Kremer noted that improvements in life expectancy in the developing world came in spite of a relative lack of increase (or even decrease) in overall income levels.[81] Sometimes the poor get lucky. As Lall, Bibile, Fatturoso, and others had done before him, Kremer listed the tools available to rectify the failures of the pharmaceutical market in the developing world: reform of patent laws, institution of differential pricing policies, reining in of irresponsible promotion, and the creation of incentives for investment in neglected tropical diseases. Seemingly without realizing it, Kremer was recapitulating a set of arguments that had already been developed nearly twenty-five years earlier.

After several decades of attempts to address the problem of differential access to the benefits of modern medicine, severe geographical disparities in access to the benefits of biomedicine persist, along with debates over their possible resolution. As this chapter has illustrated, pharmaceuticals are complex social objects. Their meanings and uses are continually contested by multiple stakeholders—physicians, policy makers, politicians, pharmaceutical executives, civil society groups, donors, and bilateral and multilateral organizations—no two of which approach pharmaceuticals as an object in exactly the same way.

The pharmaceutical is simultaneously a technology to be transferred and diffused, a commodity to be produced and circulated, and a transformative agent enmeshed in moral arguments about the role of biomedicine in the definition of human rights and humanitarian work. Yet these three understandings of the same object can lead to very different solutions to problems of access and equity, to stalemates, to multiple agencies working at cross-purposes to combat differing problems of pharmaceutical circulation, and ultimately to a perpetuation of both problem and debate.

As I have argued in this chapter, the 1979 IOM conference Pharmaceuticals for Developing Countries can be read as a logical extension of a growing policy discourse on the uneven reach of a twentieth-century "therapeutic revolution" that had been brewing since the late 1950s, and which was flexible enough to capture the attention of many stakeholders in government, industry, medicine, academia, and civil society. As the American pharmaceutical industry came under harsh public critique domestically in the late 1950s and 1960s, and internationally in the 1970s and 1980s, narratives of a twentieth-century therapeutic revolution were used by defenders and critics of the pharmaceutical industry alike.[82] From Kefauver's subcommittee in Washington to the Hathi Committee in New Delhi and the Non-Aligned Nations conference on pharmaceuticals in Georgetown, Guyana, all parties invoked in this history staked their moral claims to more or less regulation, stronger or looser patent protection, and expansion or contraction of the sphere of pharmaceutical promotion on the bedrock claim that the world after 1960 was categorically different from the world before 1940, therapeutically speaking.

All parties agreed that the uneven geographical reach of modern therapeutics was a serious problem, but they disagreed on the etiology, pathogenesis, and treatment of the condition. Did therapeutic inequalities stem from inflated prices, overhyping of expensive brand-name remedies that worked no better than older generic medicines, and lax oversight of promotional claims in both the global North and South? Were they exacerbated by the risk of receiving poor-quality generic or counterfeit medicines, or by the reduction in local investment by multinational corporations that followed attempts to reform pharmaceutical regulation, patenting, and promotional practices? These questions are still visible in early twenty-first-century struggles over therapeutic access. They run through Kremer's comments above, and through several of the other chapters in this volume.[83] And yet they are not exactly the same: the stakes and the stakeholders, have shifted in some important ways.

On the one hand, the epidemiological and technological basis for therapeutic intervention—what Walsh McDermott referred to as "technological substrate"—has changed, in some cases dramatically. The specific geography of the HIV/AIDS pandemic, the newer modes of production of biotech drugs which further complicate the transfer of drug production technology, and many other epidemiological and technological dimensions of therapeutic supply and demand have played a role in redrawing the maps of therapeutic access and inaccess. On the other hand, the institutional and economic basis for intervention has also shifted. Debates over pharmaceutical access in the 1970s invoked a field of international health in which multilateral bodies like the WHO and UNCTAD gave voice to critiques of inequity and dependency, as well as Keynesian plans to use mechanisms of the postcolonial state to address them. The articulation of pharmaceutical disparities in the early twenty-first century is taking shape in a moment in which the World Bank has outstripped the WHO in funding health-related projects; the US-dominated WTO has effectively displaced the UNCTAD as a space for articulating the economics of development; and expansive free trade compacts, from the North American Free Trade Agreement to the emergent Trans-Pacific Partnership, have become the dominant mode for speeding the transit of pharmaceuticals from one place to another.

Nonetheless, in drawing our attention to the geographical limits of the therapeutic revolutions in which both authors so ardently believed, the works of Kremer and McDermott, separated by a quarter century, illustrate the importance of thinking about the role of place in our understanding of modern biomedicine. In the situations revealed by Holloman's maps of racial and economic stratification of therapeutic access in New York City, Bibile's critiques of the pharmaceutical dependence of Ceylon soon after independence, and Silverman's indictment of cynical pharmaceutical promotion tactics across Latin America, the geography of differential pharmaceutical access was not passively produced but actively structured by specific historical processes. Conversely, this comparison draws our attention to the role of history in thinking about geographical disparities.

How can histories of development—especially technologically mediated histories of development—escape the Zeno's paradox that continuously positions the global South behind the global North in a never-ending teleology of development? One suggestion can be found in the concluding comments of the National Institute of Health's Richard M. Krause, made towards the end of the 1979 IOM conference on pharmaceuticals for the developing world. Krause described himself as an "optimist who takes the long view and believes that the developing countries will develop a science base that will influence their cultural climate 50 to 100 years from now." In Krause's view, the study of history supported his optimism because only seventy-five years earlier, the United States was "a nation almost devoid of a scientific establishment. . . . There were, at that time, no opportunities for such training in our 'developing' country."[84] Krause, too, conflated geography with chronology; developing countries were like the United States of fifty to one hundred years ago, a premodern "them" to be contrasted with a modern "us." But his reframing nonetheless offered the possibility of a world in which innovation might not always emanate from core sites in the global North and slowly diffuse to target sites in the global

South, but instead involve the development, production, reproduction, and distribution of biomedical technologies via sites of innovation more evenly distributed throughout the world.

At the very end of the 1979 conference, PMA representative W. Clark Wescoe disagreed with the argument of his "longtime friend and sometime critic" Walsh McDermott that expanding access to pharmaceuticals should be key to US efforts to improve international health and development:

It is important to recognize that the health problems of developing nations have their roots in certain fundamental social, cultural, and economic difficulties, compounded in the instance of some by vector-borne and parasitic diseases. Major components of these health problems are typically poor environments: sanitation, lack of health education, malnutrition, population pressures with explosive urbanization, and an inadequate public health infrastructure that often precludes effective delivery of health care. One or more of these problems is frequently aggravated by pressures of economic development.

The pharmaceutical industry is *not equipped* to deal with these underlying health problems.[85]

It is surprising that the lone voice of pharmaceutical modesty in this conference—suggesting that therapeutics might *not* be the revolutionary agents everyone else thought they were—should come from a pharmaceutical executive. Yet Wescoe's question merits pause: Why focus so much attention on pharmaceuticals when the social determinants of health and disease, poverty and subsistence, have been and continue to be so complicated and overlapping? Perhaps pharmaceuticals became the site of such intense debate in the late twentieth century precisely because they represented concrete sites for action. Unlike the problems of endemic poverty, failed states, and unjust social orders, the task of getting drugs into bodies seemed solvable. Different narratives of a twentieth-century therapeutic revolution shared the common vision that the modern pharmaceutical was a lever that could be used to move the world a bit closer to a better place. But the question of where to place that lever, how hard to push it, and what to push against, continues to be a subject of dispute.

NOTES

1.   Paul Beeson, *Walsh McDermott 1909–1981, a Biographical Memoir* (Washington: National Academy of Sciences, 1990).

2.   Walsh McDermott, "Historical Perspective," *Pharmaceuticals for Developing Countries* (Washington: Institute of Medicine, 1979), 9.

3.   Leighton Cluff, Introduction to *Pharmaceuticals for Developing Countries* (Washington: Institute of Medicine, 1979).

4.   Walsh McDermott, "Historical Perspective," 9.

5.   McDermott had served on a series of President's Science Advisory Committees—and had regularly spoken out in favor of the role of health technologies in international development assistance—during the formation of the US Agency for International Development (USAID) in the early 1960s. See, for example, "Research and Development in the New Development Assistance Program," memorandum, 10 November 1960, box 9, f. 8; and "The Role of Science and Technology in Development Assistance," n.d. memorandum, box 9, f. 5, both in the Walsh McDermott Papers, Medical Center Archives of New York-Presbyterian / Weill Cornell, New York.

6.   W. D. McAdams, "Three Major Marketing Problems on the Desks of Pharmaceutical Management Today," *Proceedings of the American Pharmaceutical Manufacturers' Association Midyear Meeting*, 17 December 1957; 272–280; Abraham Ribicoff, *Drug Industry and Antitrust Act* 5, 2580; Paul de Haen, "New Products Parade," *Drug & Cosmetics Industry* 90 (1962): 141–142, 222, 228, 230, 234–235, 240. This material is developed further in Jeremy A. Greene and Scott H. Podolsky, "Keeping Modern in Medicine: Pharmaceutical Promotion and Physician Education in Postwar America," *Bulletin of the History of Medicine* 83 (2009): 331–377.

7.   See, for example, Louis Lasagna, "The Pharmaceutical Revolution: Its Impact on Science and Society," *Science* 1969; 166 (3910): 1227–1233; Edmund Pellegrino, "The Sociocultural Impact of 20th Century Therapeutics," in Charles Rosenberg and Morris Vogel, eds., *The Therapeutic Revolution;* Robert Bud, *Penicillin: Triumph and Tragedy.*

8.   Osler L. Peterson, Leon Andrews, Robert S. Spain, and Bernard G Greenberg, "An Analytical Study of North Carolina General Practice, 1953–54," *Journal of Medical Education* 31 (1956): 90.

9.   Ibid.

10. John Lear, "Taking the Miracle out of Miracle Drugs," *Saturday Review* 42 (3 January 1959): 35–51.

11. The Kefauver hearings have been chronicled by several iterations of medical historians, including Richard Edward McFadyen, "Estes Kefauver and the Drug Industry," PhD diss., Emory University, 1973; Daniel Carpenter, *Reputation and Power: Organizational Image and Drug Regulation at the FDA* (Princeton, NJ: Princeton University Press, 2010); and Dominique Tobbell, *Pills, Power, and Politics: Drug Reform in Cold War America and Its Consequences* (Berkeley: University of California Press, 2012), but their thousands of pages of testimony and subpoenaed documents—and the largely unstudied research notes of Kefauver's committee preserved

in the National Archives RG 46—remain a rich trove of documentation on patenting, pricing, and promotional strategies of the pharmaceutical industry in the immediate postwar decades.

12. For more on the defensive strategies of the pharmaceutical industry under congressional scrutiny, see Dominique Tobbell, *Pills, Power, and Policy* (Berkeley: University of California Press, 2012).

13. Statement of Walter Munns, *Administered Prices in the Drug Industry: Hearings Before the Subcommittee on Antitrust and Monopoly of the Select Committee on the Judiciary*, United States Senate (hereafter *Administered Prices*) vol. 16: 8894.

14. Statement of Abraham Ribicoff, *Drug Industry Antitrust Act, Hearings Before the Subcommittee on Antitrust and Monopoly of the Select Committee on the Judiciary*, United States Senate (hereafter *DIAA*) vol. 5, 2580–2581.

15. Statement of Abraham Ribicoff, *DIAA* vol. 5: 2581.

16. Statement of Theodore Klumpp, *DIAA* vol. 4: 2265. Extended quotation: "The avowed purpose of [Kefauver's bill] is to make articles that are essential to health and life available to the public at a more reasonable price. . . . But let us look at this proposition carefully and see whether it will bring benefit to the public from a long-range viewpoint or whether it is merely another proposal to kill the goose that lays the golden eggs of progress."

17. Austin Smith, *DIAA* vol. 19: 10679–10680.

18. Statement of Roman Hruska, *DIAA* vol. 18: 10344–10345.

19. For more on the pharmaceutical industry's politically expedient use of Cold War rhetoric from the 1950s through the 1980s, see Dominique Tobbell, "Who's Winning the Human Race? Cold War as Pharmaceutical Political Strategy," *Journal of the History of Medicine and Allied Sciences* 64, no. 4 (2009): 429–473.

20. Statement of Frederick Myers, *Administered Prices* vol. 18: 10303–10305.

21. Statement of Peter Chumbris, *DIAA* vol. 3: 1346.

22. Statement of C. Joseph Stetler, *Competitive Problems in the Drug Industry, Hearings Before the Subcommittee on Monopoly of the Select Committee on Small Business*, United States Senate (hereafter *Competitive Problems*) vol. 4: 1414–1417.

23. Peterson et al., "An Analytical Study of North Carolina General Practice, 1953–54,"; Theodore Caplow, "Market Attitudes: A Research Report from the Medical Field," *Harvard Business Review* 30 (1952): 105–112; Charles C. Rabe, "The Doctor Measures the Detailman," *Medical Marketing* 11 (1952): 19–25; see also "Report on a Study of Advertising and the American Physician," in *Drug Industry Antitrust Act* (n. 39), 496; "The Fond du Lac Study," *DIAA* 698–806. For more detail, see also Greene and Podolsky, "Keeping Modern in Medicine."

24. Arthur Sackler, "Freedom of Inquiry, Freedom of Thought, Freedom of Expression: A Standard to Which the Wise and the Just Can Repair':

Observations on Medicines, Medicine, and the Pharmaceutical Industry," 11; in box 5, "William Douglas McAdams," Felix Marti-Ibañez papers, Yale University Archives. For more on Sacker and his fellow psychiatrist/advertising executive Feliz Marti-Ibanez, see Greene and Podolsky, "Keeping Modern in Medicine"; see also chapter 1 in this volume.

25. Statement of Gaylord Nelson, *Competitive Problems* vol. 4: 1287.

26. Statement of William F. Haddad, *Competitive Problems* vol. 1: 4–22. Haddad's initial research presented a forty-fold difference in the pricing of ten key drugs—meprobamate, chloramphenicol, tetracycline, streptomycine, phenazopyridine, dextroamphetamine sulphate, diphenhydramine, prednisone, sulfadiazine, and diphenylhydantoin—across a twelve-city sample. Prices in four different locations in the state of New York showed an eightfold difference in price between New York City and Albany, while prices within Manhattan varied considerably between Harlem and the Upper East Side: "The poor Negro paid more than the upper-income white for the same prescription medicine, the difference of 116 percent . . . this is not an isolated incident" (20).

27. Statement of Dr. John L. S. Holloman, *Competitive Problems,* vol. 1: 23–26. Quotation on 24.

28. "In Sri Lanka, when I proposed the creation of a Formulary Committee in 1959, I made the point that drugs used in the country was the resultant of interaction of disease and drugs and of promotional pressure by pharmaceutical industry. Thereafter we set down criteria to eliminate unnecessary drugs and it is in this manner that we have reduced the number of drugs from about 4000 to 600. . . . I believe the FDA has adopted a similar approach in their NAS/NRC Drug Efficacy Study." Senaka Bibile to Hiroshi Nakajima, 11 March 1977, E 19 81 1 folder 1, WHO archives, Geneva. The reception of the Kefauver hearings can also be seen in the development of generic pharmaceutical policy in India and Columbia, among other locales.

29. On the reading of Kefauver's transcripts by would-be therapeutic reformers in Colombia, see Victor Manuel Garcia, "Price Matters: The Introduction of Generic Drugs in Colombia," paper presented at the Society for the Social Studies of Science, Buenos Aires, Argentina, 21 August 2014.

30. See also chapter 7 in this volume; Arturo Escobar, *Encountering Development: The Making and Unmaking of the Third World* (Princeton, NJ: Princeton University Press, 1995); James C. Scott, *Seeing Like a State: How Certain Schemes to Improve the Human Condition Have Failed* (New Haven: Yale University Press, 1998); Nils Gilman, *Mandarins of the Future: Modernization Theory in Cold War America* (Baltimore: Johns Hopkins University Press, 2003); Michael Latham, "Modernization," in Theodore Porter and Dorothy Ross, eds., *The Modern Social Sciences* Cambridge: Cambridge University Press, 2003, 721–734.

31. Abdel Omran, "The Epidemiologic Transition: A Theory of the Epidemiology of Population Change," *Milbank Memorial Fund Quarterly* 49, no. 4

(1971): 510–522. This analysis builds on Paul Cruickshank's account of the convergence of health and the logics of development in the long 1970s. Paul J. Cruickshank, *The Teleology of Care: Reinventing International Health, 1968–1989*, unpublished PhD dissertation, Harvard University, 2011. See also George Weisz and Jesse Olszynko-Gryn, "The Theory of Epidemiological Transition: The Oigins of a Citation Classic," *Journal of the History of Medicine and Allied Sciences* 65, no. 3 (2010)): 287–326.

32. Bernard Soysa, "Senaka in Politics," in *Pharmaceuticals: A Third World Experience*, 40–41.

33. "Pharmaceutical Reform," in *Pharmaceuticals: A Third World Experience* (Lake House, Sri Lanka: Associated Newspapers of Ceylon, 1978), 12.

34. S. A. Wickremansinghe and S. Bibile, *The Management of Pharmaceuticals in Ceylon* (Colombo: Ceylon Industrial Development Board, 1971). S. A. Wickremansinghe and S. Bibile, "The Management of Pharmaceuticals in Ceylon," *British Medical Journal* 3 (1971): 757–759.

35. "Pharmaceutical Reform," in *Pharmaceuticals: A Third World Experience*. Lake House: Associaeted Newspapers of Ceylon, 1978, 17.

36. George Teeling-Smith, "Drugs in Ceylon" *British Medical Journal* 1971; 4(5780): 168–169.

37. "Pharmaceutical Reform" in *Pharmaceuticals: A Third World Experience*.

38. Gamani Correa to Leela Bibile, n.d., as reprinted in "Tributes to Senaka Bibile," in *Pharmaceuticals: A Third World Experience*. 53. Many of Bibile's biographers insist that the unusual circumstances point towards assassination. "There is no doubt," his colleague Osmund Jayaratne announced in his eulogy to Bibile, "that the giant multinational drug firms would have benefited enormously from his untimely death—in view of the fact that his assignment to advise thirteen Third World countries on the rationalization of drug usage—if successful—would have deprived these organisations of billions of dollars." Jayaratne, "The Humanist," in *Pharmaceuticals: A Third World Experience* (Lake House, Sri Lanka: Associated Newspapers of Ceylon, 1978), 47.

39. This was adopted in a resolution "that in the context of the revision of the industrial property systems, consideration be given to excluding pharmaceutical products from the grant of patent rights or alternatively the curtailment of the duration of patents for pharmaceuticals." United Nations Conference on Trade and Development, *Case Studies in Transfer of Technology: Pharmaceutical Policies in Sri Lanka*. Geneva: UNCTAD, 1977, TD/B/C.6/21, 10. See also "A/31/197, annex IV, resolution MAC/CONF.5/S/Res.25," cited therein.

40. Hesio Cordero, *Medicamento e Saude Publica*; Marilena Correa, Maurice Cassier, Pedro Villardi, "Pharmaceuticals and Public Health in Brazil: Copying Essential Drugs, Knowledge Acquisition and Innovation Projects in Public-Private Industrial Networks," paper presented at the Society for Social Studies of Science, Buenos Aires, Argentina, 21 August 2014.

41. In 1974 the committee, led by Shri Jaisukhlal Hathi, had been tasked to address "questions about the performance of the public sector units, multi-national firms' gaining stronghold in this field, prices of locally produced medicines, etc." *Report of the Committee on Drugs and Pharmaceutical Industry (Hathi Committee Report)* (Delhi: Ministry of Petroleum & Chemicals, 1975), 1. For more on the history of the postcolonial Indian pharmaceutical industry, see Stefan Ecks, *Eating Drugs: Psychopharmacological Pluralism in India* (New York: NYU Press, 2013).

42. Sanjaya Lall, *Pharmaceuticals for Developing Countries*, 238.

43. G. L. Monekosso, "Presidential Address," in A. F. Bagshawe, G. Maina, E.N. Mngola, *The Use and Abuse of Drugs and Chemicals in Tropical Africa* (Nairobi: East African Medical Research Council, 1974), 9. On the relation of postcolonial pharmacology and pan-Africanism, see Abena Osseo-Asare, *Bitter Roots: The Search for Healing Plants in Africa* (Chicago: University of Chicago Press, 2013).

44. Helen Tilley, personal communication; Abena Osseo-Asare, *Bitter Roots: The Search for Healing Plants in Africa.*

45. AFRO Technical Report Series no. 1, *African Traditional Medicine: Report of the Expert Committee* (Brazzaville: AFRO, 1976); World Health Organization Technical Report Series no. 622, *The Promotion and Development of Traditional Medicine: Report of a WHO Meeting* (Geneva: World Health Organization, 1978).

46. World Health Organization Technical Report Series no. 622; *The Promotion and Development of Traditional Medicine: Report of a WHO Meeting*, 9.

47. Likewise, Senaka Bibile's Department of Pharmacology at Peredinya could, on the one hand, conduct a number of studies seeking objective pharmacological verification of the validity of Ayurvedic remedies in Sri Lanka while simultaneously complaining that "there is widespread use of modern drugs by [Ayurvedic practitioners] not authorized or expected to use them. . . . over 90 per cent of those not registered to practice modern medicine are prescribing and dispensing modern drugs." United Nations Conference on Trade and Development, *Case Studies in Transfer of Technology: Pharmaeutical Policies in Sri Lanka.* (Geneva: UNCTAD, 1977), TD/B/C.6/21, 9.

48. G. L. Monekosso, "Presidential Address" in A. F. Bagshawe, G. Maina, E.N. Mngola, *The Use and Abuse of Drugs and Chemicals in Tropical Africa*, 9.

49. Andre Gunder Fran, "Sociology of Development and Underdevelopment of Sociology," *Catalyst* (1967) 3: 20–73; Fernando Henrique Cardoso, "Dependency and Development in Latin America," *New Left Review* 74 (July-August 1972): 83–95; Fernando Henrique Cardoso, "The Consumption of Dependency Theory in the United States," *Latin American Research Review* 12, no. 3 (1977): 7–24; Fernando Henrique Cardoso and Enzo Faletto, *Dependency and Development in Latin America* (Berkeley: University of California Press, 1979).

50. Richard J. Barnet, Ronald E. Muller, *Global Reach: The Power of the Multinational Corporation* (New York: Simon & Schuster, 1974).

51. Daniel Chudnovsky, "The Challenge by Domestic Enterprises to the Transnational Coporations' Domination: A Case Study of the Argentine Pharmaceutical Industry," *World Development* 7, no. 1 (1979): 45–58; "Los oligoloios intenracionales, el estado, y el desarrollo industrial en Mexico: El caso de la industria de hormonas esteroides," *Foro internacional* 17, no. 4 (1977): 490–541; "Drug Firms and Dependency in Mexico: The Case of the Steroid Hormone Industry," *International Organization* 32, no. 1 (1978): 237–286; Gary Gereffi, *The Pharmaceutical Industry and Dependency in the Third World* (Princeton, NJ: Princeton University Press, 1983).

52. Milton Silverman, *Magic in a Bottle* (New York: Macmillan, 1943); Milton Silverman and Philip Lee, *Prescriptions for Death: The Drugging of the Third World* (Berkeley: University of California Press, 1982).

53. United Nations Centre on Transnational Corporations, *Transnational Corporations And Pharmaceutical Industry* (New York: United Nations, 1979), 1–2.

54. Ibid.

55. *Pharmaceuticals for Developing Countries*, xi.

56. Edward Kennedy, "Keynote," *Pharmaceuticals for Developing Countries*, 4.

57. Ibid., 5.

58. Walsh McDermott, *Pharmaceuticals for Developing Countries*, 10–11; Edward Kennedy; *Pharmaceuticals for Developing Countries*, 4.

59. Vittorio Fattorusso, "Developing Country Perspectives: An Overview," *Pharmaceuticals for Developing Countries*, 252; Kennedy, *Pharmaceuticals for Developing Countries*, 5.

60. McDermott, *Pharmaceuticals for Developing Countries*, 13.

61. John Lesch, *The First Miracle Drugs: How the Sulfas Transformed Medicine* (Oxford, UK: Oxford University Press, 2006). The historical irony that 1937 also saw the expansion of the now-infamous Tuskegee Syphilis Study, which would eventually explicitly restrict the access of effective therapeutics such as penicillin from African-American men living with syphilis, would almost certainly not have been evident to McDermott at the time.

62. McDermott, *Pharmaceuticals for Developing Countries*, 13. For more on racial disparity and tuberculosis mortality in American cities in the early twentieth century, see Samuel Kelton Roberts, *Infectious Fear: Politics, Disease, and the Health Effects of Segregation* (Chapel Hill: University of North Carolina Press, 2009).

63. On the differential role of antitubercular therapy in the reduction of tuberculosis mortality in black South Africans, see Randall M. Packard, *White Plague, Black Labor: Tuberculosis and the Politics of Health and Disease* (Berkeley: University of California Press, 1989).

64. Kurt Deuschle, "Cross-Cultural Medicine: The Navajo Indians as Case Exemplar," *Daedalus* 115 (1986): 175–184, quotation on 176; David Jones,

"The Health Care Experiments and Many Farms: The Navajo, Tuberculosis, and the Limits of Modern Medicine, 1952–1962," *Bulletin of the History of Medicine* 76, no. 4 (2002): 749–790.

65. McDermott, *Pharmaceuticals for Developing Countries*, 13, emphasis mine. McDermott is here also referring to his earlier experiment at Many Farms, in which tertiary-level biomedical therapeutics were deployed on a Navajo reservation in Arizona in what McDermott conceptualized as a form of natural experiment, a Third-World setting inside the continental United States. The failure to show improvement in health metrics was ascribed to a mismatch in the technological substrate; unlike the situations with African Americans in New York City or Maori in New Zealand, the principal mortality burden at Many Farms was due to respiratory and diarrheal diseases of viral etiology, to which there were as yet no effective pharmacotherapeutics to offer. See Jones, "The Health Care Experiments and Many Farms," 749–790.

66. Lewis H. Sarett, "The United States Pharmaceutical Industry," *Pharmaceuticals for Developing Countries*, 130. Sarett made note of the Foundation for International Technological Cooperation recently proposed by the Carter Administration.

67. "The major diseases of the developing countries," they said simply, "are caused by a multitude of worms, fungi, protozoa, bacteria, viruses, ricketttsiae, and by a variety of ill-defined pathogens." Of the eighty-seven tropical diseases unique to developing countries, thirty-three were amenable to vaccination or specific treatment, while fifty-five had limited or no therapeutic options. Paul A. J. Janssen and Denis Thienpont, "The Pharmaceutical Industry's Perspective of Available Agents," *Pharmaceuticals for Developing Countries*, 56–57. See also Ernst Vischer and Rudi Oberholtzer, *Pharmaceuticals for Developing Countries*, 136.

68. A. O. Lucas, William S. Jordan (from the NIH), and Gilbert S. Omenn (from the White House) all detailed recent increases in federal spending on basic research and disease control related to tropical medicine. By 1978, USAID was funding nine national malaria control programs at a cost of $4.5 million per year, and it contributed close to $1 million to the new WHO TDR program. In a March 1978 speech in Caracas, Omenn detailed President Carter's recent pledge to establish an Institute for Scientific and Technological Cooperation; it was later renamed the Foundation for International Technological Cooperation.

69. John Urquhart, "Methods of Drug Administration," *Pharmaceuticals for Developing Countries*, 330–331.

70. Controversy over arbitrary division of contraceptive formulations between rich nations (which would receive them in pill form) and poor nations (which would receive them in injectable form) would roil for more than a decade. See *FDCR* August 6, 1979, T&G 6; Rachel Benson Gold and Peter D. Wilson, "Depo-Provera: New Developments in a Decade-Old Contro-

versy," *International Family Planning Perspectives* 6, no. 4 (1980): 156–160; Wendy Kline, *Bodies of Knowledge: Sexuality, Reproduction, and Women's Health in the Third Wave* (Chicago: University of Chicago Press, 2010).

71. William Foege, *Pharmacuticals for Developing Countries*, 83
72. Fatturosso, *Pharmaceuticals for Developing Countries.*
73. Vittorio Fattorusso, "Developing Country Perspectives: An Overview," in *Pharmaceuticals for Developing Countries*, 251; Sanjaya Lall, "Problems of Distribution, Availability, and Utilization of Agents in Developing Countries: An Asian Perspective," in *Pharmaceuticals for Developing Countries*, 243–246.
74. Max Tiefenbacher, "Industry Perspectives," *Pharmaceuticals for Developing Countries*, 221–223. Tiefenbacher continues, "Until the early 1970s, assaults on the patent system were on a national basis by individual countries. Since then the pharmaceutical industry has become subject to an even stronger centralized pressure from some UN agencies, in particular UNCTAD (United Nations Conference on Trade and Development) and WIPO (World Intellectual Property Organization), aiming at a drastic curtailment of the rights of the patentee, imposing heavy burdens on licensors of pharmaceuticals and related technologies."
75. Ibid., 226.
76. Jeremy Greene, "Making Medicines Essential," *BioSocieties* 6 (2011):10–33.
77. United Nations Centre on Transnational Corporations, *Transnational Corporations and Pharmaceutical Industry* (New York: United Nations, 1979), 84–94. The UNCTC report also emphasized the "revolution in drug therapy, first heralded by the advent of the sulfa drugs in the late 1930s and later confirmed by the introduction of penicillin in the mid-1940s. These medicines ushered in the 'age of wonder drugs' by establishing with certainty the fact that man-made drugs could seek out and destroy target germs without harming the living host."
78. Greene, "Making Medicines Essential."
79. For a detailed account of the links between the IOM Conference on Pharmaceutical for Developing Countries and the United Nations Conference on Science and Technology for Development, with focus on the changing roles of pharmaceuticals in bilateral and multilateral development organizations, see Heidi Morefield, "Technology for Health: Rethinking the Role of Medical Commodities and the Economics of Development at USAID, 1961–1981." Paper presented at the Colloquium in the History of Science, Medicine, and Technology, Johns Hopkins University, 31 March 2016.
80. John Urquhart, "Methods of Drug Administration," *Pharmaceuticals for Developing Countries*, 330–331.
81. Michael Kremer, "Pharmaceuticals and the Developing World," *Journal of Economic Perspectives* 16, no. 4 (2002): 67–90; Michael Kremer and Rachel Giennerson, *Strong Medicine: Creating Incentives for Pharmaceutical Research on Neglected Diseases* (Princeton, NJ: Princeton University Press, 2004).

82. Ivan Illich, *Medical Nemesis: The Expropriation of Health* (New York: Pantheon Press, 1975); Thomas McKeown, *The Role of Medicine: Dream, Mirage, or Nemesis* (Oxford: Blackwell, 1976); Archibald L. Cochrane, "Medicine's Contribution from the 1930s to the 1970s: A Critical Review," in *Medicines For the Year 2000*, ed. G. Telling-Smith and N. Wells (London: Office of Health Economics, 1979).
83. See chapters 7, 8, and 9 in this volume.
84. Richard M. Krause, "Constraints on Involvement of the US Government and Academic Research Community, including Manpower Consideration," *Pharmaceuticals for Developing Countries*, 192.
85. W. Clark Wecoe, *Pharmaceuticals for Developing Countries*, 183.

SEVEN

After McKeown: The Changing Roles of Biomedicine, Public Health, and Economic Growth in Mortality Declines

PAUL FARMER, MATTHEW BASILICO, AND LUKE MESSAC

What saves lives? At least since the revolutionary ferment in Europe during the 1840s, a debate has raged over the sources of population-level mortality decline. The debate has often been framed as a contest between techno-logical intervention and broader social transformation. Perhaps the most famous opening salvo came from the nineteenth-century Prussian pathologist and politician Rudolf Virchow, widely hailed as the father of social medi-cine, who declared, "The improvement of medicine may eventually prolong human life, but the improvement of social conditions can achieve this result more rapidly and more successfully."[1] More than a century later, as heroic narratives of mid-twentieth-century pharmaceutical in-vention gave way to concerns over rising costs and unethi-cal experimentation, British physician Thomas McKeown refuted claims that medicine was responsible for the sec-ular decline in mortality rates in Western Europe.[2] His seminal monograph, *The Modern Rise of Population,* argued that the decline in mortality in England and Wales since the start of registration in 1837 owed little to medical therapeutics.

Virchow and McKeown both spoke deep truths, but the lessons of their work must be understood in light of evolving circumstances. Almost eighty years after the first sulfonamide marked the advent of a new era in antimicrobial therapy, and over a decade into the global scale-up of antiretroviral therapy for AIDS, the theses posited by Virchow and McKeown are ripe for reexamination. What were McKeown's motivations for pitting access to the fruits of biomedicine against broader social and political change? Should we continue to reiterate this dichotomy? Or does the recent past afford reason to believe the two are not in opposition, but are instead mutually integral to the prevention of premature death and unnecessary suffering?

This chapter refigures narratives of the mid-twentieth century "therapeutic revolution" as a geographically uneven process. Walsh McDermott, a physician-scientist who played a key role in the development of antibiotics in the early 1950s—and in the interpretation of McKeown in global public health in the early 1980s—bestowed the moniker "the golden decade" upon the years 1941 to 1951. During that time, he noted, the medical armamentarium grew from little more than "quinine, Atabrine, the arsenicals, and the [recently introduced] sulfonamides" to "the penicillins, the streptomycins, the tetracyclines, chloramphenicol, and isoniazid."[3] And yet, for decades after the golden decade, billions of the world's people lived and died without access to these new therapies.

McDermott's struggle to make sense of the limits of the therapeutic revolution is described by Jeremy Greene in chapter 6 of this volume. And yet, as we explore in this chapter, an equity platform—that is, an actionable plan for chemotherapeutics, preventives, and other tools of "the youngest science"— has recently begun to take shape as a result of social and political processes that escaped McKeown's analysis and which McDermott would not live to see. Though it has been realized unevenly, this platform has demonstrated that the delivery of medical and public health interventions can contribute to significant population-level declines in mortality. The physician's imperative to treat the patient need not conflict with the broader project of fashioning a healthier future.

McKeown's findings for eighteenth- and nineteenth-century Britain and Wales are not timeless truths. In light of recent experience and evidence, we posit a new hypothesis: In any setting in which the heaviest burden of disease is caused by ailments amenable to effective medical therapies, population-level mortality declines need not await secular increases in standards of living. Since this describes most settings of dire poverty in the 2010s, and since the appropriators and agenda setters in global health had never—before the past two decades—engaged in a sustained attempt to deliver effective therapy to those in greatest need, the data to make this claim have previously been absent. But that is no longer the case.

Between 2000 and 2011, Rwanda achieved the steepest decline in mortality of children under age five in recorded history, alongside precipitous drops in death during childbirth and dramatic increases in life expectancy. In addition, a growing body of literature in econometrics has demonstrated a significant contribution of medical and public health interventions, even in the face of continued material deprivation. In sum, the McKeown thesis is not applicable to the global present, as the social realities and technological possibilities of the United Kingdom during the industrial revolution or during the immediate post–World War II era no longer obtain.

This chapter will proceed in three parts. First, we sketch McKeown's thesis, examine the intellectual and political circumstances of its production, and pick out those portions of the argument with special salience for global public health in the following decades. Second, we interrogate the continued relevance of McKeown's claims about the contribution of medicine to population-level health outcomes. Here we review recent data demonstrating that medicine has, in fact, contributed to mortality declines, particularly in the nations of the Global South. We conclude with a discussion of the continuing need for a biosocial "science of delivery" to help ensure that our sizable and growing armamentarium for health (including but not limited to diagnostics, chemotherapeutics, and preventives) reaches the global poor.

The McKeown Thesis and Its Discontents

In a series of papers[4] and monographs[5] written between the mid-1950s and the 1970s, the British physician Thomas McKeown weighed the relative importance of different possible causes of the rise in population in England and Wales since the late eighteenth century. The first of McKeown's conclusions was that the rise in population was due mainly to a decline in mortality from infectious disease. He then discounted the possibility that there had been a meaningful decline (at the population level) in the virulence of pathogens, though he allowed for the possibility of such a decline in the case of the causative bacteria in rheumatic fever. McKeown allowed some causal significance for sanitary interventions in the mortality decline, but he claimed that the effects of public health interventions were primarily felt with waterborne diseases like cholera, which accounted for a smaller portion of mortality than did airborne diseases.[6]

McKeown was most forceful in his claim that medicine had *not* played a crucial role in the overall mortality decline in England and Wales. Effective interventions for major causes of mortality—in particular, effective drugs against tuberculosis and other airborne diseases—were not available until the mid-twentieth century. McKeown's most famous graph showed that most of the decline in mortality from tuberculosis in England and Wales preceded the introduction of streptomycin, the first effective antitubercular agent, in 1947 (see Figure 7.1). Those diseases for which there were earlier effective interventions, namely smallpox and diphtheria, were in his data fairly unimportant causes of mortality at a population level. In the end, McKeown settled on improved nutrition—a function, he believed, of a steady rise

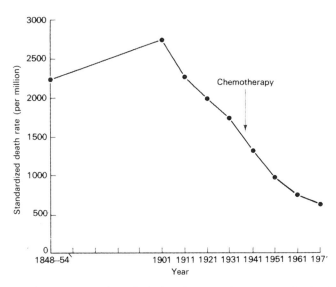

FIGURE 7.1 Thomas McKeown's graph of tuberculosis mortality rates in England and Wales. From *The Modern Rise of Population* (New York: Academic Press, 1976).

in standards of living—as the main reason for the decline in mortality.[7] Improved nutrition, he said, had increased resistance to airborne pathogens like the tubercle bacillus.

McKeown's argument became the focus of a fierce and prolonged historiographical debate. Historical demographer Simon Szreter composed one of the most comprehensive critiques in a 1988 article where, among other objections, he argued that McKeown's "positive explanatory thesis," which gave such pride of place to nutritional improvements derived from a rising standard of living, was decidedly immodest. Szreter argued that public health interventions had played a larger role than McKeown allowed, noting, for instance, that respiratory tuberculosis (TB) was often a sequela to waterborne illnesses that reduced intestinal absorption of nutrients. The decline in respiratory TB, Szreter claimed, was in many cases a downstream consequence of sanitation infrastructure that brought clean water and better sewage disposal to the people of England and Wales.[8] More recently, an analysis by David Cutler and Grant Miller estimated that clean water interventions alone could account for nearly one-half of the mortality reduction in the United States between 1900 and 1936.[9]

Still, Szreter remained convinced by McKeown's demonstration "that those advances in the science of medicine which form the basis of today's conventional clinical and hospital teaching and practice, in particular the immuno- and chemo-therapies, played only a very minor role in accounting for the historic decline in mortality levels." Charts in McKeown's *Modern Rise of Population* showed that mortality in England and Wales from respiratory TB, bronchitis, whooping cough, measles, and scarlet fever declined long before the introduction of effective chemotherapeutics or immunizations. McKeown's subsequent work used similar methods to demonstrate the irrelevance of biomedical intervention to the mortality declines in Sweden, France, Ireland, and Hungary.[10]

This argument about medicine was the linchpin in McKeown's broader political project. A professor of social medicine at Birmingham University, McKeown argued against contemporary planners who sought to shunt the limited resources of the National Health Service to high-tech curative medical interventions rather than to preventive efforts to modify environment and behavior, which he called "the predominant determinants of health."[11] McKeown did not even veil the political valence of his historical scholarship. He wrote that history was "essentially an operational approach which takes its terms of reference from difficulties confronting medicine in the present day."[12] His accounts of the past were explicitly intended to inform action in the present.

It is crucial, though, to remember the ways in which McKeown's present differed from the early decades of the twenty-first century. He wrote during the central decades of the postwar industrial boom and the consolidation of the British cradle-to-grave welfare state system, when few assumed that social spending would ever face severe budget cuts. In this context, pointing to the centrality of living standards in improving population-level health—over high-tech medical intervention or locally administered Victorian-era public health schemes—was an argument for an egalitarian politics.

Later critics noted that McKeown's dismissal of the role of sanitary measures ignored the possible role for political movements, state intervention in social provision, and human agency more generally.[13] Some also objected to his failure to specify that he analyzed Western European experiences almost entirely before the "golden decade," experiences that might not be applicable to the post-antibiotic era or in different parts of the globe.[14] But the underlying assumptions of McKeown's argument—a secular rise in general living standards, the universalization of British industrialization as a developmental trajectory, the capacity of aggregate statistics to approximate individual experiences, and the primacy of economic growth—were rarely questioned by politicians or social scientists of his era.

The idea that growth must precede spending on health and education was common to the writings of modernization theorists of the 1950s and 1960s. For example, Walt Rostow's influential modernization text, *The Stages of Economic Growth: A Non-Communist Manifesto* (1960), presented a universal five-stage trajectory of economic growth that hinged on the reinvestment of profits in production for decades before the investment in social services. Only when nations reached an "era of high mass-consumption" (a state which Rostow saw occupied only by the United States, Western Europe, and Japan in 1960) would the prudent politician see fit to invest in social welfare and social security programs.[15] Even social democrats like Swedish economist Gunnar Myrdal and African socialists like Tanzanian President Julius Nyerere echoed this admonition that government spending on welfare programs must be curtailed in favor of investment in pursuit of maximal production.[16] Thus, McKeown's conflation of a neatly schematized British history—where rapid growth long preceded concerted efforts at social provision—with a prescription for the present was not a unique perspective.[17]

To be sure, there were among McKeown's contemporaries in historical demography those who recognized that the nineteenth-century British experience might not so neatly inform the contemporary moment. As Jeremy Greene explains in chapter 6 of this volume, by the early 1970s the epidemiologist Abdel Omran theorized that the "epidemiologic transition" would be effected by "imported medical technologies in much of Africa and Asia."[18] Unlike the work of either Rostow or McKeown, Omran's model was comparative: "The transition in the now developed countries was predominantly socially determined, whereas the transition in the 'third world' is being significantly influenced by medical technology."[19]

In some ways, Omran's faith in technology was becoming marginal in American and European public discourse by the 1970s. Revelations of the teratogenic consequences of thalidomide in the early 1960s and the carcinogenic consequences of DES in the early 1970s, the scandal of the Tuskegee syphilis study (halted by the Centers for Disease Control in 1972), and the ever-increasing costs of medical care led to increasingly critical views on physician culture and biomedical technology.[20] Still, Omran's confidence in diffusionist technology would presage the faith that would help justify the pursuit of vertical interventions during the 1980s, as well as the World Bank's focus on cost-effective technologies in the 1990s.

But in contrast to McKeown, Omran, and medicine's critics, the dominant strain in international public health in the 1970s saw social change and technological intervention as equally necessary. Halfdan Mahler, director-general of the World Health Organization (WHO) from 1973 to 1988, was a lead proponent of the Primary Health Care (PHC) movement that culminated in the 1978 Declaration of Alma-Ata, proclaiming, "Health for all by the year 2000." The definition of primary health care set out at this conference included medical chemotherapeutics and preventives, trained staff, basic sanitation, nutrition, and health education.[21] Though Mahler would criticize those who advocated technical "fixes" to social ills, he was the champion of the WHO's Action Program on Essential Drugs, which sought to make commercial pharmaceutical products part of a "public health commons" susceptible to human rights claims.[22] In the 1970s, the idea of a delivery revolution that included health among a panoply of social and economic rights was at least rhetorically possible, though insufficiently funded on a global scale.

Public Health Nihilism and Possibilities:
McKeown in the Postmodern Moment

To some later commentators, McKeown's explanation that his historical demography "takes its terms of reference from . . . the present day"
would prove prescient in ways he had not predicted. Reflecting on the
legacy of the McKeown thesis in the late 1990s, the historians Amy
Fairchild and Gerald Oppenheimer argued that his devaluation of
medicine and public health had the unintended consequence of jus-
tifying draconian budget cuts during the neoliberal era, beginning in
the 1980s and 1990s:

> In rapidly changing ideological circumstances, McKeown's findings were gratefully
> absorbed by the rising ideology of the New Right, which radically questioned the
> value of the whole welfare-state system and its associated policies of full employ-
> ment, income redistribution, and free public services.[23]

The supposed "fit" between McKeown's argument and the political
imaginary of market fundamentalism would have been anathema to
McKeown himself. In *The Role of Medicine* (1979), McKeown argued that
medicine had a clear role in contemporary society even if its claims to
effectiveness were often immodest. He called, in the name of public
health, for increased public investments in the subvention of food for
children and the elderly in Britain, redoubled efforts to improve family planning and agricultural production in developing countries, and
broader training of health professionals (especially physicians) in the
environmental determinants of health.[24] At no point did McKeown sub-
scribe to the laissez-faire philosophies embodied by the rhetoric of British Prime Minister Margaret Thatcher and US President Ronald Reagan.

Nevertheless, later public health scholars singled out McKeown as
an adherent of "public health nihilism," the idea that "improvements
in health status can only come with the amelioration, if not the radical
transformation, of adverse social conditions."[25] In *The Modern Rise of
Population,* McKeown posited that the major reasons for the improvement in general nutritional status were the introduction of more efficient property regimes (particularly enclosure) and technological innovation (e.g., crop rotation, new farm implements, canals for transport).[26]
McKeown did not mention workers' movements for higher wages and
better housing in nineteenth-century Britain and Wales. Despite his social democratic sympathies, McKeown's work inspired others, particu-

193

larly the economist and Nobel laureate Robert Fogel, to attribute secular population-level mortality declines entirely to increases in aggregate production.[27] Fogel's *The Escape from Hunger and Premature Death, 1700–2100*, published in 2004, stressed the links between aggregate agricultural output, average nutritional intake, and physiological resistance to disease ("robustness and capacity of vital organ systems").[28] As the economist Angus Deaton observed, Fogel was "at some pains to emphasize the close tracking of health and income."[29] Through such reductionist logic, McKeown's intellectual heirs would obscure many of the steps toward equity that he had endorsed during his lifetime.

But McKeown's work was not the only—or even the main—reason for this change in focus. He did not spur the transition within orthodox economics from a pragmatic and socially embedded Keynesian consensus to a utopian vision of efficient markets.[30] And the global dismantling of state services would take hold only after increases in oil prices and interest rates, decreases in the prices of agricultural exports, and the withdrawal of commercial credit and the draconian demands of international financial institutions following Mexico's 1982 default on its sovereign debts.[31] The decline of government medical and public health services in the last quarter of the twentieth century owes more to these intellectual currents and global forces than to (mis)interpretations of McKeown.

Another even more disturbing argument against biomedical intervention in settings of poverty is unrelated to McKeown's work but merits mention here, if only for its historical persistence and deadly consequences. At the height of the neoliberal enthusiasm for disinvestment in public services, leading global public health journals published articles by experts who claimed that reducing child mortality would only increase populations to ecologically unsustainable levels and doom them to famine or other forms of ruin. In a 1990 article published in *The Lancet*, the physician Maurice King counseled against "such desustaining measures as oral rehydration . . . since they increase the man-years of human misery," and urged the realization that "health services may not be a priority for [poor] communities."[32] This was little more than another resuscitation of an oft-repeated prophecy, embodied in the "population bomb" foreseen by prognosticators in the 1960s[33] and by Thomas Malthus's injunctions against feeding the poor at the turn of the nineteenth century.[34]

Health care provision has nowhere been shown to be a threat to demographic stability, nor is the denial of it an effective means of population control. Studies of regions with high levels of AIDS mortality find inconsistent effects on population-level fertility rates.[35] Meanwhile, nations that have achieved substantial fertility declines have often done so not by permitting mass die-off, but by expanding human capabilities. For instance, an analysis in India found that female literacy accounted for approximately three-quarters of fertility decline in regression models, while the relationship between fertility and per capita income was not statistically significant.[36] Too often, neo-Malthusian claims have functioned as thinly veiled justifications for exploitative policies; Megan Vaughan documented how officials in British colonial Africa who clamored about the perils of overpopulation did so to obscure a recent expropriation of arable land.[37] Demographic entrapment, like public health nihilism, retains an allure that cannot be explained by its (thin) evidentiary basis.

Luddite Traps

Yet if McKeown has had a lasting legacy beyond the maelstrom of critique he drew from fellow travelers in social medicine and his inspiration of scholars like Fogel, it was his humbling of medicine. Our major aim here is to temper McKeown's claims about medicine in light of the history of global health, particularly the experiences of the last three decades. While Szreter believed McKeown's claims about medicine to be the most amply proven portion of his thesis, other commentators have joined Omran in arguing that the rather small part played by medicine in Britain's mortality decline should not automatically discount the possibility that medicine might have had a much larger role in averting premature death during the post-antibiotic era. In the eyes of these writers, the aspect of the McKeown thesis that has withstood criticism the longest—the irrelevance of medicine in Britain's mortality decline—should not be used as a rhetorical tool in defense of global inaction in the present.

One prominent proponent of this argument was Walsh McDermott, a physician-scientist whose involvement in debates over the value of pharmaceuticals in developing countries is detailed by Jeremy Greene in chapter 6 of this volume. In a 1980 commentary in *Science*, McDermott argued that population-level declines in mortality could be achieved using chemotherapeutics and vaccines that fell into one of two categories. First were those like the sulfonamides that were effective against many fatal diseases. And second were those drugs that were effective against important and closely monitored causes of death like tuberculosis. McDermott constructed a graph of the US death rate in the twentieth century and noted the sudden acceleration in its decline starting in 1937, the same year in which he claimed "an adequate supply of sulfonamide was put into the hands of the personal-encounter physicians all over America."[38]

McDermott argued against those academics who "fall into the trap of assuming that diseases of multifactorial origin cannot be cured by a specific technology unless the multifactors, or at least a number of them, are also appropriately controlled." To support his case he pointed to tuberculosis treatment, which, following the introduction of isoniazid (a drug McDermott had played a crucial role in developing), was found to be curable in all manner of social settings. "When introduced to poverty areas, the rate of fall in the death rate has been identical with that elsewhere."[39] McDermott's was an argument not only for the population-level effects of health interventions, but for what we have described elsewhere as "pragmatic solidarity" with the destitute sick.[40]

With respect to the contrast between the abrupt substantial fall in death rate following the introduction of tuberculosis chemotherapy and the long, steady previous decline in death rate, one could say as a rough calculation that without the drug technology we would have arrived where we are now by about the year 2020, say 40 years from now. To the historian of the public good, 40 years is not really too long to wait; but for the thousands of individuals whose survival came by technology through the encounter physician system of the past 30 years, the equation comes out differently.[41]

As a physician first and a "historian of the public good" second, Mc-Dermott was unwilling to accept the untimely demise of patients when he knew effective treatments were available. To be sure, he admitted, a concerted attack on all the ills—including poor housing, unemployment, war, racism, and malnutrition—that contribute to poor health outcomes was the most preferable strategy, but he was unwilling to forestall pragmatic steps on behalf of the sick in the present to wait for yet-unrealized visions of a just society.

Nonetheless, there was an element of pessimism in McDermott's prognosis for health services in poor countries. At times he even fell into the same Luddite traps he criticized in others.[42] Specifically, Mc-Dermott doubted that pharmaceuticals would reach most people in developing countries in the absence of vast increases in per capita health spending. Without the personnel or resources to reach populations through "personal-encounter physicians," poor countries needed some other method to access biomedicine. "The appropriate technology for control of diseases of economically underdeveloped countries happens to be mainly that applicable to groups as a whole," he explained.[43] Though McDermott did not specify what form such widely distributed technologies would take, he did say they would require a new kind of biomedical research infrastructure to overcome existing delivery challenges. McDermott's willingness to dichotomize population-level intervention and "personal-encounter" care was simply another framing of the "either/or" approach he claimed to reject.

Indeed, before the twenty-first century no one seriously tested medicine's ability to spur population-level mortality decline. McDermott saw a set of interventions he helped conduct on the Navajo Reservation in Arizona at mid-century as such an experiment. His clinics failed to stem many prominent causes of morbidity and mortality; pneumonia, diarrhea, ear infections, measles, and impetigo remained constant. But his interventions were neither advanced nor intensive, even for their day. McDermott admitted that additional interventions, such as home-based nursing care, could have helped lower diarrheal death rates.[44] Other examples of nationwide mortality decline during the mid-twentieth century can be attributed only partly to medicine. Improvements in life expectancy during Mao Zedong's rule in China (from forty-one years in 1949 to sixty-two in 1976) and during Fidel Castro's rule in Cuba (from sixty-three years in 1959 to seventy-four in 1979) were achieved in large part through public investments in housing, sanitation, and education alongside rural medical infrastructure.[45]

Other efforts to improve health outcomes through expanded access to biomedicine have been cut short by the rise of contrary political ideologies. In South Africa, an ambitious plan designed by physicians Sidney and Emily Kark to extend comprehensive health services to the country's poor—both black and white—came to an abrupt end when the Afrikaner National Party imposed apartheid in 1948.[46] And, as noted above, in the wake of the neoliberal dismantling of public health systems, the Alma-Ata Declaration's promise of "health for all by the year 2000" was never seriously pursued. In its place, the Selective Primary Health Care counseled ministries of health to focus on "minimum packages" that would not even attempt to treat tuberculosis and other major global causes of mortality. Between the advent of the antibiotic era and the turn of the millennium, there was no sustained equity-based platform to address the major killers of the global poor with high-quality tools.

Viewed from 2015, McDermott's two major points of pessimism can be seen as assuredly outdated, disproven by decades of experience. First, his despondency over the cost of therapeutic interventions failed to acknowledge that these costs were not immutable truths, but rather were the product of political institutions (such as intellectual property regimes) that were susceptible to social forces. Second, his assumption that medical chemotherapeutics (in pill form) would never be widely accessible to poor people in poor countries revealed another too-constrained understanding of the possible. Both of his doubts will be answered by our revisiting some of the developments of the recent history of the global AIDS pandemic.

In short, times have changed. The technologies McDermott had on hand could not achieve the sort of revolution he wished to see in the delivery of biomedical interventions. Take, for instance, the tools available for averting deaths from diarrheal disease, long a major killer of infants and children. In 1980, McDermott doubted the value of existing therapeutics to help solve this problem:

Our medical technology has relatively little to offer infants in a sanitarily unprotected home environment. . . . A disease pattern of great importance in developing societies—the pneumonia-diarrhea complex of infants—largely disappeared from our society without the use of today's technology, but in a setting of widespread economic uplift; and attempts to substitute the drugs effective in U.S. society for a complete lack of sanitary barriers in the home may have quite limited value in developing societies.[47]

While McDermott's pessimism may have had some merit in 1980, it was soon to be refuted by a simple new intervention. Oral rehydration therapy (ORT), a mixture of sodium, a carbohydrate, and water, was introduced in 1979, but McDermott died in 1981 and did not live to see its successful implementation. By 2000, the WHO estimated that ORT was given during most episodes of childhood diarrhea worldwide. Alongside other interventions, including the promotion of breastfeeding and female education, ORT helped decrease the number of deaths attributable to diarrhea among children under five from 4.6 million in 1980 to 1.5 million in 1999.[48] Rotavirus, the most important cause of severe gastroenteritis among children globally, became a preventable illness with the introduction of two vaccines in 2006; since 2009 the WHO has recommended that rotavirus vaccine be included in all national immunization programs.[49]

The Delivery Decades: Equity Platforms since 2000

In the years since the turn of the millennium, the limits of medicine posited by McDermott and McKeown have come to appear even more outdated. There has been an unprecedented rise in public attention to and funding for HIV/AIDS and other problems in global health. Development assistance for health from public and private institutions rose from $8.7 billion in 1998 to $21.8 billion in 2007. The number of people around the world on antiretroviral therapy reached 8 million by the end of 2011, a twentyfold increase since 2003. Still, approximately 1.7 million people die each year from AIDS, a preventable and treatable disease, while at least 6 million people currently eligible for treatment are not receiving it. But, in large part as a result of increased access to therapeutics, deaths from AIDS-related causes in sub-Saharan Africa fell by 32 percent between 2005 and 2011.[50]

Increasing funding for curative and preventive medicines in the twenty-first century has spurred similar declines in mortality for other major infectious killers. Between 2000 and 2012, mortality rates from malaria decreased by 45 percent globally and by 49 percent in Africa.[51] Preventive measures have helped reduce incidence of malaria by 29 percent globally and by 31 percent in Africa. Between 1990 and 2012, tuberculosis mortality rates fell by 45 percent worldwide.[52]

Nothing about this expansion in access to lifesaving therapy was foreordained. For decades, the paltry appropriations for foreign aid budgets and the high prices of brand-name pharmaceuticals for lifesaving interventions in poor countries faced few serious challenges. But an initially slow response to AIDS and other major infectious killers proved susceptible to social forces.[53] Drug costs for antiretroviral drugs plummeted after transnational activism challenged the inviolability of intellectual property regimes, and after innovative market coordination allowed for the harnessing of newfound economies of scale and generic production. The lowest available annual per-patient price of the most common first-line ART regimen in the developing world fell from $10,000 in the late 1990s to $300 in 2002, and to $87 in 2007. This precipitous decrease in drug prices created a new opportunity to scale up AIDS treatment programs.

When foreign aid administrators in the United States declared lifesaving interventions like antiretroviral treatment too complex to be delivered in resource-poor settings, pilot programs in rural Haiti and urban South Africa demonstrated treatment adherence and immune reconstitution. Low levels of international donor funding, long seen as an immutable reality, changed drastically as an unlikely coalition of activists, academics, cultural figures, and political leaders from across the political spectrum advocated for increased funding for AIDS care and treatment. The Global Fund for AIDS, TB, and Malaria (GFATM) and the US President's Emergency Plan For AIDS Relief (PEPFAR), the latter initiated—to the surprise of many—by President George W. Bush, made global health equity more of a reality than Alma-Ata or all previous declarations. These developments were made possible by technical innovation and medical therapies, but just as crucially by social and political processes involving diverse actors in many countries.

Ours is not an unquestioned belief in the inviolable benevolence of chemotherapeutics. Indeed, the historical record is littered with medical and public health campaigns in the global South that used *cordons sanitaires* as thin veils for racial segregation and promoted pills as panaceas for rapacious systems of colonial rule. In her history of sleeping sickness outbreaks and control strategies in the Belgian Congo between 1900 and 1940, the social historian Maryinez Lyons contended that the disruptive and coercive policies of King Leopold's colonial regime (rubber quotas, taxes, forcible dislocations) were a primary cause of the sleeping sickness epidemics. During the outbreaks, even more overbearing policies—including passports for internal travel, further dislocations, and massive chemotherapy with dangerous medicines that often caused blindness—advanced the prerogatives of the state. Lyons documented how the response by colonial health officials simultaneously advanced "constructive imperialism" in the minds of European publics and wholesale social engineering of African societies. Germ theory was used as a justification for sleeping sickness control programs that focused on isolation and chemotherapy over ecological and socioeconomic interventions.[54]

The need to include both the technological and the social in critical analysis is equally important today, if for different reasons. The introduction of novel chemotherapeutics into settings of great poverty can be simultaneously miraculous and ambivalent for patients. In chapter 8 of this volume, Julie Livingston explores patients' experiences of cancer therapy in Botswana's central hospital, where oncologists have intermittent access to chemotherapeutics but scarce supplies of morphine or antiemetics. And while the number of African patients on antiretroviral treatment (ART) continues to increase, some have been spared death from AIDS only to be tormented by hunger. When patients—left

desperately poor and malnourished by the virus and the resulting inability to earn a living—begin ART, they can be brought back from the
edge of death. But this recovery may start an unrelenting hunger in
patients who had been starving to death without feeling it. Physicians
and activists (and pharmaceutical companies) may have succeeded in
putting "pills into bodies," but those same bodies are now starving. In
the words of one Mozambican patient recently resuscitated by highly
active antiretroviral therapy (HAART), but unable to quench his renewed appetite without a job or farming inputs, "All I eat is ARVs."[55]

Yet, more often than not, poor people in poor countries have been
the ones calling for more equitable access to the fruits of modern medicine. In July 2000 at the International AIDS Conference in Durban,
South Africa's Treatment Action Campaign led a march of more than
five thousand demonstrators to demand lower prices for antiretrovirals
and increased access to treatment. One reason the Treatment Action
Campaign could bring out so many people was the level of organization among the heavily infected black communities. In South Africa
many of the most active organizers, like Zackie Achmat, were veterans
of the antiapartheid movement. But they were often on the outskirts of
the new South Africa. Zackie Achmat was a radical gay man who had
been a sex worker, and most members of TAC were black South Africans
living in slums that were a legacy of apartheid rule. South Africans living with AIDS were fighting not only pharmaceutical companies, but
also their own post-apartheid government, led by Thabo Mbeki, who
had denounced antiretrovirals as harmful to health and refused to de-
vote significant public funding toward treatment. Though Achmat had
AIDS and needed the medicines, he went on a heavily publicized "drug
strike," refusing to take medicines himself until poor South Africans
had access to them.[56] What anthropologist Adriana Petryna has called
biological citizenship has become a central locus of the sociopolitical
struggles alongside age-old battles over other political, civil, social, and
economic protections conferred by the state.[57]

The Role of Medicine: Evidence for a New Hypothesis

In recent decades, economists and demographers have produced evidence of mortality declines in settings of poverty wrought in significant measure—though certainly not solely—by improved access to biomedicine. Demographer Samuel H. Preston's 1975 study, suggestively
titled "The Changing Relation between Mortality and Level of Eco-

nomic Development," found that only 10 to 15 percent of the improvement in global life expectancy between 1930 and 1960 was driven by income growth, while technological change—improvement in health interventions—was the major factor in this transformation.[58]

In the next few decades, the debate over the contribution of public health and medical interventions in population-level mortality declines continued in the econometric literature. An influential paper by Deon Filmer and Lant Pritchett using 1990 data found that "cross-national differences in public spending on health account for essentially *none* (one-seventh of 1%) of the differences in health status." Yet Andrew Sunil Rajkumar and Vinaya Swaroop responded with the finding that public sector health spending did much more to decrease under-five mortality in those countries with high scores on indices of "good governance."[59] In a 2001 paper, Dean Jamison, Martin Sandbu, and Jia Wang pointed out that cross-national studies often assumed (contrary to the historical record) that the rate of technological progress was constant across countries. After allowing for country-specific technology levels, they found that only 5 percent of the reduction in the infant mortality rate was driven by income improvements, while the largest contributions came from technological change.[60] A review of the literature by Michael Kremer in 2002 contended that the weight of evidence favored the view that "the role of pharmaceuticals and medical technology in improving health in developing countries stands in contrast to the historical experience of the developed countries. . . . Modern medical technologies allow tremendous improvements in health even at low income levels."[61]

Cross-national studies emphasizing the contributions of public health and medicine to health improvements in poorer countries during the second half of the twentieth century were supported by country-specific studies with similar findings. Demographer Olukunle Adegbola argued that the deployment of biomedicine in Nigeria helped increase life expectancy by fourteen years between 1963 and 1980.[62] Working with health and demographic data from a small area of rural Senegal, French demographer Gilles Pison and colleagues attributed a reduction in under-five mortality of more than 75 percent between 1967 and 1992 to the introduction of basic medical services ("a dispensary and a maternity clinic") as well as to growth surveillance, health education, vaccination, and malaria programs. The authors argued that, at least in this region, changes in women's educational levels and transportation improvements were quite limited and could not account for the decline.[63] In the Indian state of Kerala, a universal health care scheme and close access to medical facilities allowed 97 percent of expectant mothers to deliver in hospitals or other biomedical institutions. The state spent $28 per capita on health in 2000 while recording an infant mortality rate of fourteen per thousand live births and life expectancies of sixty-seven years for women and seventy years for men. In contrast, in the same year the US spent $4,703 per capita on health while achieving an infant mortality rate of seven per thousand births and life expectancies of eighty for women and seventy-four for men.[64] None of these successes are attributable to pills alone. Scholars seeking to determine the causes of mortality decline face the challenge of colinearity; investments in health infrastructure often occur alongside broader social and economic changes. Yet these studies demonstrate that biomedicine has made a discrete and measurable contribution to population-level mortality decline.[65]

The single most dramatic and most richly evidenced example of biomedicine's ability to contribute to population-level mortality decline is Rwanda in the twenty years since the 1994 genocide. This country—the poorest on the planet in the wake of the genocide—has experienced some of the steepest declines in mortality in recorded history. The fall in Rwanda's under-five mortality rate between 2000 and 2011 has far outpaced those of England and Wales and other countries in Western Europe at any point in the nineteenth and twentieth centuries, and even of "Asian tigers" like South Korea during their initial takeoff in the 1950s and '60s (see figure 7.2). Life expectancy rose from twenty-eight years in 1994 to fifty-six years in 2012. More than 97 percent of the population has health insurance coverage through community-based, civil service, military, or private plans. The national vaccination program has achieved 93-percent coverage for each of nine vaccine-preventable illnesses, up from 25-percent coverage for five illnesses in 1999. In a single decade, mortality associated with HIV disease fell by 78.4 percent, while mortality from tuberculosis fell by 77.1 percent. Between 2005 and 2011, mortality from malaria dropped by 87.3 percent. Between 2000 and 2010, the maternal mortality ratio declined by 59.5 percent. Between 2000 and 2011, the probability that a child would die by age five decreased by 70.4 percent. The crude death rate (deaths per thousand people per year) fell from thirty-three in 1995 to fourteen in 2000 to seven in 2012. The decline in the crude mortality rate since 2000 represents the single fastest decline during the same period of any country in the world.[66] Taken together, the pace of Rwanda's improvement in health outcomes is altogether unprecedented—not just for this period, but for any period in global history.

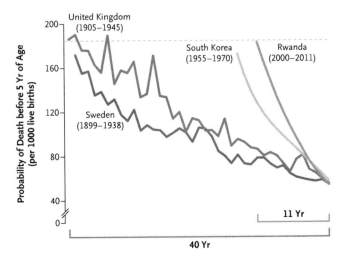

FIGURE 7.2 Probability of death before five years of age during various periods in the United Kingdom, Sweden, South Korea, and Rwanda. From Paul Farmer, "Shattuck Lecture: Chronic Infectious Disease and the Future of Health Care Delivery," *New England Journal of Medicine* 369 (2013): 2424–2436. Accessed at http://www.nejm.org/doi/pdf/10.1056/NEJMsa1310472.

To be sure, improvement in health services was coterminous with a rapid increase in aggregate economic output and a substantial decline in poverty (from 77.8 percent in 1994 to 44.9 percent in 2010). Widespread political violence has been absent from Rwanda since 1997. Chronic malnutrition—McKeown's favored determinant of mortality rates in England and Wales—saw a modest decline in Rwanda, from 51 percent in 2006 to 43 percent in 2012.[67] These rapid social transformations have surely contributed to Rwanda's impressive improvements in health outcomes. But recent research in Rwanda points to the causal importance of medicine and public health in the improvement in economic indices. The ecologist and economist Matthew Bonds and his colleagues have used the Rwandan experience to demonstrate how investments in comprehensive health systems can facilitate improvements in health indices and reductions in poverty rates alike.[68] Rwanda after the genocide gave the lie to demographic entrapment and other theories that ignored the virtuous cycle between improved health and economic growth.

But what role, we might ask, did medicine and public health play in this mortality decline? Though claims of causality are manifold, we point to the establishment of a strong public sector health system as a significant contributor. By harnessing enthusiasms for vertical programming among global health funding agencies to efforts to strengthen health systems more generally, planners in the Ministry of Health "resisted pitting prevention against care, public sector against private, and vertical programmes against primary care." Instead, the ministry sought to "identify and address the leading causes of mortality and morbidity, while expanding access to basic health services to the poor and strengthening the health system."[69] This allowed for vast improvements in all manner of health problems, including less prominent problems of global health such as maternal morality and vaccine-preventable illness.

In 2004, Partners In Health's sister organization, Inshuti Mu Buzima (IMB, Kinyarwanda for Partners in Health), started working in Rwanda at the request of the government and the William Jefferson Clinton Foundation. In Rwanda, IMB sought to prove that comprehensive primary health care (including care for AIDS, TB, and malaria) could be delivered effectively and equitably in two rural resource-poor districts. These districts were home to a half-million residents and zero doctors before IMB's arrival. New and renovated hospitals and clinics—with new laboratories, larger and well stocked pharmacies, and, in some cases, operating rooms—would belong to the Ministry of Health and be staffed by its employees. The goal was for the ministry to gradually assume full control of IMB-supported facilities.

Over the next decade, IMB built a long-term partnership to accompany the Rwandan government in the design and implementation of its District Health System Strengthening Framework. Pilot programs at IMB facilities were considered for national scale-up if proven effective. One such program involved the use of community health workers to deliver care to patients in their homes. Before IMB's arrival, fewer than one hundred patients were on antiretroviral treatment at the six facilities in its catchment area. In its first year, IMB enrolled more than one thousand people on ART. Every patient was paired with a paid *accompagnateur*, usually a neighbor, who made daily or twice-daily home visits to observe ingestion of medications and ensure that patients had access to food (each patient receiving ART was entitled to food packets for the first six months of therapy), housing, transportation, schooling for children, and other forms of psychosocial assistance. All services were provided at no cost to patients. By 2012, the Ministry of Health had commenced a large-scale national recruitment and training program for community health workers and deployed 45,000 such workers across the country.[70]

A nationwide system of health insurance initiated by the government of Rwanda has also been central to increasing access to care. In 2006, the Ministry of Health announced the national implementation of a community-based *mutuelle* health insurance scheme, mandating that every citizen purchase health insurance. Annual premiums vary by region, but in the rural districts in which IMB works, membership costs one thousand Rwandan francs (slightly less than US$2) per year. For members, the copayment for most primary care visits is 150 RWF (about US$0.27); for hospitalizations, patients pay 10 percent of the cost of drugs, consultations, and procedures. Patients presenting for HIV counseling and testing, antiretroviral therapy, tuberculosis care, or prenatal visits are not charged any copayments. The Global Fund pays premiums and copayments for people deemed by local leaders to be too poor to pay. IMB contributes additional funds to the coffers of *mutuelle* accounts in its catchment area to support free care for children under five years old and for patients presenting for malaria diagnosis or care.[71] By the end of 2008, the Ministry of Health reported that 95 percent of Rwandans were enrolled.

Considered alongside the wealth of cross-national data reviewed above, the details of country-specific cases—with Rwanda as the best-illustrated example—beg a reconsideration of the McKeown hypothesis. The role of medicine in spurring mortality declines and demographic transition is not what it was in the nineteenth and early twentieth centuries. The years since 2000 represent the first time that delivery of a panoply of high-quality therapeutics has been sustained on any appreciable scale. The "child survival revolution" of the 1980s and early 1990s demonstrated the capacity for "minimal packages" including simple interventions such as oral rehydration and vaccines to spur declines in disease-specific mortality. The subsequent "delivery decades" have proven that treating ailments amenable to effective biomedical therapeutics can effect population-level mortality declines far more rapidly than long-term increases in general living standards.

Conclusion: Unfinished Business

Even with the weight of this evidence, access to effective biomedical intervention has been expanded unevenly. In many of the "economically underdeveloped countries" invoked by McDermott as settings inimical to the purchase and delivery of biomedical interventions, the therapeutic revolution has rarely and only recently been felt. While the problem of delivering effective therapies to patients in resource-poor areas is old, the "know/do gap" has received renewed attention in the last decade. Multidisciplinary research endeavors, such as the Global Health Delivery Project at Harvard University, have attempted to advance a science of delivery through systematic analysis of strategies of medical and public health implementation. Among the recurring principles that have emerged through this exercise is the need to link health delivery and economic provision. "Patients who cannot afford transportation to a local clinic, for example, may need additional resources to borrow a car or, in remote rural areas, rent a donkey; the destitute sick may also need help with child care or food for their families."[72]

The emerging "science of delivery" is a framework through which the potential for health improvements can be realized in almost any economic environment. Delivery takes as a starting point the observation that almost any medical service has been delivered in resource-limited settings, yet vast heterogeneity in access to these services persists. From this foundation, the guiding questions become constitutively practical: How can interventions be delivered in a certain time and place?[73] Delivery researchers investigate the conditions under which health care access can improve dramatically. They necessarily undertake an interdisciplinary study of these conditions, what has been termed a *biosocial* approach,[74] using methods of observation from fields such as epidemiology, political science, anthropology, economics, history, and clinical medicine. Delivery approaches can be standardized (e.g., protocols for the management of Burkitt's lymphoma) or tailored to community aspirations, political institutions, and local understandings of illness. Modern improvements in health without concomitant improvements in income have demonstrated a path toward improving health, regardless of large-scale structural change. Delivery science embraces the complexity and wide range of problems that prevent beneficial health intervention while pointing out that these problems are solvable.

The potential for improvements in health outcomes even in the absence of broader structural change does not obviate the desire to link health with more fundamental transformations in the structures of production and resource allocation. While this aspiration has a long history, the empirical record of purposive structural change is inconsistent at best. Indeed, many leading development scholars have concluded that universal levers of economic development (including growth that is inclusive of the poor) remain elusive.[75] More ambitious and ideologically extreme programs made grand claims about structural transformation that ultimately rang hollow. Instead of waiting for a holy grail of etic formulas for economic development, a delivery approach acknowledges the transformations that can be attained in human development, even if the process of structural change and poverty reduction has yet to be fully elucidated.

Equitable delivery platforms for therapeutic interventions—including diagnostics, preventives, and treatments—continue to demand new financing mechanisms. "When treatments are easily administered, convenient, and likely to result in cure or excellent clinical response, there will be great demand for them. But when such need is seen as demand only if there is an established market for these innovations, it is fair to talk about market failure."[76] In the absence of concerted political action, recent biomedical innovations such as bedaquiline for drug-resistant tuberculosis or sofosbuvir for hepatitis C may not reach the places where the burden of disease is highest for decades. Solutions to these problems are readily apparent in the recent histories of AIDS and drug-resistant tuberculosis, both of which have seen declines in the price of therapies of more than 97 percent over the last two decades.[77]

Beyond infectious disease, other global health challenges can be addressed by improving the delivery of effective therapies. In certain clinical areas, little progress was made in expanding access to treatment even during the 2000s, an era of what science journalist Laurie Garrett called "marvelous momentum."[78] Poor people in poor countries in the 2010s received little more care for cancer, diabetes, heart disease, and mental health than they had a decade earlier. In these areas the same rhetoric of doubt—that treatments can be delivered in poor settings, that funds can be raised, that anything other than gradual improvements in palliation and prevention can be achieved—has pervaded global health discourse. A 2006 article in the *Annals of Oncology* explained, "Palliative care . . . should be given a high priority in every country. This is especially true in poor countries where . . . the majority of cancer patients will remain uncured in the coming decades."[79] Yet even as this judgment was rendered, off-patent chemotherapeutic agents capable of curing many malignancies had already been included in the WHO list of essential medicines, and pilot programs in poor countries have shown good patient outcomes, even in the absence of trained oncologists.[80] Expanded access to such therapies need not await the recapitulation of tired doubts about feasibility in resource-poor settings.

The long and complex legacy of Thomas McKeown in the fields of medicine and public health led us to consider carefully the temporal and geographic specificity of his findings. A careful empiricist can note at once the substantial evidence that an improvement in social conditions contributed more to particular health improvements in Great Britain during the nineteenth and early twentieth centuries, as well as the sizeable contributions of medical therapeutics to the reduction in mortality globally in the last two decades. There is no either/or choice between addressing the social determinants of disease and ensuring good clinical care based on adequate clinical protocol.

McKeown and McDermott debated the significance of the twentieth-century therapeutic revolution, but neither would live to see the "delivery decades" of the twenty-first century. The dichotomies of twentieth-century discourse—the Luddite traps and neoliberal prescriptions that insisted that effective medical interventions had to await the end of poverty and the fears of demographic entrapment that made lifesaving therapeutics the culprit for ecologic disaster—need not be repeated. Care in interpreting data and a healthy skepticism of the claims of medical omnipotence and nihilism alike can quell the recurring urge to pit economic growth, public health intervention, and medical delivery against each other.

NOTES

1. Erwin Ackerknecht, *Rudolf Virchow: Doctor, Statesman, Anthropologist* (Madison: University of Wisconsin Press, 1953).
2. Claims that biomedical pharmaceuticals were responsible for the secular decline in mortality could be quite bold. Take, for instance, a print advertisement by the pharmaceutical manufacturer E. R. Squibb and Sons from 1936 (even before the first antibiotic drugs, the sulfonamides, reached the market), which extolled: "Man's ancient enemy, disease, is everywhere. But now his ranks are thinning, before the light of medical science— the light that shines for all. . . . The House of Squibb is dedicated to the service of scientific medicine. We shall go on working with the medical profession, guardian of the public health." Quoted in Nancy Tomes, "Merchants of Health: Medicine and Consumer Culture in the United States, 1900–1940," *Journal of American History* 88, no. 2 (2001): 519–547.
3. Walsh McDermott, "Pharmaceuticals: Their Role in Developing Societies," *Science* 209, no. 4453 (1980): 240–245.
4. Thomas McKeown and R. G. Brown, "Medical Evidence Related to English Population Changes in the Eighteenth Century," *Population Studies* 9 (1955): 119–141; Thomas McKeown and R. G. Record, "Reasons for the Decline of Mortality in England and Wales during the Nineteenth Century," *Population Studies* (1962) 16: 94–122; Thomas McKeown, R. G. Brown, and R. G. Record,

"An Interpretation of the Modern Rise of Population in Europe," *Population Studies* 26 (1972): 345–382; Thomas McKeown, R. G. Record, and R. D. Turner, "An Interpretation of the Decline of Mortality in England and Wales during the Twentieth Century," *Population Studies* 29 (1975): 391–422.

5. Thomas McKeown, *The Modern Rise of Population* (London: Edward Arnold, 1976).

6. This distinction between water- and airborne diseases was significant because McKeown considered the development of sewage the only meaningful new intervention in public health since the start of registration. Thomas McKeown, *The Modern Rise of Population*.

7. McKeown's precise wording of this primary cause of the mortality decline was "a rising standard of living, of which the most significant feature was improved diet." McKeown, *The Modern Rise of Population*.

8. Simon Szreter, "The Importance of Social Intervention in Britain's Mortality Decline, c.1850–1914: A Re-interpretation of the Role of Public Health," *Social History of Medicine* (1988).

9. David Cutler and Grant Miller, "The Role of Public Health Improvements in Health Advances: The Twentieth-Century United States," *Demography* 42, no. 1 (February 2005): 1–22. This analysis relied on a difference-in-difference approach that used "plausibly exogenous" differences in the timing of installation of water filtration and chlorination technologies among American cities to measure the causal role of these technologies in mortality declines.

10. McKeown, Brown, and Record, "An Interpretation of the Modern Rise of Population in Europe," 345.

11. Thomas McKeown, *The Role of Medicine: Dream, Mirage, or Nemesis?* (Oxford: Basil Blackwell, 1979), xvi. McKeown certainly saw an important role for medicine; he argued that resources for preventive and curative interventions should be based on sound evidence, and that more resources should go toward symptom relief for the chronically ill. See Robert Aronowitz, *Risky Medicine: Our Quest to Cure Fear and Uncertainty* (Chicago: University of Chicago Press, 2015), 220.

12. Thomas McKeown, "A Sociological Approach," in *Medical History and Medical Care* (London: Oxford University Press, 1971), 4. For a discussion of McKeown's approach to history, see Amy Fairchild and Gerald Oppenheimer, "Public Health Nihilism vs. Pragmatism: History, Politics, and the Control of Tuberculosis," *American Journal of Public Health* 88, no. 7 (1998): 1105–1117.

13. See, for instance, Anne-Emmanuelle Birn, "Gates' Grandest Challenge: Transcending Public Health Ideology" *Lancet* (2006): 514–519. See also Allan Brandt and Martha Gardner, "Antagonism and Accommodation: Interpreting the Relationship between Public Health and Medicine in the United States during the 20th Century," *American Journal of Public Health* 90 (2000): 707–715.

14. James C. Riley, *Rising Life Expectancy: A Global History* (Cambridge: Cambridge University Press, 2001), 10.

15. Walt W. Rostow, *The Stages of Economic Growth: A Non-Communist Manifesto* (London: Cambridge University Press, 1960).

16. Take, for instance, this typical passage from one of Nyerere's speech to the Tanzanian Parliament: "We have to increase our production of goods if we are to enable everyone to live in conditions of human dignity." Julius K. Nyerere, *Freedom and Socialism: A Selection from Writings and Speeches* (Dar Es Salaam: Oxford University Press, 1974), 9. For Myrdal's views on the imperative for growth before social provision, see Gunnar Myrdal, *Rich Lands and Poor: The Road to World Prosperity* (New York: Harper and Brothers, 1957).

17. Later commentators have noted not only that modernization theorists relied on schematized versions of British history, but that these histories elided centuries of British history in an attempt to draw a clear transition from feudalism to capitalism. See Ian Roxborough, "Modernization Theory Revisited," *Contemporary Studies of Society and History* 30, no. 4 (1988): 753–762.

18. Abdel Omran, "The Epidemiologic Transition: A Theory of the Epidemiology of Population Change," *The Milbank Memorial Fund Quarterly* 49, no. 4 (1971): 510.

19. Omran, "The Epidemiologic Transition," 521.

20. An exemplar of this critical moment was Ivan Illich, whose *Medical Nemesis*, published in 1970, helped spur the reevaluation of medicine's beneficence. Ivan Illich, *Medical Nemesis: the Expropriation of Health* (New York: Pantheon, 1970).

21. Edward Kennedy, "Keynote," *Pharmaceuticals for Developing Countries*, 4. In comparison to the use of "primary care" in popular American parlance, Alma-Ata's definition of primary health care was a broad one: "Education concerning prevailing health problems and the methods of preventing and controlling them; promotion of food supply and proper nutrition; an adequate supply of safe water and basic sanitation; maternal and child health care, including family planning; immunization against the major infectious diseases; prevention and control of locally endemic diseases; appropriate treatment of common diseases and injuries; and provision of essential drugs." For more on Mahler and Alma-Ata, see Matthew Basilico, Jon Weigel, Anjali Motgi, et al., "Health for All? Competing Theories and Geopolitics," in Paul Farmer, Jim Y. Kim, Arthur Kleinman, and Matthew Basilico, eds., *Reimagining Global Health* (Berkeley: University of California Press, 2013).

22. Jeremy A. Greene, "Making Medicines Essential: The Emergent Centrality of Pharmaceuticals in Global Health," *Biosocieties* 6, no. 1 (2011): 10–33. And as Greene notes elsewhere in this volume, Senator Edward Kennedy, the US representative at this conference, used the opportunity to call for

greater global access to lifesaving biomedical therapeutics. He pointed to the injustice that the majority of the world's poorest nations had "virtually no access to them at any time."

23. Fairchild and Oppenheimer, "Public Health Nihilism vs. Pragmatism," 1105–1117.

24. McKeown, *The Role of Medicine.*

25. Ronald Bayer, "Public Health Nihilism Revisited," *American Journal of Public Health* 90, no. 12 (200): 1838.

26. McKeown, *The Modern Rise of Population*, 130–133.

27. McKeown, *The Modern Rise of Population,* 129–133. Szreter focuses on this issue, explaining: "The form and economistic terminology of McKeown's argument by exclusion has resulted in the suppression of any explicit consideration of the independent role of those socio-political developments which were responsible for such hard-won improvements as those in working conditions, housing, education, and various health services. . . . McKeown's interpretation of the epidemiological evidence has, therefore, been crucially misleading in suggesting that these social, cultural, and political dimensions can quite properly be conceived merely as the automatic corollary of changes in a country's per-capita real income. . . . It is precisely the importance and necessity of this slow dogged campaign of a million Minutes, fought out in town-halls and the local forums of debate all over the country over the last quarter of the nineteenth century, which has been missing in our previous accounts of the mortality decline." Szreter, "The Importance of Social Intervention," 16.

28. Robert Fogel, *The Escape from Hunger and Premature Death: Europe, America, and the Third World* (Cambridge: Cambridge University Press, 2004), 21.

29. Angus Deaton, "The Great Escape: A Review of Robert Fogel's *The Escape from Hunger and Premature Death, 1700–2100,*" *Journal of Economic Literature* 46 (2006): 106–114. Deaton objects: "While the synergism is surely there, it is far from automatic. . . . In a country by country examination of mortality decline, whether in Europe in the nineteenth century or in the third world since World War II, we sometimes find that economic growth is correlated with improvements in longevity, but just as often not." For more evidence questioning the primacy of economic growth in explaining secular mortality declines, see David Cutler, Angus Deaton, and Adriana Lleras-Muney, "The Determinants of Mortality," *Journal of Economic Perspectives* 20, no. 3 (2006): 97–120.

30. Daniel Rodgers chronicles the rise to prominence of the highly abstracted, ahistoric models favored by Chicago School economists (led by Milton Friedman). In this account, market fundamentalists obscured the obstinate realities of monopoly power and ignored histories of exploitative policies. Daniel Rodgers, *The Age of Fracture* (Cambridge, MA: Harvard University Press, 2011).

31. Basilico, Weigel, Motgi et al., "Health for All?"

32. Maurice King, "Health Is a Sustainable State," *Lancet* 336, no. 8176 (1990): 664–667.

33. See Paul and Anne Erlich's *The Population Bomb* (New York: Sierra Club/ Ballantine Books, 1968). Although this work shares King's fears of demographic collapse, its policy prescriptions focus mostly on family planning and mass sterilization campaigns rather than abstention from lifesaving measures for sick children. In 1994, Luc Bonneux published an essay on "demographic entrapment" in Rwanda as a cause of that country's genocide. Like the Ehrlichs, Bonneux argued for family planning. Using a McKeownian argument about the relative unimportance of medicine to population figures, he explicitly argued against King's counsel to let dying children expire: "The argument that we have no time to lose, and that desperate times need desperate solutions, does not hold. The time we gain by maintaining high child death rates is trivial. . . . The good news is that there is no agonizing moral dilemma for doctors, public health, and primary health care—and there [never] is and never was a reason to let children die an easily preventable death. The bad news is that the international community stood by and watched Rwanda die from excess fertility." Luc Bonneux, "Rwanda: A Case of Demographic Entrapment," *Lancet* 344 (December 17, 1994): 1689–90.

34. Thomas Malthus, *An Essay on the Principles of Population*, 1798.

35. In a recent empirical study using survey data from twelve sub-Saharan African countries, Fortson shows evidence that the HIV/AIDS epidemic has had a negligible effect on fertility rates. See Jane G Fortson, "HIV/ AIDS and Fertility," *American Economics Journal: Applied Economics* 1, no. 3 (2009): 170–194.

36. Jean Drèze and Mamta Murthi, "Fertility, Education, and Development: Evidence from India," *Population and Development Review* 27, no. 1 (2001): 33–63.

37. Megan Vaughan, *The Story of an African Famine: Gender and Famine in Twentieth-Century Malawi* (Cambridge: Cambridge University Press, 1987).

38. Walsh McDermott, "Medicine: The Public's Good and One's Own," *World Health Forum* 1 (1980): 128.

39. McDermott, "Medicine: The Public's Good and One's Own," 128.

40. See Paul Farmer, *Partner to the Poor: A Paul Farmer Reader* (Berkeley: University of California Press, 2010).

41. Walsh McDermott, "Medicine: The Public's Good and One's Own," 129.

42. "Addressing the social roots of disease is sometimes held to be incompatible with advocating the delivery of high-quality, high-tech care—an opinion often voiced by critics of private-sector medicine. . . . Nothing is wrong with high-tech medicine, except that there isn't enough of it to go around. It is, in fact, concentrated in precisely those areas where it will have the most limited effects. We need more and better clinical services for those marginalized by poverty and discrimination. . . . We are

living in a time when double standards of care must be questioned." Paul Farmer, *Infections and Inequalities: the Modern Plagues* (Berkeley: University of California Press, 2001), 14.

43. Walsh McDermott, "Pharmaceuticals: Their Role in Developing Societies," *Science* 209, no. 4453 (1980): 240–245.

44. Even in the area where McDermott claimed success—tuberculosis, where incidence among children declined from 50 to 6 percent, and no deaths occurred between 1956 and 1962—the Navajo were given isoniazid monotherapy. Streptomycin, which required daily injections, was deemed too difficult to deliver in the sparsely settled and impoverished reservation. See David Jones, "The Health Care Experiments at Many Farms: Tthe Navajo, Tuberculosis, and the Limits of Modern Medicine, 1952–62," *Bulletin of the History of Medicine* 76, no. 4 (2002): 749–790. The long-term consequences of monotherapy, in particular the development of drug resistance, are not mentioned in this article, but can be found in Paul Farmer, *Pathologies of Power: Health, Human Rights, and the New War on the Poor* (Berkeley: University of California Press, 2004).

45. Karen Eggleston. "Health Improvements under Mao and Its Implications for Contemporary Aging in China," http://healthpolicy.stanford .edu/events/health_improvement_under_mao_and_its_implications_for _contemporary_aging_in_china/.

46. Shula Marks, "South Africa's Early Experiment in Social Medicine: Its Pioneers and Politics," *American Journal of Public Health* 87 (1997): 452–459.

47. Walsh McDermott, "Pharmaceuticals: Their Role in Developing Societies," *Science* 209, no. 4453 (1980): 240–245.

48. Cesar G Victora, Jennifer Bryce, Olivier Fontaine, et al., "Reducing Deaths from Diarrhea through Oral Rehydration Therapy," *Bulletin of the World Health Organization* 78, no. 10 (2000).

49. Shabbir Madhi, Nigel Cunliffe, Duncan Steele, et al., "Effect of Human Rotavirus Vaccine on Severe Diarrhea in African Infants," *New England Journal of Medicine* 362, no. 4 (2010): 289–298.

50. Luke Messac and Krishna Prabhu, "Reimagining the Possible: The Global AIDS Response," in Paul Farmer, Jim Y. Kim, Arthur Kleinman, and Matthew Basilico, *Reimagining Global Health.*

51. World Health Organization, "World Malaria Report 2013," http://www .who.int/mediacentre/news/releases/2013/world-malaria-report-20131211/ en/.

52. World Health Organization, "Global Tuberculosis Report 2013," http:// apps.who.int/iris/bitstream/10665/91355/1/9789241564656_eng.pdf.

53. Paul Farmer, "Chronic Infectious Disease and the Future of Health Care Delivery," *New England Journal of Medicine* 369 (2013): 2424–2436.

54. Maryinez Lyons, *The Colonial Disease: A Social History of Sleeping Sickness in Northern Zaire, 1900–1940* (Cambridge: Cambridge University Press, 2002).

55. Ippolytos Kalofonos, "All I Eat Is ARVs: The Paradox of AIDS Treatment Interventions in Central Mozambique," *Medical Anthropology Quarterly* 24, no. 3 (2010): 363–380.

56. Messac and Prabhu, "Redefining the Possible." Also see Jean Comaroff, "Beyond Bare Life: AIDS, (Bio)politics, and the Neoliberal Order," *Public Culture* 19, no. 1 (2007): 197–219 .

57. Adriana Petryna, *Life Exposed: Biological Citizens after Chernobyl* (Princeton, NJ: Princeton University Press, 2003).

58. Samuel Preston, "The Changing Relation Between Mortality and Level of Economic Development" *Population Studies* 29, no. 2 (1965): 231–248.

59. Andrew Sunil Rajkumar and Vinaya Swaroop, "Public Spending and Outcomes: Does Governance Matter?" *Journal of Development Economics* 86 (2008): 96–111.

60. Dean Jamison, Martin Sandbu, and Jia Wang, "Cross-Country Variation in Mortality Decline, 1962–87: The Role of Country-Specific Technical Progress," *Commission on Macroeconomics and Health Working Paper Series* 4 (April 2001): 3–16.

61. Michael Kremer, "Pharmaceuticals and the Developing World," *Journal of Economic Perspectives* 16(4) (2002): 67–90, quotations on 67. Kremer noted that life expectancy in Vietnam increased to sixty-nine years by the year 2000, even though per capita income was less a tenth of that of the United States in 1900, while life expectancy grew by 10 percent in low-income sub-Saharan African nations, even as economic growth was anemic or even negative.

62. Olukunle Adegbola, "The Impact of Urbanization and Industrialization on Health Conditions: The Case of Nigeria," *World Health Statistics Quarterly* 40 (1987): 74–83.

63. Gilles Pison et al., "Rapid Decline in Child Mortality in a Rural Area of Senegal," *International Journal of Epidemiology* 22 (1993): 72–80. Also see James Riley, *Rising Life Expectancy: A Global History* (Cambridge: Cambridge University Press, 2001).

64. Basilico, Weigel, Motgi, et al., "Health for All?" By the mid-1980s, the public health literature and foundation reports alike began to feature numerous examples of rapid improvement in health outcomes in settings of poverty. These examples highlighted the importance of integrated approaches to health and social welfare more broadly, approaches that included immunizations and basic health services alongside female education and farm inputs. For a famous and paradigmatic example, see the Rockefeller Foundation's *Good Health at Low Cost* (1985), featuring case studies of China, Sri Lanka, Costa Rica, and Kerala.

65. This literature is not without its own immodest claims. In 1975, Orubuloye and Caldwell published a study arguing that statistically significant differences in under-five mortality between two otherwise very similar areas of rural Nigeria could be attributed almost entirely to the presence

of a hospital in only one of the towns. But anthropologist Steven Feierman noted that the authors "ignored several possible explanatory factors," including higher rates of male migration and of well water contaminated with feces in the town without the hospital. See Steven Feierman, "Struggles for Control: The Social Roots of Health and Healing in Modern Africa," *African Studies Review* 28, no. 2–3 (1985): 73–147.

66. "Death Rate, Crude (per 1,000 People)," World Bank, http://data.world bank.org/indicator/SP.DYN.CDRT.IN/countries/1W?page=2&display= default. Rwanda's decline of 46.9 percent between 2000 and 2012 is greater than that of any of the other 232 countries and territories for which the World Bank has data.

67. John Paul Sesonga, "Rwanda: Report Indicates Some Improvement in Food Security," 13 March 2013, World Food Programme. http://www.wfp.org/stories/rwanda-report-indicates-some-improvement-food-security.

68. Matthew H. Bonds, Andrew P. Dobson, Donald C. Keenan, "Disease Ecology, Biodiversity, and the Latitudinal Gradient in Income," *PLoS Biology* 2012; 10: e1001456.

69. Paul Farmer, Cameron Nutt, Claire Wagner, et al., "Reduced Premature Mortality in Rwanda: Lessons from Success," *British Medical Journal* 346 (2013): f65.

70. Peter Drobac, Matthew Basilico, Luke Messac, et al., "Building a Rural Health Delivery Model in Haiti and Rwanda," in Matthew Basilico, Paul Farmer, Jim Kim, and Arthur Kleinman, eds., *Rethinking Global Health: An Introduction*, (Berkeley: University of California Press, 2013).

71. Ibid.

72. Jim Y. Kim, Michael Porter, Joseph Rhatigan, et al., "Scaling up effective delivery models worldwide," in Paul Farmer, Jim Y. Kim, Arthur Kleinman, and Matthew Basilico, editors, *Reimagining Global Health* (Berkeley: University of California Press, 2013), 186.

73. Jim Yong Kim, Paul Farmer, Michael E. Porter. "Redefining Global Health-Care Delivery," *Lancet* 382, no. 9897 (September 2013): 1060–1069.

74. Farmer, *Infections and Inequalities*.

75. Dani Rodrik, "Goodbye Washington Consensus, Hello Washington Confusion? A Review of the World Bank's *Economic Growth in the 1990s: Learning from a Decade of Reform*," *Journal of Economic Literature* 44 (2006): 973–987. Also see Gerald M. Meier, *Biography of a Subject: An Evolution of Development Economics* (Oxford: Oxford University Press, 2005).

76. Paul Farmer, "Chronic Infectious Disease and the Future of Health Care Delivery," *New England Journal of Medicine* 369 (2013): 2424–2436.

77. For more on market failures and coordinated financing in pursuit of price reductions, see Rajesh Gupta, Jim Yong Kim, Marcos Espinal, et al., "Responding to Market Failures in Tuberculosis Control," *Science* 293 (5532): 1049–1051.

78. Laurie Garrett, "The Challenge of Global Health," *Foreign Affairs*, January/February 2007.

79. Twalib Ngoma, "World Health Organization Cancer Priorities in Developing Countries," *Annals of Oncology* 17, suppl. 8 (2006): viii9–14.

80. Paul Farmer, Julio Frenk, Felicia Knaul, et al., "Expansion of Cancer Care and Control in Countries of Low and Middle Income: A Call to Action," *Lancet* 376 (2010): 1186–1193. Also see Felicia Knaul, Julie R. Gralow, Rifat Atun, et al., *Closing the Cancer Divide: an Equity Imperative* (Cambridge, MA: Harvard University Press, 2012).

EIGHT

Chemotherapy in the Shadow of Antiretrovirals: The Ambiguities of Hope as Seen in an African Cancer Ward

JULIE LIVINGSTON

Is the golden age of pharmaceuticals over in southern Africa? Or is it just beginning? Perhaps it is still to come? This chapter asks how we might think about the therapeutic revolution from southern Africa. More specifically, it uses ethnographic and historical research from Botswana to call into question a series of universalizing or normative assumptions upon which understandings of biomedical therapeutics and their histories often rely.[1] Southern Africa, as a site of colonial and postcolonial political economy, of local biology, and of public health, provides a critical context that undermines any implicit universality and progressive telos of our narratives of therapeutic development.[2] Over the past century there, consecutive, linked, and recursive waves of epidemic disease, along with the uneven politics of access to life-extending therapies, have shaped a therapeutic revolution that reminds us of the older circular temporality of revolutions subsumed under their modern telos of rupture and transformation.[3] At the same time, the southern African context necessitates an expanded definition of efficacy in order to grapple with the uneven outcomes of shifting pharmaceutical regimes. In other words, from Botswana we can see and question aspects of the therapeutic revolution that are present in the "West" but are sometimes obscured by the revolutionary rhetoric surrounding drugs.[4]

Pharmaceuticals are sometimes imagined in essentialized ways—as universal technologies and goods that circulate through a global marketplace, carrying with them a biological efficacy that transcends the specificities of place and time. There is some truth to this image, and yet drugs, like all aspects of biomedicine, only take shape and meaning, only "work" or don't, as they are deployed and applied within particular historical contexts. In other words, a significant portion of therapeutic efficacy is contained by the biochemical properties of a given drug, and a portion is distributed across infrastructure, markets, social context, and so on. And efficacy grows more complicated if we are to take into account "side effects," "drug resistance," and the dynamism of epidemiological patterns.[5] Southern Africa, and specifically Botswana, offers a critical context in which to consider these historically situated questions of efficacy. The combination of epidemiologic patterns and the politics of therapeutic access have long given an added urgency to these issues in the region, and they become more complicated over time as the revolutionary drugs that enter the region shift from antibiotics to antiretrovirals to chemotherapy.

What we see clearly revealed in Botswana is something of a seesaw that alternates between despair and hope. In part, this is because of the distinct timing and scale of regional epidemiological histories. Thomas McKeown's famous graph (see chapter 7) illustrating how far tuberculosis (TB) rates had fallen before the introduction of antibiotics and vaccination in the United Kingdom doesn't quite fit in Botswana (previously Bechuanaland).[6] As Randall Packard has shown, TB was already on the decline in Europe and North America by the time it began to spread through southern Africa.[7] Colonial labor policies in the late nineteenth and early twentieth centuries introduced and then disseminated the disease, which reached epidemic proportions in places like Bechuanaland by the 1930s.[8] Decades after antibiotics were developed, endemicity remained *extremely* high until the 1970s and 1980s, when the government of a newly independent Botswana began expanding access to antibiotics, successfully routing the disease from the country. The annual risk of infection declined from 5.8 percent in 1956 to 0.1 percent in 1989.[9] TB notification rates declined from 506 out of every 100,000 inhabitants in 1975 to less than half that figure by the end of the 1980s.[10] Yet this golden moment was short-lived. Within a decade, despite the availability of drugs, tuberculosis—reignited by the presence of HIV—was back, and its incidence was rising rapidly. By the 2000s, Botswana had one of the highest TB notification rates in the world. Since 2000, Botswana has consistently reported in excess of 590 cases per 100,000 annually. As of 2007, tuberculosis was responsible for 13 percent of all adult deaths in the country.[11] In other words, the revolution was only temporary.[12]

The therapeutic revolution in southern Africa also takes on its particular character because of the political economy of access to pharmaceuticals. Perhaps the most obvious example is with antiretrovirals (ARVs). Since the mid-1990s, when ARVs were first developed, southern Africa (including Botswana) has had the highest incidence of HIV in the world. Yet until 2002, access to ARVs, revolutionary in their ability to extend the lives of those with the virus, was limited to the handful of patients with enough resources to acquire these drugs through private channels. In Botswana, again, in a repeat of the pattern with TB, there was a truly massive epidemic, with widespread loss of life and pervasive debility that went on for years despite the existence of effective therapies, because of problems of access. Through the work of activists and shifts in pricing strategies and corporate philanthropy and new configurations in what has come to be called global public health, prices came down. And so in Botswana, antiretrovirals became publicly available starting in 2002, and then gradually scaling up.[13] With both TB and HIV there are epidemiological issues of trajectory and scale to consider, wherein massive epidemics continue to accelerate *after* the development of revolutionary therapies, alongside issues of access.

Botswana as a case study helps to highlight that access is more complicated than a straightforward or simple relationship between poverty and disease.[14] As noted above, the nation has suffered from lags in access, and even common and inexpensive therapies regularly go in and out of stock in the country's hospitals due to bureaucratic and institutional problems. Still, once new therapies enter the health system, they are available to citizens regardless of income. Botswana has in place a robust system of universal care that it has steadily built since independence. Unlike in other countries throughout the region, or in certain wealthy countries like the United States, drugs in Botswana are provided through the national health care system to all patients who meet the biological criteria for their use. So, with ARVs on the ground in Botswana, another therapeutic revolution was at hand.

But even with access to revolutionary pharmaceuticals like antibiotics for TB or antiretrovirals for HIV, miracles quickly give way to yet new problems, epidemiologically speaking. One disease is not separate from another, as we saw with the resurgence of TB in the wake of HIV. In Botswana there is now a rapidly emerging cancer epidemic, which to some extent is related to HIV and the presence of these revolutionary ARVs. Many of Botswana's cancer patients are suffering from virus-associated cancers that are facilitated by their HIV-related immunosuppression. A minority, but a significant number, of HIV patients will develop a virus-associated cancer before being initiated on antiretroviral therapy, or during the process of partial immune reconstitution.

Before ARVs were available, many of Botswana's patients died *with* a cancer, but *from* other AIDS-related infections. Since 2002, when Botswana's ARV program began, however, many have been surviving their HIV only to grapple with virus-associated cancers made all the more aggressive and difficult to treat by HIV (and in many cases tubercular) coinfection. In Botswana, where nearly a quarter of all adults have the HIV virus, ARVs are critically necessary, yet those drugs have a secondary effect of exposing this deadly relationship between cancer and HIV. At the same time, the establishment of oncology services to assist patients with the new HIV-related cancers has helped to identify a significant population of patients with cancers not necessarily related to HIV who previously might have gone undiagnosed and untreated.

This grim underside to the otherwise impressive success of Africa's first national ARV program coupled with the significant burden of other cancers already prevalent in the population to create a situation of overwhelming proportions. And so, conjoined to the politics of HIV and the AIDS industry, new questions of access emerged around cancer treatment, prompting Botswana to establish an oncology service in 2001 in anticipation of the cancers that would follow from the revolutionary ARV program it was about to begin.

But the politics of access is not straightforward, and it is indeed related to the sense of revolution, of miracle. We might assume that access to drugs in the face of ever-shifting epidemics bears a straightforward relationship to cost. If one disease is attached to another and cannot be so easily cleaved apart (as we saw with TB and HIV or HIV and, say, cervical cancer, and as is also the case for symptoms and the underlying disease), the logics of global health facilitate a therapeutic revolution that focuses on mortality and disease prevention rather than on the experience of illness itself. Somehow off-patent chemotherapy is a life-or-death access issue and thus available—and yet off-patent morphine is not always easily accessed by those same patients after chemotherapy has failed, and they now lie in agony, dying slowly in their homes. Patented Gleevec (a very expensive drug meant to be taken for life) is available through the good efforts of Botswana's lone oncologist to tap into possibilities for corporate philanthropy, while patented Kytril (a less expensive but still costly antiemetic meant to be used for very short periods of time) is not.[15] And yet it is profound nausea and vomiting that keeps many patients from returning to the cancer clinic to complete their chemotherapy cycles.

In other words, these political and economically shaped conditions —epidemiology and access—do indeed map a therapeutic revolution of sorts, one experienced on the remarkably wide scale of profound epidemics. But they also condense and intensify the sense of despair and doubt that shadow all revolutions. There is a narrative of access here, but just what is it that patients are clamoring to get? Incredibly important, powerful, life-extending drugs. At the same time, these drugs are not necessarily the miracles for which one might hope, as any patient on ARVs undergoing chemotherapy for their lymphoma will tell you.[16] Longing for antibiotics is not the same as longing for ARVs is not the same as longing for chemotherapy is not the same as longing for morphine or antiemetics.

In the rest of this chapter I shift scale, to offer a close-up look at the next moment of this therapeutic revolution as it is unfolding for patients in a small cancer ward in Botswana, where I have done extensive ethnographic research. It is my hope that this kind of fine-grained view of a single site might reveal some of the intellectual, existential, and moral stakes of the pharmaceutical future for patients in middle-income places like Botswana. So while I acknowledge Botswana's uniqueness within the southern African region, I also want to suggest that it is facing certain health care issues (including a rapidly emerging cancer epidemic) that are shared by many other middle-income countries such as Barbados, Chile, or Iran.

Botswana has a small population, about two million citizens (though there is a substantial population of undocumented immigrants), with a per capita annual income of nearly US$13,000. Meanwhile, the gap between rich and poor has grown and continues to harden over time, as it appears to be doing in most of the world right now. The country has tarred roads, clean water, a sanitary infrastructure, and a good system of telecommunications, and the government spends nearly 10 percent of the gross domestic product on health care. Given that reaching this economic and infrastructural status is the goal of many countries in Africa and throughout the global South, where a cancer epidemic is emerging rapidly, a close look at Botswana's cancer ward reveals the dimensions of a pharmaceutical horizon for so-called developing nations. For the region in particular, I argue that cancer is the critical face of African health *after* ARVs.

A look at chemotherapy in the shadow of ARVs enables us to contemplate the ambiguities of hope that abound in such a context. The dynamics of the therapeutic revolution described above are, as noted, really more cyclical or like a seesaw than like a forward march of progress. This cyclical nature of both profound and serialized epidemics and a very public politics of access serve to condense and highlight the relationship between despair and hope that lies at the center of any therapeutic revolution. In the cancer ward of Princess Marina Hospital (PMH), Botswana's central referral hospital, which at the time of my research was the only dedicated cancer ward in the country, this cycle is played out in miniature among individual patients. It is further complicated by a third characteristic that shapes the nature of this most recent phase of what Joao Biehl has called the "pharmaceuticalization of public health" in Botswana: southern Africa's location on the periphery of the cancer industry.[17] Cancer drugs are not developed in Botswana for its particular biological or technological context. Instead, doctors in Botswana must tailor drugs created elsewhere to the unique biological and technical field in which this new cancer epidemic is emerging.

Together, the cyclical nature of miracle and epidemic, remission and recurrence, alongside this necessarily improvised oncology raises critical questions about the nature of efficacy. There is urgent need for well-funded and well-run *public* oncology settings across Africa. Yet I want to suggest that while political and economic hopes for improved care are crucial, developmental fantasies that hinge on improved technological access, whether in the form of ARVs or chemotherapy, will not allow Batswana (as citizens of Botswana are called) to avoid the contradictions and dilemmas that accompany contemporary revolutionary biomedicine, pharmaceuticals included.

My focus here on the limits of efficacy in cancer medicine is not to deny the many patients in Botswana who survive their cancers or persist in the face of grueling pain and loss, thanks to PMH oncology. Nor is it to deny the potential benefits of improved health infrastructure, including a more comprehensive oncology service. Clinical care at PMH oncology is hampered by technological challenges that are the necessary partner to effective pharmaceuticalized therapies.[18] The hospital has no cytology lab, MRI machine, endoscopy, or mammography. Histology is a dicey prospect, tumor markers are unavailable, genetic screening is impossible, machines break regularly, and various drugs (like all supplies) go in and out of stock unpredictably.

Nonetheless, in the shadow of the palpable, collective redemption brought by ARVs, cancer patients, their relatives, and clinical staff use oncology as a technological field in managing and shaping their existential angst. Novel technologies like chemotherapy burst onto complicated political and economic landscapes, generating new desires and hopes. But as they become normative and embedded in infrastructural contexts that are not equipped to handle them, like this national referral hospital, their ambiguities are revealed and the challenges of practice become more burdensome, spawning both political critique and individual creativity.[19] In the case of tuberculosis, laboratories, X-ray machines, technicians, and clinics were all necessary to create and enhance the biological efficacy of antibiotics. In the case of HIV, laboratories, rapid test kits, clinics, and CD4 machines (notoriously prone to breakdowns) are all needed to transform the politics of access to ARVs into the extension of life. With cancer, the situation grows ever more complex.

In other words, Batswana are learning rapidly about the power of certain biomedical technologies like chemotherapy to ease suffering and to potentially stave off death. And yet these same goods are proving more complicated and less miraculous than hoped. Cancer illnesses and deaths expose the ironies and problems lying at the intersection of therapeutic revolution and international development progress narratives, whereby Batswana, having seized hold of their futures through ARVs, now long for political economy in which they can inherit the power of clinical oncology. I want to suggest that developmental fantasies that hinge on improved technological access, whether in the form of ARVs or chemotherapy, will not allow Batswana to avoid the contradictions and dilemmas that accompany revolutionary therapeutics. In one sense, Botswana is a more "utopian" or "revolutionary" setting for oncology in that the simple provision of chemotherapy is in itself a form of rapid progress, and yet the ambiguities of this progress are rapidly felt. In many cases these technologies, powerful as they are, cannot forestall a cancer death. Nor are such technologies neutral. The side effects of chemotherapy can entail intense misery, even as it is often the sole vehicle of therapeutic hope for the otherwise terminally ill. Cancer's recent and rapid emergence as an illness category in Botswana imbues oncology with a sense of progress and hope, but also exposes patients to a backlash of uncertainties and often crushing disappointments.

Cancer in Botswana differs from cancer in the global North, where it lies at the heart of cutting-edge, highly capitalized biotechnical research and public narratives about loss, heroism, and hope. The rapidly emerging field of global oncology promises to extend the geographic reach of oncology, and to thereby introduce new research questions and concerns. But for now, the problem of cancer in places like Botswana, whose biological, epidemiological, and technological context does not match those of North America or Europe, remains far outside the center of gravity in oncological research, which is still mainly based in the North and tends to focus on the development of ever newer drugs and technologies. As a result, key questions remain unresearched. Few studies address the chemotherapeutic challenges of simultaneous HIV and tubercular coinfections in treating cancer.[20] Yet because of drug interactions and compounded side effects, and because some of Botswana's cancer patients are too immunocompromised to withstand standard chemotherapeutic regimens, charting a treatment course for patients who are coinfected with HIV and cancer is difficult. Newer "smart" drugs like herceptin are too expensive to consider, and important support interventions like neupogen (for treating the neutropenia that is a common side effect of chemotherapy) are too costly to use in any but the most compelling of circumstances (though Gleevec is available for some patients through a corporate philanthropy program). Nursing conditions are also different in PMH, such that the necessary support care to enable, for example, concurrent radiotherapy and chemotherapy (the standard of care for many cancers) is not possible. Evidence-based oncology protocols published in the leading medical journals do not say what to do when etoposide, 5FU, or cisplatin (all core chemotherapy drugs in PMH's stripped-down arsenal) suddenly go out of stock, as each did for some time during my research. While chemotherapeutic drugs are universal goods circulating through biomedical nodes in a global market, to make use of them to help Botswana's cancer patients, their oncologist must borrow, adjust, and even deny, but never simply import metropolitan knowledge.

How does this look in practice? Let us think about it across different moments of the illness trajectory. Because oncology (and, therefore, diagnosing and counting cancers) is a novelty in Botswana, there is little capacity for screening within the national health system, and necessary laboratory and diagnostic tests are concentrated in PMH. This means that patients have to be referred to oncology by a doctor. Village clinics are staffed by nurses, so even seeing a general practitioner, a specialist in one of the hospitals, or a doctor in one of the HIV clinics is already the product of serious effort in pushing upwards through an overcrowded public health care system.

Doctors who staff the primary hospitals and nurses who staff the clinics have had almost no training in cancer. A nurse in a local clinic told Sekgabo, a mother of four who was in her mid-thirties at the time, that the painful hard lesions on her face were the result of using skin lighteners. On another visit, her laborious breathing was attributed to possible TB infection, and she was put on TB treatment. After several such visits, her gums and upper palate had become extremely painful, so she finally bypassed her clinic and local primary hospital and made her way to the dental clinic at PMH, where the maxillofacial surgeon immediately suspected Kaposi's sarcoma (KS) and referred her for HIV testing and a consultation with Dr. P., the oncologist. By the time she arrived in oncology to confirm her cancer diagnosis and begin chemotherapy, the cancer was in her lungs and all over her legs, destroying the lymphatic system and causing massive swelling in her knee that needed to be drained. The pain was intense, as was her disfigurement. Two years later the KS was in remission, but her left knee was a fixed joint and she could only walk short distances. As a result, she lost her job as a cashier and lived in chronic pain. Chemotherapy and ARVs saved her life, but her body was greatly diminished.

By the time they arrive in oncology, many of these patients are quite desperate. Many are experiencing a new if fragile faith in biomedicine born of the recent, widespread experience of ARVs and a desperation rooted in their bodily anguish. But this faith is tenuous; it coexists with a cynicism, a wariness born of the failures of the referral system to tend to patients in a timely manner. For those with HIV, this cynicism is furthered by the way in which cancer and HIV are all clotted together—in other words, the promised miracle of ARVs has not fully materialized for these patients, and they are still quite sick. Perhaps they have KS in their lungs, which has been treated repeatedly as TB, getting worse rather than better while on antibiotic therapy. Perhaps they have cervical cancer, originally mistaken for and treated as a routine STD. In other words, these patients have already taken plenty of drugs, and those drugs have failed them.

Upon arrival in PMH oncology, after diagnosis is confirmed, treatment begins. Initial diagnostic conversations are usually quite brief: "You have kankere. We will now treat you with injections. This injection might make you vomit, it might make you lose your hair, it might make you dizzy. This is temporary. Your hair will grow back. You must come for six injections. First today, then three weeks later, then three weeks, three weeks, three weeks until we reach six injections. When you come for injection come early in the morning. Get your blood drawn at the lab, bring the lab result here and then take a number and join the queue. Do you understand?" Or, for a new inpatient: "You have cancer of the mouth. We will give you a bed and shrink it with medicine."

From this stage onwards, the oncology team works to shore up the boundaries of cancer as a distinct disease, one that is separable from the patient's HIV, hypertension, or tuberculosis. Chemotherapy is central to this process of disease ontology and to the renewal or creation of therapeutic optimism. As demonstrated by the clinical conversation described above, it is not explanation of "how" chemo works, but phenomenological experiences of the extreme potency of chemotherapy that work to instantiate its power. Because the health system does not yet enable screening for early detection, by the time most patients arrive in the cancer ward and its clinic, they already have advanced disease. These are not asymptomatic cancers; these are patients in real pain and discomfort, and those with solid tumors usually have big, hard, palpable growths and often necrotic, suppurating wounds. Patients bring the fullness of bodily experience and affliction to their chemotherapeutic trajectories. Given this, chemotherapy is central to disease ontology and to therapeutic optimism for its ability to rapidly and often visibly shrink tumors. Yet chemo, for all its power, is a highly aversive experience for many.

Over time, as the patient returns for subsequent treatment cycles, embodied knowledge is sedimented and a chemotherapeutic habitus of sorts is built. Patients hold their arms out and pump their fists for blood taking and chemo injections. Those awaiting chemotherapy with "anticipatory nausea" developed through previous experience learn to sit where they cannot see or smell the chemo room. Patients might wait for several hours or even all day for their turn in the chemo room, bored and anxious about how they will manage the bus journey (and then the walk) home. They chat with one another (and their accompanying relatives) on the bench, trading information and encouraging and commiserating with one another; or they sit silently, a nervous anticipation building all the while. They complain loudly at anyone who tries to jump the queue. But despite their desire to make their way to the front of the line and then home, entering the chemo room is a thoroughly loaded experience. Let me illustrate with an entry from my field notes.

In the treatment room, the woman with the mastectomy and the lump we are watching finishes her injection. While the old man with lung cancer lies on the table getting his one-hour drip, she is in the chair. She is finished, but she looks green and doesn't stand up to go. Dr. A. says she looks nauseous. I take her to sit next to the sink to wait. Then I give her some damp paper towels to hold, and I wipe her forehead with another towel. After a minute she says, "OK, I can wait outside." She rejoins the bench, and the other patients ask after her. She just says one word: "sebete" (nausea).

I call the next patient in: a woman with breast cancer. "My God," she says, and then in Setswana, "Modimo o ntusa" (God help me), and braces herself to enter the room. She cringes as the cannula is put in; it is painful. It has to be replaced from the crook of the left arm to the right hand, because it has infiltrated on the first vial. Immediately after the cannula is all into her right hand, she goes to the sink and vomits—as I am ripping the tape for her arm. I put the tape on the gauze as she vomits—and then I get her paper towels. She washes her mouth, collects herself, and leaves.

The kind of learning I've just described underscores the stakes here. These are patients with a very serious and often tremendously painful disease, who are now undergoing a highly aversive therapy, one that requires a totalizing commitment. Most patients cannot work during the first week after chemotherapy, which is usually administered every three weeks. Many spend several days or even a week or more with overwhelming nausea and vomiting once they arrive home after their chemo injections. Patients who are regularly employed are entitled to paid sick leave, but those who make their living through casual labor, domestic service, petty trade, and agriculture must manage the economic consequences, as must those patients whose employers cheat them out of sick pay. Each appointment means countless hours spent waiting in queues for a turn with the phlebotomist, the oncologist, and then for the chemo itself.

Because of intense pressures for available space on the ward, chemo is given as a push injection whenever possible. But those whose cancers require three- or five-day inpatient chemo treatments must surrender themselves totally to the ward. In May 2009, as Rra P. lay in the ward attached to a chemo drip, he received a call on his cell phone informing him that his younger brother had just died. Despite his grief and his considerable responsibilities as eldest brother in making important family arrangements at this critical time, he was not allowed to leave the hospital earlier than expected. He needed to finish his chemo course and the postchemo hydration IV. Only then, three days later, could he return to his home village, several hours away by bus, and begin the mourning process.

Amid all this, and because of all this, efficacy takes on a compounded urgency. Yet efficacy in this case is far narrower than patients might initially hope. For Dr. P., efficacy means shrinking a tumor and preventing or halting the process of metastatic spread. For patients, efficacy means feeling better and enacting a socially and physically meaningful future. It means seeing one's children safely to adulthood. Bridging these two horizons of hope necessitates a further consolidation of cancer as a biomedical object, one meant to be separate from and foreign to the patient's self. By getting patients with solid tumors to focus on the mass, different horizons of efficacy held by the patient and by the clinician could be partially merged and patient hopes reinvested in biotechnological modalities.

But cancer is not a stable entity; its phenomenology is porous. After cancer is established, perhaps in only skeletal and shadowy form, it continually threatens to overspill its delicate etiological boundaries. And so patients come with shrunken tumors but are suffering from infections and other complications. Dr. P. attempted to marginalize these complaints, shoring up the boundaries of cancer and the efficacy of chemotherapy even as he treated the side effects of disease or therapies when possible. "Yes, yes, but that doesn't matter," he would say to someone who came in miserable from an infection, and then point to their shrunken tumor. "Is *this* smaller or larger than when we started?" he would ask, even as he wrote the order for the antibiotics needed to treat the infection. The side effects of the chemotherapy were agreed to be awful, but they had to be constituted as side effects: unpleasant distractions from the central work of chemotherapeutic healing.

A middle-aged man came into the clinic, sat down, and showed his arm, which was in pain. The chemo had infiltrated the vein, which is quite destructive since it is cytotoxic. It had produced an ugly black mark, but fortunately no open sore. Dr. P. joked with him: "But the tumor is much better, right? You are like someone who comes in with a big huge tumor that we fix and then complains, 'But Doctor, I now have this little tiny itch on the sole of my foot.'" The patient laughed: "No, the tumor is much better, but the medicine that cured it did this to my arm."

Yet efficacy of this sort is nonetheless often disappointing. I wrote in my notes for March 2007: "Dr. P. has a debate with a patient who came yesterday. He says he is not better. Dr. P. looks in his mouth, and the tumor is much smaller. But it has left a big hole." What is efficacy here?

Many patients emerged from oncology disfigured and debilitated. For others, focusing on a tumor meant that they would refuse further intervention once the tumor appeared shrunken. Women with breast cancer whose tumors had been downsized with chemotherapy might then refuse the planned mastectomy. For others, therapeutic optimism failed early. After a miserable cycle or two of chemotherapy, they would return to the *ngaka* (Setswana doctor) or the healing prophet, but not to the ward. "Whatever happened to the woman from Barclays Bank with the KS in her lungs?" we would ask one another. "Can she still be alive?" Often such people would suddenly reappear after some months, now suffering from a massive recurrence of their disease. Having earned the new label of defaulter, they incurred Dr. P.'s wrath for having vacated the efficacy of his treatment.

These patients return with a thoroughly embodied desperation, saturated with existential angst. In this state they join those for whom progress, relief, and remission have proven fleeting, patients who are now entering the final phase of their illness. Amid the relative scarcity of tertiary care and the tremendous pain and angst of a cancer illness, it is not surprising that just as the end draws into sight, many patients and their relatives seek to secure a future through more revolutionary therapies. For them, progress after ARVs is about pursuing more high-tech medicine, including chemotherapy. And yet, in Botswana, where the majority of patients are diagnosed with already advanced disease and where their treatment might be further complicated by an HIV coinfection, prognosis is often poor. One of the difficulties facing the oncologist is the knowledge that even if there were more drugs and more machines, many of the patients in the PMH cancer ward would still be terminally ill. Dr. P. must help the growing population of patients and relatives to maintain hope in chemotherapy as a techno-medical pursuit that can extend life and ease suffering. But he must also regularly acknowledge individual instances of therapeutic futility, especially since that acknowledgment is critical to the rationing of care.

In 2007 Mothusi, a young man with end-stage cancer of the naso-pharynx, lay shivering and sweating, the feeding tube exposed on his bare abdomen. He cringed in pain as Dr. A., the medical officer, pushed the four tubes of chemotherapeutic drugs into the central intravenous line implanted in his chest. He lay back, exhausted and shaking, and prepared for the onslaught of nausea and fatigue that would soon follow. Mothusi had arrived in the ward on a hot, crowded day by ambulance from South Africa, where he had been hospitalized while attending university. The doctors there had sent him home to Botswana as a terminal case. The tumor was so large it blocked his throat entirely, so that he could not even swallow his own saliva. He was anxious that he might be given a tracheostomy, something he very much did not want.

In the counseling session immediately after his arrival, Mothusi's parents wanted everything possible done for their son—couldn't Dr. P. at least "give chemo?" Because of the incomplete medical records from South Africa, there was at least a small chance that Mothusi, despite the advanced stage of his disease, might respond to treatment. So Dr. P. acceded to the parents' insistence that Mothusi receive an aggressive course of chemotherapy, and that the oncology team not give up on him. In various counseling sessions over the next several weeks, Dr. P. began to lay out where the road would end.

Chemo was quite a miserable experience for Mothusi, but it did at least initially provide him with some relief, even as he suffered its side effects. Dr. P., Dr. A., and I all knew there was little chance that the chemo would significantly extend Mothusi's life, but it was impossible not to hope along with his parents. And so, with the CT scan machine broken, we clung to and debated the ambiguities of his X-rays: were the metastases in his lungs shrinking, or was the exposure of the film different? Life on chemotherapy was at times agonizing. For three solid days after the painful injections, Mothusi would face total nausea, dizziness, and intense exhaustion. Then, five or six days later, when his white cell count would plummet, he would succumb to a series of nasty infections in his chest, intestinal tract, or ears.

This is not to say that Mothusi found no pleasure in life. He listened to music. Friends and relatives came to visit him. He read the newspapers. His mother and father were there every day, and on some weekends he was even allowed to go home on a hospital furlough. He joked with his doctors, with his nurses, and with the ethnographer who followed them around. For his parents, he embodied the emergence of an aspirational ethos of patient care in Botswana, where "First World" high-tech medicine hovers as an imagined promise against which Batswana evaluate risks and imbue value in the lives of patients. Then, several weeks after his arrival in the ward, Mothusi choked to death in the middle of the night as Dr. P., called in from home, watched the surgical officer gave him the emergency tracheotomy he had feared. Did it matter that Mothusi died with a scalpel jutting out of his throat? Did it matter that his mother would survive him, knowing that she hadn't given him up without a fight? Was this a charade of therapeutic futility or a necessary exercise in hope? Would his mother have charted a different course if chemotherapy were not so novel in Botswana? Would Mothusi himself done so if he'd been the one to choose? Did chemotherapy fail here? Did clinical ethics fail?

This stance of therapeutic optimism around chemotherapy surprised me at first. I had worked as an ethnographer in Botswana in the 1990s, when most people I met were highly cynical about biomedicine. For more than a decade, people heard relentless messages about AIDS being a death sentence, and then watched many people die ugly deaths. So too they were living through the ubiquitous return of tuberculosis. But with ARVs, although deaths continue, many people are seeing their own bodies and those of their neighbors, coworkers, and relatives reconstituting—and drugs are the conduit to this renewal. Increasingly, one sees relatives pushing for more highly technologized interventions, including new rounds of chemo for the supposedly dying, and allowing little bits of hope to seep into their decisions to medicate.

Oncology, as a highly pharmaceuticalized endeavor, emphasizes "hope" as a vital force orienting and animating biotechnical research, patient narratives, and practices of care.[21] For oncologists, researchers, fundraisers, and cancer patients, hope emerges as a mantra that discursively anchors the center of a vast and complicated enterprise. Oncology constantly and simultaneously produces knowledge and uncertainty, therapy and futility, and hope provides much-needed ballast for well-meaning and sometimes desperate people. In the dark moments that shadow all cancer wards, the hope that patients, knowledge, and techniques will improve is crucial to this often brutal and violent, if well-meaning, domain of technoscientific practice. Indeed, so much hope is wielded so often that it seems impossible that so little improvement has actually occurred in survival rates for many high-profile cancers over the past century.[22] At its most cynical, hope and the repetition of its name provide a fig leaf for an enormous multibillion dollar drug industry. As Sarah Lochlann Jain cautions us, our focus on the atomized hope of individuals distracts us from a broader oncological politics of publics living in a toxic and capitalistic world.[23] Botswana's position—as an African nation that against all odds prioritizes universal health care and corporate capitalism simultaneously, and as a place where the cancer epidemic itself is in some part an outgrowth of a philanthropic project by Merck & Company to extend the lives of those with HIV through the provision of antiretrovirals—suggests that a somewhat different but no less compelling politics is afoot.

One way to understand all of this is through a narrative of difference. Africa is poor. Africa is sick. Africa is broken. Such images have a long history, and are foundational to the salvific narratives that lie at the heart of the pharmaceuticalization of public health in an era of extensive privatization.[24] And indeed, Botswana's cancer ward throws the politics of access, the existential imperatives of therapy, the seesaw of hope and despair, and the ambiguities of efficacy into stark relief. But these issues are not specific to the region. They lie at the heart of our collective pharmaceutical future and whatever revolutionary promises it might contain.

NOTES

1. For a more detailed account of my methods and the research from which this essay is drawn and from which some of the narration is taken, see Julie Livingston, *Debility and the Moral Imagination in Botswana* (Bloomington: Indiana University Press, 2005); idem, *Improvising Medicine: An African Oncology Ward during an Emerging Cancer Epidemic* (Durham, NC: Duke University Press, 2012).

2. On local biology, see Margaret Lock and Vinh-Kim Nguyen, *An Anthropology of Biomedicine* (Oxford: Wiley-Blackwell, 2010). On the therapeutic revolution, see, for example, Jane S Smith, *Patenting the Sun: Polio and the Salk Virus* (New York: William Morrow and Company, 1990); David Oshinsky, *Polio: An American Story* (New York: Oxford University Press, 2006); Siddhartha Mukerjee, *The Emperor of All Maladies: A Biography of Cancer* (New York: Scribner, 2011); Chris Feudtner, *Bittersweet: Diabetes, Insulin, and the Transformation of Illness* (Chapel Hill: University of North Carolina Press, 2003).

3. Reinhart Koselleck, *Futures Past: On the Semantics of Historical Time* (New York: Columbia University Press, 2004), chapter 3.

4. For more on how the pharmaceutical industry shapes this rhetoric, its effects on the definition of health and disease, and the constitutive economic logics of health as market, see Jeremy Greene, *Prescribing by Numbers: Drugs and the Definitions of Disease* (Baltimore: Johns Hopkins University Press, 2007) and Joseph Dumit, *Drugs for Life: How Pharmaceutical Companies Define our Health* (Durham, NC: Duke University Press, 2012).

5. On side effects see, Joe Masco, "Side Effect," on Somatosphere: http://somatosphere.net/2013/12/side-effect.html (accessed 6 August 2014).

6. Thomas McKeown, *The Modern Rise of Population* (New York: Academic Press, 1976); Thomas McKeown, *The Role of Medicine: Dream, Mirage, or Nemesis?* (London: Nuffield Provincial Hospitals Trust, 1976). Of course, as Farmer, Basilico, and Messac review in chapter 7 of this volume, McKeown's argument has been subject to great scrutiny, but the central narrative concerning the timing of the decline of tuberculosis stands. Amy Fairchild and Gerald Oppenheimer, "Public Health Nihilism vs. Pragma-

tism: History, Politics, and the Control of Tuberculosis," *American Journal of Public Health* 88 (1998): 167–187; James Colgrove, "The McKeown Thesis: A Historical Controversy and Its Enduring Influence," *American Journal of Public Health* 92 (2002): 725–729; Simon Szreter, "Rethinking McKeown: The Relationship Between Public Health and Social Change," *American Journal of Public Health* 92 (2002): 722–725.

7. Randall Packard, *White Plague, Black Labor: Tuberculosis and the Political Economy of Health and Disease in South Africa* (Berkeley: University of California Press, 1989).

8. Julie Livingston, *Debility and the Moral Imagination in Botswana* (Bloomington: Indiana University Press, 2005).

9. Republic of Botswana, Ministry of Health, *National Tuberculosis Programme Manual*, sixth edition (2007), 11.

10. Ibid.

11. Ibid.

12. Again it resonates with Koselleck's account of revolution before it took on its modern linguistic character. Koselleck, *Futures Past: On the Semantics of Historical Time* (New York: Columbia University Press, 2004), chapter 3.

13. Botswana was the first in the region, in fact on the continent, to do so. Other countries are attempting to follow suit, but with uneven results.

14. See also Vinh-Kim Nguyen, *The Republic of Therapy: Triage and Sovereignty in West Africa's Time of AIDS* (Durham, NC: Duke University Press, 2010); Joao Biehl, "The Activist State: Global Pharmaceuticals, AIDS, and Citizenship in Brazil," *Social Text* 22 (2004): 105–132; Susan Reynolds Whyte et al., *Second Chances: Surviving AIDS in Uganda* (Durham, NC: Duke University Press, 2014).

15. For more on the politics of Gleevec pricing, see Stephan Ecks, "Global Pharmaceutical Markets and Corporate Citizenship: The Case of Novartis' Anti-Cancer Drug Glivec," *Biosocieties* 3 (2008): 165–181.

16. And, as is well documented, ARVs pose a tremendous dilemma for patients who suffer as a side effect the ravenous hunger associated with immune reconstitution in households where food is already in short supply. Ippolytos Kalofonos, "All I Eat Is ARVs: The Paradox of AIDS Treatment Interventions in Central Mozambique," *Medical Anthropology Quarterly* 24 (2010): 363–380; Susan Reynolds Whyte et al., *Second Chances: Surviving AIDS in Uganda* (Durham, NC: Duke University Press, 2014).

17. Joao Biehl, "Pharmaceuticalization: AIDS Treatment and Global Health Politics," *Anthropological Quarterly* 80 (2007): 1083–1026.

18. These observations are based on ethnographic work undertaken between 2006 and 2009. Technology, infrastructure, and staffing are not stable, and there have, no doubt, been significant changes since then.

19. Sherine Hamdy, "When the State and Your Kidneys Fail," *American Ethnologist* 35 (2008): 553–569; Sharon Kaufman, *And a Time To Die: How American Hospitals Shape the End of Life* (Chicago: University of Chicago

Press, 2005); Brian Larkin, *Signal and Noise: Media, Infrastructure, and Urban Culture in Northern Nigeria* (Durham, NC: Duke University Press, 2008).

20. In the late 1980s and early 1990s, there was clinical research and interest in AIDS-related malignancies, particularly KS and lymphoma, among US-based researchers, but cases of cancer began to decline after the introduction of protease inhibitors in the mid-1990s. A greater flurry of work on AIDS-related malignancies (particularly non-Hodgkins Lymphoma, HPV-related genital cancers, and KS), including treatment recommendations, followed in the late 1990s and early 2000s. See for example, several of the articles collected in Feigal, Levine, and Biggar, eds., *AIDS-Related Cancers and Their Treatment* (New York: Marcel Dekker, 2000). But the therapeutic guidance in this volume ranged from pronouncements of the difficulties of treating cervical cancer in HIV-positive individuals to discussions of cytokines, which are not relevant to the PMH context. In 1995 the National Cancer Institute established the AIDS Malignancy Consortium (AMC), with fifteen main clinical trials sites to develop therapeutic protocols for AIDS-related cancers. But while the AMC focused on KS, lymphoma, and HPV, in Botswana there are also high rates of head and neck tumors associated with HIV, and many other patients suffering from breast, lung, esophageal, leukemia, bone, and other cancers who happen to have HIV coinfection. For a brief overview of the AMC to 2003, see Jeffrey Shouten, "The Rise and Fall of the AIDS Malignancy Consortium," *Research Initiative Treatment Action,* (2003): 30–31. More recently, the AMC has begun research on non-AIDS-defining cancers, and has a new mandate to "expand their capacity to conduct trials in Africa"; http://oham.cancer.gov/oham_research/programs/consortium/, accessed 26 July 2011. A new body of work interested in coinfection is beginning to emerge. Meredith S. Shiels et al., "Cancer Burden in the HIV-Infected Population in the United States," *Journal of the National Cancer Institute* 103 (2011): 753–762; Sandro Vento and Massimiliano Lanzafame, "Tuberculosis and Cancer: A Complex and Dangerous Liaison," *Lancet Oncology* 12 (June 2011): 520–522; Scott Dryden-Peterson et al., "Malignancies among HIV-Infected and HIV-Uninfected Patients in a Botswana Prospective Cohort," paper presented at the 19th Conference on Retroviruses and Opportunistic Infection, Seattle, 2012; G. Suneja et al., "Cancer in Botswana: A Prospective Cohort Study of Cancer Type, Treatment, and Outcomes," *International Journal of Radiation Oncology* (2013).

21. Mary Jo Delvecchio Good, Byron Good, Cynthia Schaffer, and Stuart Lind, "American Oncology and the Discourse of Hope," *Culture, Medicine, and Psychiatry* 14 (1990): 59–79; Barbara Ehrenreich, "Pathologies of Hope," *Harpers Magazine,* 1 February 2007.

22. Sarah Lochlann Jain, "Living in Prognosis: Towards an Elegiac Politics," *Representations* 98 (2007): 77–92; Robert Aronowitz, *Unnatural History: Breast Cancer and American Society* (Cambridge: Cambridge University

Press, 2007); Robert Proctor, *Cancer Wars: How Politics Shapes What We Know and Don't Know About Cancer* (New York: Basic Books, 1995).

23. Sarah Lochlann Jain, "Cancer Butch," *Cultural Anthropology* 22 (2007): 501–538.

24. Jean Comaroff, "The Diseased Heart of Africa: Medicine, Colonialism, and the Black Body," in Shirley Lindenbaum and Margaret Lock, eds., *Knowledge, Power, and Practice: The Anthropology of Medicine and Everyday Life* (Berkeley: University of California Press, 1993); Megan Vaughan, *Curing Their Ills: Colonial Power and African Illness* (Palo Alto, CA: Stanford University Press, 1991).

NINE

Volatility, Speculation, and Therapeutic Revolutions in Nigerian Drug Markets

KRISTIN PETERSON

On the north end of Lagos Island, there is an enormous wholesale market for which millions of pharmaceuticals await distribution to Nigerian as well as west and central African private drug markets. The market resides within an old, historic neighborhood called Idumota. The residents are descendants of former freed slaves from Brazil who re-patriated themselves to this island in the mid-nineteenth century and live alongside Yoruba indigenes, including traditional rulers whose old, dilapidated palaces are inter-spersed among the wholesale pharmaceutical trade. The drug market comprises just one section of a much larger market that is home to locally made goods such as fabrics, Nigerian music, and Nollywood films, but it mostly sup-plies imported goods from the Middle East and Asia such as kitchen wares, spare car parts, second-hand clothing, second-hand computers, packaged food, and other goods essential to life and living. On any given day the market is crowded with sellers wooing customers, cars and mo-torcycles pushing an impossible path through the hustle and bustle, and traders selling drinks, snacks, and mobile phone credit, earning bare amounts for their families. Yet in the pharmaceutical section, millions of dollars worth of mostly antimalarials, nutritional supplements, over-the-counter analgesics, and antibiotics pass through this market *each and every day.* And while it may seem like an out-of-the-way place—out of the way and distinct from the high-earning drug markets of North America, Europe, and Japan—Idumota is essentially tied into the manufacturing and distribution chains of the brand-name drug industry.

Idumota as a massive commercial pharmaceutical market did not exist until the 1980s. Its very formation was the result of two major political and economic events that restructured the Nigerian state and the international pharmaceutical markets. The first was the structural adjustment programs (SAP) administered by the World Bank and the International Monetary Fund, which privatized national African economies in the face of the collapse of continent-wide commodity markets. SAPs had devastating consequences, including mass-induced societal poverty, job loss, basic subsidy removals, food insecurity, and an enormous accumulation of state debt.[1] The second was the restructuring of the pharmaceutical industry. At that time, the drug companies were experiencing a profitability crisis, which was generated by expiring drug patents, a lack of products in the research and development pipeline, and new competition in the global generics market.[2] As the Reagan administration flushed the life sciences industry with new financing, pharmaceutical and especially biotech companies also pursued equity financing, which was primarily obtained via the NASDAQ stock market and venture capital.[3] That is, the US drug industry got tied to the speculative marketplace as a mode of survival. In the process, a new therapeutic revolution was promised by investors, for which new biotechnologies, undergirded by trade, patent, and technology transfer laws, would resurrect a troubled industry.

At the close of the 1970s, Nigeria was home to a thriving brand-name pharmaceutical market for which very few generic products even existed. The convergence of structural adjustment in Africa and the remaking of the pharmaceutical industry produced a violent dispossession and remaking of the Nigerian pharmaceutical market. These events, combined with others taking place locally in the Nigerian drug market, dramatically crashed Nigeria's brand-name market by the end of the 1980s. In the immediate aftermath, drugs became scarce and a new market had to be built. Idumota represents the aftermath of such events, where new wholesaling systems transferred from formal to informal trading spheres, and the global circulation of pharmaceuticals in Nigeria shifted from North America and Europe to mostly Asia and the Middle East. Brand-name drugs were replaced with low-end, low-cost high-selling therapeutics that are largely inefficacious and often faked—that is, made with intentionally falsified ingredients.

Drawing on ethnographic research I conducted with pharmaceutical wholesale traders, retail and industry pharmacists, marketers, industry journalists, and regulators in Nigeria between 2005 and 2010, I argue that it is critical to place the question of a supposedly new "second pharmaceutical revolution,"[4] promised during the 1980s by investors in the speculatively capitalized field of biotech, in the context of African dispossessions.[5] That is, the promise of biotech converged with the global disparity in access to safe, effective forms of essential medications as described in chapters 6, 7, and 8 of this volume. In explaining what made this convergence possible, I first describe the brand-name drug industry's arrival in Nigeria in the 1960s and 1970s, and its subsequent abandonment of this market, which took place in the 1990s. I then describe how the Nigerian market was remade into a generic market that experiences enormous volatility. Many speculative practices are now employed to cope with the uncertainty. Speculative practices in the drug industry (such as massive mergers and investments in high-risk biotech companies) must be understood alongside lateral arbitrage strategies that speculated on wild currency fluctuations in the Nigerian pharmaceutical market, because these are two reverberations in a system of movements in a supply chain that must always anticipate market volatility.

I conclude by arguing that the 1980s therapeutic revolution only promised the remaking of its highest-earning markets while producing disastrous results for the West African market. In this sense, it was successful from the point of view of Northern-based speculative capital, which helped to create renewed profitability for the companies that survived this era of frenzied mergers and acquisitions, characteristic of new-market making. While such activities secured high rates of economic growth as demanded by the investment industry, they also impeded the development of actual therapeutic breakthroughs. At the same time, these events violently remade new markets and social orders in what was once a significant West African drug market. If there was any therapeutic revolution to be found in Nigerian drug markets in the aftermath of brand-name industry abandonment, it was innovation in new ways to fake drugs, and not the realization of new and much-needed therapeutics.

Histories of Market Making in Nigeria

During the 1950s and 60s, brand-name pharmaceutical companies had pioneered and marketed key new drugs such as birth control pills, blood pressure medication, tranquilizers, and heart and psychiatric drugs (see chapters 2, 3, and 6 in this volume). During that period, brand-name drug companies such as Bayer, Boots, Ciba, Hoechst, Imperial Chemical Industries, Parke-Davis, Roche, and Wellcome had established a presence in Nigeria, and indeed among many colonial territories of several European empires. By the early 1970s, companies had extended distribution to manufacturing in many countries outside their home territories, and foreign markets made up over one-half of industry revenue.[6] Just ten years after its independence in 1960, Nigeria experienced a huge oil boom, which attracted more companies manufacturing and marketing products within Nigeria for the West African region. The oil boom promulgated a rather buoyant middle class and Nigeria's brand-name market. By 1980, the pharmaceutical industry's earnings in Nigeria were estimated to be N400 million (about US$400 million), and it employed nearly ten thousand Nigerians.[7]

The arrival of the multinationals led to an increase in the professionalization of pharmacists, especially those who worked as marketers for brand-name pharmaceutical companies. Industry pharmacists were enticed by much bigger salaries than their community and hospital pharmacist counterparts received, and the companies provided them with perks such as housing allowances and chauffeur-driven cars.[8] They received extensive marketing training and were flown around the world to attend conferences and seminars. Eventually, many of the pharmacists ventured out on their own, maintaining their relationships with their former company employers as retailers and distributors of company products. Pharmacists turned retailers identified specific clients they could work with over time, establishing shops to cater to those clients' needs. They might also have secure contracts with oil companies or federal hospitals. Several pharmacists working in Nigeria during this period told me that once a big contract was established, much of one's business was essentially secured. In such situations, drug companies provided generous credit relationships and bulk discounts,[9] and pharmacists routinely recouped a price markup on imported drugs, which was as high as 33 percent at that time.[10] The link between industrial labor and Western capital and the lifestyles that echoed oil boom extravagance meant that all pharmacists in the private and public sector had high prestige and were well regarded in society. The oil boom facilitated lasting relationships between first-generation postcolonial pharmacists and global companies that would shape future definitions of quality in drugs and in the distribution systems that moved the drugs from manufacturer to consumer.

But the prestige of professional pharmacy and the circulation of high-end pharmaceuticals were soon threatened by Nigeria's oil bust at the end of the 1970s. The bust occurred alongside a global recession, and a series of events took place that amounted to what was commonly known as "foreign exchange scarcity"—a term used by almost every industry pharmacist who discussed the period with me. It refers to the drying up of money for imports due to a lack of oil output, but also to the massive flight of industrial capital from Nigeria. Indeed, there was a huge strain on all business in Nigeria, including brand-name drug companies.

At that point the Nigerian government responded in late 1983 by implementing what was known as the import license policy. The intention was to allocate foreign exchange to manufacturing companies, trading companies, and distributors, in order to import what became classified as "essential commodities," which included pharmaceuticals. However, holding an import license opened new opportunities to engage in what are often referred to as "corrupt" money-making practices.[11] Licenses were awarded to a select few firms and individuals, many who had never imported drugs in the past.[12] For example, in 1984—the first round of distribution—only N55 million out of N100 million was used to import drugs; the remaining balance was diverted to "fake" companies that funneled the money to other means, according to *Pharmanews*, the leading monthly newsmagazine of the Nigerian drug industry, which cited the minister of health's findings on diversion.[13]

Many health care services collapsed as a result.[14] With each round of foreign exchange administration, pharmacists and smaller-scale drug companies were increasingly edged out of the importation and distribution of drugs.[15] Indeed, between 1983 and 1984 the pharmaceutical labor force was cut nearly in half as a result of the policy, falling from thirteen thousand workers in 1983 to seven thousand in 1984.[16] These figures were matched by an acute shortage of drugs all over the country.

By 1986, the economic situation worsened—national debt had increased and national revenue decreased. The World Bank and the International Monetary Fund insisted on a structural adjustment program that removed fuel, agricultural, and pharmaceutical subsidies. The crisis also devalued Nigeria's currency, the naira, which had been on par with the US dollar and the British pound sterling during most of the 1970s. The ever-widening exchange rate meant that drug companies could not sell their stock and continued to lose profits. A marketer who had been working at the time for Ciba Geigy, the Swiss pharmaceutical company, explained:

I know most companies left the country during the IMF SAP policy, which devalued the naira. And at the point of devaluation you cannot meet up with turnover. Before devaluation of the naira, Ciba Geigy Nigeria was number three worldwide [in terms of sales] because the naira was strong, at times challenging the dollar. With heavy deregulation, Ciba sold off their plants. You need to sell a lot more to meet turnover in dollars. But was the economy able to support this? It was not. So companies started forming decisions on how to do business in the country. I could remember my company sponsored a meeting here in the Sheraton [Hotel] on how to do business in Africa, and they started looking through each therapeutic area and what the population could afford. And of course price still remains an issue when you have a population where 70 percent live below the poverty line. And this was a result of IMF SAP policy.

Pharmaceutical companies had to calculate new circuits of accumulation, recognize a new increase in standards of living outside of Africa, and at the same time rapidly identify sites of dispossession that global recessions and structural adjustment had generated. They adapted quickly, and pharmaceuticals were entirely rerouted and aligned to these new logics of labor and capital. Some former medical representatives told me that managers in the parent companies were concerned about losing the entirety of the Nigerian market, and advocated for staying put. But these long-term partnerships, and the social and economic investments that had developed those relationships over time, were structurally severed—a decision taken by managers at companies' headquarters. Such decisions had to do with being smart about flexibility in an environment that was changing the game of global competitiveness. The companies understood what needed to shut down, what needed to move, and which ties needed to be cut. They responded to the new inducements to move from one country to the next. Every foreign drug company that had been manufacturing in Nigeria packed up and moved abroad (many to Asia), shut down, or expanded their operations at existing sites and manufacturing plants in Europe and the United States. The companies that divested or ended their operations in Nigeria included Imperial Chemical Industries, Bayer, Upjohn, Parke-Davis, Boots, Wellcome, and Hoechst. Pfizer divested after a forty-year presence in Nigeria.

Just as the Nigerian market was rapidly disappearing, the pharmaceutical industry was undergoing immense restructuring. These efforts were a response to an economic crisis experienced in the United States beginning in 1973. At the end of the 1970s, the United States implemented several policies to deal with its financial crisis. Namely, its government stopped pumping the economic system with liquidity. Instead, it aggressively competed for capital by increasing interest rates; lowering taxes for corporations, speculators, and the wealthy; and lifting restrictions on capitalist enterprise. Crucially, these measures provoked an appreciation of the US dollar, which attracted capital back into the United States. In effect, the direction of capital flows—that is, the economic gain of Nigeria (and other oil-producing states in the global South) and the economic contraction of the United States—was reversed.[17]

But beyond these aspects of US economic financialization, the US drug industry turned to speculative capital as a way out of its own economic downturn. That is, it pursued equity financing via the NASDAQ, and also wooed venture capital for biotechnology start-ups.[18] Capturing the volatility of the 1990s pharmaceutical and genomics stocks, the science studies scholar Mike Fortun asks what value is: "How do you tell a real genomics company from a counterfeit one?"[19] A similar question about the value of drugs came to haunt Nigerians in the same period. It was at this time that Wall Street's "takeover movement" began.[20] As the anthropologist Karen Ho explains, Wall Street accomplished this takeover by "putting corporations 'in play' . . . where all the largest corporations were up for grabs to the highest stock-price bidder, thus forcing them to be immediately responsive to the exigencies of the stock market."[21] As a result, the standard of pharmaceutical value became entirely set by the investment community. Specifically, investors valued drug companies not simply by the amount of profit earned but also, and more importantly, by high rates of growth.[22] As Wall Street orchestrated companies' market valuations, there emerged a highly competitive scramble in the realm of drug discovery.[23] In short, the industry was completely reshaped through financial speculation, new property regimes, new molecular technologies, and new entrants into the field that manage the industrialization of life itself—a field in which, sociologist Melinda Cooper argues, the rise of biotechnology became inseparable from the rise of neoliberalism.[24]

Wall Street dictates led the brand-name drug industry into a frenzy of consolidations during the 1990s, which made international news. But less attention was paid to an equally important strategy to survive competition and unrealistic investor expectations: asset dumping in foreign markets. A Nigerian senior manager at that time, whom I will call Mr. Adebayo, explained to me that his company, Upjohn, was facing very low sales in Nigeria, and operations were scaled down to minimal levels. By the mid-1990s, all the brand-name multinationals decided to divest themselves of holdings in Africa, a development to which Adebayo referred as a "voluntary collapse" of a significantly large foreign market.

The impact of voluntary collapse or market abandonment was substantial. For example, Mr. Adebayo marketed Togamycin, a brand-name version of spectinomycin that was used in Africa in the late 1970s and early 1980s as a second-line antibiotic for the treatment of gonorrhea and other infections.[25] Togamycin had become standard treatment in the United States after strains of such infections had developed resistance to penicillin.[26] In Nigeria two million units of the drug were sold per year; according to Adebayo, that amounted to 15 percent of Upjohn's global sales, making Nigeria its number-one market in the world at the time. Upjohn decided to simply stop marketing Togamycin in Nigeria. A demand for Togamycin did not emerge in other places, which meant that 15 percent of its worldwide sales could not be absorbed elsewhere. Moreover, the company had also just launched Unicap M, which within one year had become the top-selling oral multivitamin in Nigeria. It was designed to be the first drug or supplement wholly produced in that country, from manufacture through packing to marketing. But one year after establishing this marketing plan, the import license system was imposed, followed by structural adjustment—and Unicap M was never manufactured in Nigeria.

As these events were happening in Nigeria, Upjohn's worldwide earnings were on the rise. But its products faced a great deal of marketing challenges. Motrin, prescribed for arthritis and menstrual cramps, lost out to other competitors, and Xanax, used for anxiety treatment, faced sluggish sales once its addictive qualities were discovered. Another drug, Halcion, a sleep-inducing agent that earned seventeen million prescriptions per year, got bad press for causing memory lapses and addiction, which provoked over a hundred lawsuits. By 1992 Upjohn had the lowest sales in the brand-name drug industry.[27] The company responded in several ways: it sped up its research efforts, worked to develop high-selling over-the-counter drugs, and, like other companies, increased its pharmaceutical prices. But drug companies were merging, consolidating, and getting bigger. Upjohn merged with Pharmacia, a Swedish company, and the new entity became the ninth largest company in the world, with seven billion dollars in annual sales. However, the merger did not bring in the expected revenue or huge projected growth.[28] So in 2000 Pharmacia and Upjohn merged with Monsanto and Searle, becoming Pharmacia. Three years later, Pfizer bought Pharmacia.

Pfizer, a global giant, began operations in Nigeria in 1954, and by 1974 it had established a full-fledged manufacturing plant in Lagos.[29] During the early 1980s its global sales, like those of other companies in the industry, were stagnant. Pfizer's coping strategy was to increase its research and development budget by 100 percent. In addition to acquiring Warner-Lambert,[30] which made Pfizer the highest-earning pharmaceutical company in the world, Pfizer also initiated a licensing program with foreign companies, paying them royalties in exchange for marketing rights on their newly developed drugs.[31]

At the same time, Pfizer's Nigerian sales began to slow, also due to currency devaluation and the high cost of raw materials in the country. During the oil boom Pfizer, like other multinational companies, had charged high prices for its products because many Nigerian customers had the capacity to pay them. But after earning power was depleted and currency devaluation set in, the company was no longer earning upward of five million dollars per year in Nigeria. Like other European and US pharmaceutical companies doing business in the country, Pfizer was unwilling to lower prices even for a society that was plummeting into ever-expanding poverty. By 1997, ten years after the implementation of structural adjustment, Pfizer had halted all manufacturing in Nigeria and divested itself of its holdings in the country, and its local subsidiary, Neimeth, bought out the parent company. Yet, at the same time these events were occurring in Nigeria, Pfizer released thirty-seven new products, eight of which were blockbusters. Less than a year after the company left Nigeria, it also sold off several subsidiaries around the world for a total of $4.35 billion. By 2000 it had earned an unprecedented $29 billion. Upjohn disappeared through mergers, and Pfizer became the largest drug company in the world. But both firms, along with all the other foreign drug companies that had been operating in Nigeria, abandoned the Nigerian market.

These dramatic changes in the industry exacerbated some existing problems that Adebayo and other high-level managers identified in the structure of the Nigerian drug market. As a product manager, Adebayo attempted to fill the gap between available drug products and health care needs that had received less attention. Certainly some products were marketed as ways to prevent or treat diseases with high morbidity, like malaria, and those products sold well. However, drugs for very prevalent diseases from worm infections, such as schistosomiasis and leishmaniasis, were far less available, due to the fact that pharmaceutical companies did not prioritize such drugs, then—and do not do so now. But there were also less prevalent diseases, such as hypertension and cancer, for which there were few available products. Although companies were selling millions of bottles of antidiarrheal medication, the only hypertension drug on the market, Minoxidil, was "coming in trickles" to Nigeria, according to Adebayo. As Nigerian pharmacists knew at the time, morbidity related to malaria and diarrhea was very high, and the tendency was to overlook a broader disease landscape that also included cancer, hypertension, and diabetes, to name just a few diseases. The mismatch between disease burden and the products available in the market began in the moment that foreign companies began exporting drugs to West Africa. And although Adebayo and others tried to remedy the problem, divestment halted their efforts, and a long-term pattern of mismatch was established. By the time the economy had fully contracted, it was difficult to introduce new drugs; all companies found it less expensive to import drugs than to manufacture them in Nigeria.[32]

Just four years after the SAP's implementation, Nigeria—a country that had once been a great consumer and producer of brand-name drug products—accepted a loan from the World Bank for twenty million dollars to purchase essential drugs to make up for the pharmaceutical shortfall in hospitals and for the high cost of drugs.[33] By 2000, after the bulk of consolidation and asset dumping had taken place, the top five brand-name companies' wealth amounted to twice that of the gross domestic product for all of sub-Saharan Africa.[34] As drug companies first invested in and then pulled out of Nigeria, hauling offshored assets out of much of the African continent, the Nigerian market dramatically opened for new forms of South-South investment and trade.

Birth of Unofficial Drug Markets and New Market Structures

By the time brand-name companies divested out of Nigeria, there was a huge drug shortage problem, as multinationals' brand-name products had constituted more than 90 percent of Nigeria's drug market. The market was subsequently rebuilt in unpredictable and unanticipated ways. Igbo traders who hailed from the eastern part of the country stepped in to take control of a collapsed national and, indeed, West African regional private drug distribution system. Traders began to import drugs manufactured in Asia, eastern Europe, the Middle East, and South America. Currently, Indian companies command more than 50 percent of the drug market, and Chinese companies control nearly 100 percent of the medical technologies market. And so the drug trading patterns shifted from North America and Europe to primarily Asia.

Separately and quite significantly, the capital used to attract new generic drugs was made possible by the international narcotics trade. As structural adjustment meant a severe decline in available jobs, some Nigerians turned to the narcotics trade at a moment when Nigeria became an international transitional transfer point between two important sites of narcotics production—Latin America (coca) and Asia's Golden Triangle (opium). Narcotics traders found clever ways of repatriating earned cash from narcotics deals, as was explained to me by former narcotics dealers and by those involved in present-day pharmaceutical distribution. Narcotics dealers purchased legitimate pharmaceuticals (as well as personal computers and luxury cars) on a massive scale at cheap bulk rates in Europe or elsewhere, and shipped them to Nigeria, dumping them into newly emerging and expanding markets.[35]

Idumota Market on Lagos Island became the main wholesale market that imported directly from manufacturers and distributed to large clients as well as other wholesale markets in the West and Central African regions. The market grew in this neighborhood after many people lost their jobs at the end of the civil war (1969), and it continued to grow with the onset of structural adjustment (1986). The growth is significant. Kunle Okelola, executive secretary of the Pharmaceutical Manufacturers Group of the Manufacturers Association of Nigeria, estimates that in 2009 the entirety of Nigeria's national drug market amounted to more than two billion dollars in sales.[36] The chairman of the Lagos State Medicine Dealers Association, the union representing over hundreds of members operating pharmaceutical shops in Idumota, estimated to me that "billions of naira" (the equivalent of hundreds of thousands to millions of US dollars) pass through this drug market *every day*; and the Pharmaceutical Manufacturers Association estimates that this market grows by up to 15 percent every year.[37] There are more than seven hundred traders working here between different levels of official and unofficial business, selling to hospitals, clinics, corporations, government institutions, and the oil industry.

What kind of drugs are traded? While the 1970s drugs were primarily brand-name and mostly patented, the new drugs entering unofficial markets were almost all generics meeting the needs of a population newly impoverished by structural adjustment. By the early 1990s, however, the structure of the pharmaceutical market appeared to be out of control, as far as Nigerian government officials were concerned. Research studies and reports indicated that just a few years after the implementation of structural adjustment, fake drugs comprised 30 to 70 percent of the entire national drug market in Nigeria.[38] For this period, reports also indicated that fake drugs were sold in tens of thousands of illegal places in Lagos state.[39] While current reports indicate that Nigeria's fake drug problem has declined since the 1990s, the United Nations recently declared that West Africa has the worst fake drug problem in the world.[40] Presently, fake drugs comprise anywhere between 30 to 50 percent of the entire West African regional market.[41] These numbers and declarations are epistemologically blurred because, while fake drugs are perceived as prolific, their numbers cannot actually be counted or ascertained. But they do catch the priority attention of regulatory officials, thus leaving other important issues rather obscured. These latter issues include the problem of substandard drugs and high levels of drug resistance. Substandard drugs are not intentionally faked, but contain too little or too much of their active ingredients as a result of shortfalls in the Nigerian or other manufacturing processes. Moreover, the most commonly sold drugs, such as older-generation antibiotics, often encounter the highest levels of drug resistance—up to 100-percent resistance for some older-generation antibiotics in certain parts of the country.[42]

Critically, the structure of the market that includes fake, substandard and non-efficacious drugs has a tendency to consistently reproduce itself. In the 1970s, many drugs found in the Nigerian market were simple antibiotics, antimalarials, and analgesics. But though these drugs were simple, many of them, such as Togamycin, were also effective and widely sold in North American and European markets. As the biologies of numerous bacteria and malaria-causing parasites changed over time, however, the imported drugs meant to tackle these infectious agents did not change to meet new medical needs. Instead, an intensive competition for market share had and still has a tendency to encourage the manufacturing, importation, and sale of nonefficacious and often low-quality pharmaceuticals approved for market in the 1970s.

Although elements of this structure were in place at the inception of Nigeria's pharmaceutical market, market divestment and devaluation accentuated that structure. Specifically, when global drug markets were restructured in the 1980s and 1990s, two critical events took place. The first is that manufacturing sites were reorganized. The brand-name drug industry was already well established outside of the markets of middle- and high-income countries, but with market crashes and currency fluctuations, many companies relocated to—or consolidated their operations in—Asia, especially South Korea, Singapore, Taiwan, India, and China.[43]

A former Nigerian worker for Abbott Laboratories, a US-based company, illustrated the situation for me. He invoked the example of erythromycin, an antibiotic that remains relatively effective in parts of Nigeria. Abbott manufactures erythromycin not in the United States but in Pakistan, among other places. The lower labor costs involved in doing so give the company higher profit margins. Abbott then sells the product in Nigeria at rather inflated US prices, because the company can claim that it is a US drug. This pricing strategy recognizes the buying-and-selling culture in Nigeria. Abbott was one of the first companies to bring erythromycin to Nigeria. It was years before any generic drug manufacturer began to market erythromycin, and as of 2010 there were no more than five small companies distributing generic erythromycin in Nigeria. Product recognition and prescribing patterns that do not often substitute high-quality generic drugs for brand-name products mean that the original version of the drug can still command high sales.

The second critical event was that while markets were restructuring, labor was rendered cheap; but at the same time, manufacturing became too costly in Nigeria. Moreover, an array of long-term SAP-imposed taxes on imports, along with the state's retreat from providing the very basic infrastructure (such as electricity) needed for industrial manufacturing, made it far more lucrative to import and trade drugs than to make them. For example, Nigerian pharmaceutical importers and distributors who do not work in the market, and who distribute to retailers or clinics instead, gave me several different scenarios for ideal importation strategy. If an importer wants to distribute several drug products, she must first establish a product line that always maintains high sales, such as one that includes antimalarials, antibiotics, and over-the-counter painkillers—the most common and fastest sellers in the market. If she is in business for the long term, she looks to drugs made for smaller but possibly emerging markets, such as statins (anti-cholesterol products) or antihypertensives, for which there are fewer competitors because they cater to consumers with higher purchasing power. If she works with a US or European company that manufactures brand-name drugs offshore, she may assume that the drug quality will remain high, which in turn ensures that she will have a steady income stream.

One of the wealthier Idumota traders told me that he is very reluctant to buy products that are new in the market, because they are not a sure thing in terms of sales. He always waits to see how such new products initially perform—allowing others, usually distributors, working outside Idumota, to take the risk first. If he thinks a product is doing well, then he will consider a partnership of sorts with retail clients. Traders pay very close attention to how fast products move in the market. If, for instance, customers—especially those who buy in bulk—come to the market asking for a product that is not available there, traders are known to drop everything, figure out where the drug is selling in West Africa, and travel all night to a market outside Nigeria to buy the product. Or, if the price of one over-the-counter generic drug crashes, traders in the market seek out new high-earning products instead; they may alter what stock they carry according to the market's boom-and-bust dynamics. Because the vast majority of drugs are imported via trading networks, the future of drug products is located not in specific antihelmintics, cardiovascular drugs, or new drug product breakthroughs, but in market price fluctuations and, certainly, the public's recognition of and ability to pay for certain drugs. Thus, existing market practices and market structure are continually reinforced as massive numbers of cheaply priced and largely ineffective (due to drug resistance) antibiotics and analgesics outpace the pharmaceutical needs resulting from other diseases.

These pricing logics not only drive the proliferation of older and inefficacious drugs for the West African market, but also drive the "chemical arbitrage" that leads to fake drugs. The principle remains the same as in "price arbitrage," which is the primary mechanism that moves drugs from manufacturer to end user. That is, drugs are routinely priced differently across national markets due to regulatory regimes that include price caps on drugs, and to the manufacturer's discernment of an appropriate price for that market. Wholesale distributors take advantage of the price difference, attempting to buy low in one market and sell high in another.[44] Chemical arbitrage means that instead of capitalizing on price differentials, the distributors intentionally deviate drug chemistry and drug dosages from their standard ranges. These practices provide a wider markup margin, and enable further arbitraging activities. For example, instead of the usual two-hundred-milligram dose for paracetamol, a trader or distributor can negotiate with a manufacturer to make pills containing only one hundred milligrams of paracetamol, but label them as containing two hundred milligrams. Or a drug that may not sell well in one market (such as an anti-inflammatory) can be renamed and relabeled to sound like a high-selling drug (like an antibiotic), and then be exported to the Nigerian market, usually at a very high price markup.[45]

Within the market, these products are exchanged via arbitrage practices that directly coincide with traders' attempts to "hustle the day" or "make it now,"[46] in the context of new and unfamiliar goods hitting rebuilt markets and new forms of risk that change alongside these dynamics. Indeed, these practices of exchange, pricing, credit, and labor interact with, and indeed often drive, large-scale as well as nuanced microlevel market dynamics. For example, pricing strategies and price wars not only create uncertainty over a drug's reliability and point of origin, but also present numerous ways of hedging risk against business practices that allow one to derive additional cash from exchanges in the distribution chain. Moreover, credit practices are tied to labor and high-risk entrepreneurialism. For example, a trader may get drugs on credit from a marketing representative and then be unable sell them because of a sudden currency devaluation, or because the drug's market price has suddenly dropped drastically, or because an order of drugs arriving from abroad is too close to its expiration date and cannot be sold. Any one of these scenarios, as well as many others, could mean the end of someone's business. As a result, uncertainty always undergirds valuation, and volatility is always anticipated.

These arbitrage dynamics and market practices in the Idumota market are not isolated. They represent one point of many in the drug manufacturing and distribution processes. They rely on similar arbitrage strategies that move pharmaceuticals of all qualities across continents. The imported drugs that travel to markets in Nigeria and elsewhere are conceptualized, manufactured, and distributed on the basis of competition emerging from ever-downward pricing pressures, the regulatory regimes of nation-states, and the worldwide porousness of the international borders through which pharmaceuticals must travel.

Downward Pricing Pressures and Offshoring after Market Restructuring

When I first encountered the problem of fake and substandard drugs in Nigeria, the people producing these products were portrayed as operating in the shadows, and their identities remained largely unknown. How could anyone responsible for producing this enormous supply of drugs remain unknown? But there are several avenues that enable local and transnational operations to remain veiled. Perhaps the most important factor driving the hidden nature of fake drugs is not illicit activity, but the more transparent activity of offshore manufacturing.

In the 1990s, a speculative wave of consolidations and asset dumping in the pharmaceutical industry converged with the creation of new sites of offshored and outsourced manufacturing in response to increased pressures to reduce cost. Outsourcing is the procurement of goods or services under contract with an outside supplier, and offshoring is the practice of moving or basing a business operation abroad. For example, brand-name companies offshore their research and production to companies abroad, which in turn outsource to smaller local manufacturers. There is a great deal of licensing activity and partial merging—up and down the pharmaceutical value chain—from preclinical chemistry to clinical trials.[47] Opportunities to outsource and offshore pharmaceutical manufacturing and raw material production (materials that are either unprocessed or minimally processed, such as chemicals) became available in the Chinese and Indian economies, both of which were growing rapidly.[48] The rise of these markets and drug economies was key to the survival of brand-name drug manufacturers. It was also important to the development of a new Nigerian drug market, with Chinese and Indian companies taking the largest share of the pharmaceutical, pharmaceutical raw material, and medical technologies market.

The promise of a therapeutic revolution, which gave rise to drug industry consolidation in the 1990s, did not produce many hoped-for new products. Now there are even fewer drugs in the pipelines, with more upcoming expiring patents, and rising R&D costs.[49] As European and North American companies move to Asia, they are closing down plants or dumping assets in both their home markets and foreign markets, with higher costs. They are also acquiring or licensing to a number of national firms based in Chinese and Indian home markets, and this helps to grow these industries.[50] One outcome of increased industry consolidation is that the wealthiest Indian and Chinese companies have been acquiring American and European firms. The industry literature refers to this as "reverse offshoring"[51]—a misnomer, if we understand these activities through the impetus of capital rather than via US and European hegemonic trading power.

The more recent patterns of company merging and acquisition differ slightly from strategies prevalent in the 1990s. At that time, consolidation in the drug industry led to companies dumping their less productive assets while strengthening their existing product lines and adding already well-earning products to their profiles. In this more recent scenario of Chinese and Indian companies acquiring American and European firms and vice versa, the focus of expansion is on reduced or abstracted stages of manufacturing. Active pharmaceutical ingredients (API)—the key chemical or biological ingredients in drugs—are made in the primary manufacturing stage. China and India are the world's first and third highest producers of API, which is a multibillion-dollar industry (Italy is second). They produce API for drug companies around the world, including Nigeria. Then comes secondary manufacturing, which is the production of pharmaceuticals in their final form. The third stage is the tableting or packing of drugs for distribution to wholesalers. A possible intermediary step within these stages could, for example, involve outsourcing some of these manufacturing stages to a local company that makes API within the offshored site.

These activities are further complicated in China, which has a large chemical industry with more than eighty thousand companies.[52] Chemical companies can make API, or can cross completely into drug production itself. But the Chinese Food and Drug Administration, has no jurisdiction to inspect chemical companies.[53] If it did, regulation would be especially difficult, simply due to the sheer size of both chemical and pharmaceutical industries. This is a problem that all national regulatory agencies face. Regulation in any country is designed to oversee manufacturing on the basis of national regulatory laws. But regulatory bodies and their legal mandates are not well designed to oversee the crisscrossing of prolific offshored and outsourced manufacturing, and this makes it nearly impossible to inspect, or sometimes even locate, manufacturing premises.[54] Even though Nigerian, US, European, and other regulatory authorities have offices in overseas markets just as China does, none of those countries, including China, has the capacity to actually inspect and regulate all of the facilities under its purview in any rigorous way.[55] It is therefore difficult for any drug regulatory agency to guarantee the safety of a national drug supply.

The fake-drug industry relies on offshoring to Asia, outsourcing within offshored sites, and the impossibility of regulation. For example, a major fake-drug scandal involved Scientific Protein Laboratories (SPL), a US company based in Wisconsin. It was the primary owner of a Chinese company, Changzhou SPL, which manufactured heparin, an anticoagulant (blood thinner) derived from pig intestines. In March 2008 the FDA claimed that 81 deaths and 785 adverse effects occurred across eleven countries due to heparin contaminated with a much cheaper raw material.[56] The *New York Times* investigators David Barboza and Walt Bogdanich reported that the Changzhou plant "was certified by American officials to export to the United States even though neither government [Chinese nor US] had inspected it. The plant has been exporting heparin to [the American health care company] Baxter [International] since 2004. . . . Some experts say as much as 70 percent of China's crude heparin—for domestic use and for export—comes from small factories in poor villages. One of the biggest areas for these workshops is . . . in coastal Jiangsu Province, north of Shanghai, where entire villages have become heparin production centers."[57]

The heparin case shows how chemical arbitrage opportunities have become available in offshoring activities, as well as in the outsourcing that occurs in offshored sites. These manufacturing processes feed into distribution channels as well. When producers and distributors work together (most often in ways that, unlike in the heparin case, do not draw attention to themselves), they first ascertain the regulatory capacity of a drug's destination. Different regulatory regimes have different capacities to monitor the various aspects of fake drugs (from chemistry to packaging), and fake drug producers and distributors take this into account. Once producers ascertain these regulatory constraints, they calculate the lowest amount of API and the cheapest amount of inactive ingredients needed to create a drug that will make it into the destined market without raising the suspicions of regulators. There is often far more deviation outside the standard API range in difficult-to-regulate markets than in markets that are more rigorously regulated.[58]

Finished drug products, as well as raw materials for pharmaceutical ingredients that are manufactured in Asia, move laterally among multiple distributors. At this stage, they can pass through as many as six trading companies before they reach the pharmaceutical manufacturer or wholesaler. As the journalist Katherine Eban has discussed, the lateral moves at this stage are made within an extensive network in the wholesale drug trade.[59] The network includes distributors, intermediaries, secondary wholesalers, and a vast array of businesses that run the gamut between the official and unofficial, the licit and the illicit. Just as is the case with arbitrage conducted in Idumota, where multiple lateral exchanges make gains in sales, global pharmaceutical distribution chains are driven by price differentials set by the manufacturers. These traders—diverters or arbs, as they are commonly called—

take advantage of the price differences by buying discounted drugs and reselling them at marked-up prices to other distributors and wholesalers.[60] The distribution chain constitutes many people and companies across global regions, with each link in the chain involving a new price markup. In Europe, the arbitraging of pharmaceuticals is allowable via parallel import laws, which makes it an especially pervasive practice there.[61] The international trade lawyer Donald deKieffer has pointed out that "the major distributors operate at very thin profit margins, rarely exceeding 5%. If, however, they can purchase inventory at 10% or more below the price offered by the manufacturer, the result goes directly to the bottom line. This has traditionally been too tempting to resist for even the most ethical of companies."[62] The multiple lateral movements, many of which take place in Europe, do the work of obscuring manufacturing origins. One may never know that drugs or raw materials have come from an unregulated or unregistered company, or from a company that is registered but only part of whose manufacturing chain is regulated.[63]

Distributors draw almost entirely on the regulatory gaps, price differentials, and gray trade links to facilitate the global fake-drug trade, which uses the same distribution routes as does the trade in legitimate pharmaceuticals. The distribution chain for both intentionally faked and legitimate products relies on free-trade zones around the world, like those in Dubai or the Panama Canal, which are not subject to rigorous inspection, and then moves on quickly to sites of sale or manufacture.[64] Indeed, counterfeiters use free-trade zones to hide the origins of pharmaceuticals and chemicals as well as to make, resell, market, or relabel fake drugs.[65] In Dubai, where many fake drugs stop in transit to West Africa, the usual requirement for local ownership of companies is waived, and there are no import and export fees or income tax.[66] As authorities catch on to the regular use of one free-trade zone in this way, counterfeiters simply move on to new sites that are not so well surveilled.

These examples from the global distribution chain highlight a number of gray areas in which breakdowns in regulation are driven by global drug economies. The massive dispersals in the production and distribution chains make it difficult to discern the difference between intentionally faked or unintentionally substandard drugs, because regulatory inspection—of everything from raw materials to finished products—has difficulty distinguishing between them. When they reach markets like Idumota, they are lost to regulatory oversight.

Asian Drugs to West African Markets

How do fake and substandard drugs enter Nigeria? Mr. Kumar (not his real name), who has been importing Indian pharmaceuticals since 1980, asserted in an interview over dinner:

There is one basic community from Nigeria, and most of the Indian third-grade companies export all sorts from India, where they may not even have a pharmaceutical factory. So these are the two people who are involved in bringing these fake and substandard products . . . There has [sic] been some good Indian companies which have been here, like Ranbaxy has been here for now twenty, twenty-five years. They have put up a manufacturing unit here. Vitabiotics, this is another good company here. May Organics, another good company here. There are good companies that are here, but the point is that there are also these trading ulcers which exist.[67]

The transnational alliances about which Kumar spoke materialize in several ways. Tony—a former officer of the Pharmaceutical Society of Nigeria, the main Nigerian professional organization for pharmacists—told me that he had received an e-mail message from a company in China informing him that the firm took orders for any specifications, including not only drug chemistries but also tablet coloring and packaging. Manufacturers such as this are largely small- to medium-scale companies that are linked to the entire supply and distribution chains across continents.

As the anthropologist Yi-Chieh Jessica Lin points out, the distribution of fake products out of China entails moving these products by boat or air to neighboring southeast Asian ports, from which they are then shipped to intermediary countries in places like Europe, where new documents or relabeling can take place.[68] Given that parallel importation is legal in Europe as mentioned above, much of the actual faking occurs there. Critically, like the money-laundering strategies of Nigerian businessmen who have created and relied upon capillary networks, the total embedding of local networks to move fake goods out of China is equally important.[69] Certainly, these networks employ or draw on key actors in customs and transport to move fake products across borders.

After fake drugs travel out of Asia by sea or over land, they pass through free-trade zones or other porous regulatory depots, and arrive in West Africa. As both Tony and regulatory officials explained to me, drugs entering Nigeria can be smuggled directly into the country. But more often they are first directed to neighboring West African countries. If they are traveling by boat, they are usually shipped to the Republic of Benin, whose main port is in the city of Cotonou, only one hundred kilometers west of Lagos. The port authority in Benin inspects only shipments that are to remain in that country, which is a major hub for used-car imports and sales. Goods that are simply passing through Cotonou in transit to other West African countries are not inspected (although the cargo may be scanned for explosives or searched for narcotics). Rather, the "acquits" system, which allows transporters to cross borders within the region without global customs clearance,[70] provides documentation that certifies inspection at the point of embarkation and is enough to pass goods through customs. As the scholars Carine Baxerres and Jean-Yves Le Hesran note, "Consequently, trade of specific goods, such as pharmaceuticals, can escape customs statistics."[71] Smugglers take advantage of these gaps in the regulatory apparatus by packing fake drug products in the middle of a shipping container and surrounding them with legitimate products. Once the drugs have cleared customs in Cotonou, they are loaded onto trucks and move to their final destination in Nigeria. When the trucks reach the Nigerian border, it is up to the border officials to decide what to do about the cargo. A driver may insist that only Nigeria's drug regulatory agency can inspect it, and payments made to officials can ease the truck's way across the border.

Much of what is known about fake drugs results from the joint work among national, regional, and international agencies, including members of the United Nations, federal governments, nongovernmental organizations, and brand-name drug companies. While there are regional task forces that cooperate across the fifteen West African states, there are no harmonized laws. One reason is that the term "counterfeit" has no similar juridical definition within the region. Moreover, coordination among police, courts, legislators, and regulatory bodies is not well executed.[72] However, regional coordination is nascent and has been supported by recent joint statements and action plans. These global statements on fake drugs always acknowledge problems with consumer safety and public education, but responses to these problems mostly involve appeals to commercial interests that match war-on-drugs strategies. These include punishment mechanisms, such as stringent jail time, rather than policies to prevent the selling of fake drugs, such as price regulation.

Nigeria's Narcotics and Counterfeiting Federal Task Force is responsible for surveilling fake drugs that enter the country. It also cooperates with other West African agencies to surveil the West African region, due to pervasive fake drug smuggling over West African borders. A high-ranking member of the task force explained to me that it provides extensive coordination among West African state drug regulatory authorities to identify and stop transnational business chains that facilitate the movement of products into West Africa. For example, the same official told me about the arrest of a Nigerian businessman who was discovered with a drug product that he claimed had been manufactured in India even though the airport manifest stated that it had come from China. The official took the matter to Interpol in Europe, which notified the police in China. The police located the Chinese company, which had faked an Indian company's drugs (something that may be increasingly common, given the trading relationships between China and India). The Chinese police retrieved the phone number of the Nigerian importer from whom the arrested Nigerian businessman had purchased the drugs, and the importer turned out to be a clearing agent at the Lagos airport. The task force official was later informed that the fake drug manufacturers in China were executed. I sensed that the official was slightly horrified but also pleased that some punishment had been implemented.

Idumota market has its own unofficial drug task force. Chidi, a market trader, is one of its members. With other members who are also pharmaceutical traders, he goes on daily sweeps checking for fake drugs in market stalls. He told me:

[Fake drugs] will give us a bad image. If we catch you selling fake drugs firstly, the union [Lagos State Medicines Dealers Association] will first of all apprehend you . . . the union will now alert the police. You will be in police custody before we now [tell] NAFDAC [the federal drug regulatory agency], [which] will now go and pick you up. NAFDAC will subject [your products] to a chemical test. You pay heavily—105,000 naira for the test, for each product. Then after the test, you will pay if you didn't pass the test, and you will go to court and from there go to jail. . . . During the time of investigation, the shop will remain closed. . . . So we don't tolerate any type of fake drugs in this market.

Drug traders like Chidi make it very clear that if there is even a rumor of fake drugs, much less their actual presence in Idumota, it is bad for everyone's business. Regardless of whether the actions Chidi described really go as smoothly as he reported, he and others claimed to me that the internal market surveillance has actually prevented the dumping of fake drugs in Idumota. Like global distribution networks that seek out porous transit sites, dealers of fake drugs turn to new destinations in Nigeria if a market like Idumota applies strict internal control mechanisms. Opinions about what drives and fuels the distribution of low-quality drugs vary widely. Tony, the former Pharmaceutical Society of Nigeria official, said:

It is only the [fake drug importers] that have been put in place by these market people, and they have their businesses dotted all around. If you put a continual load of fake drugs in Idumota this evening, by tomorrow morning you wouldn't see it because they would have distributed it, put it on the buses that are doing night travels, [and] they are off! The one going to Kano [a city in the northern part of the country] has gone [*snaps fingers*], the one going to [the airport after they] land, someone picks it [*snaps fingers*] [and] dispatches it to the north. It goes off like that! That is why we are saying, the society is saying, "No to drug market," because if you have well organized drug distribution you don't allow them to prosper together because when they prosper they are able to do more evil. . . . So what I am saying in essence is that the drug market is the bedrock of fake drug distribution. So things should be taken off and because of that you can't do a recall. So the recall system is not there because they are not bothered, they are all interested in the money.

Ikenna, a wealthy pharmaceutical trader, insisted to me: "It is the system. It is the system. There are people working [to facilitate the sale of fake drugs] at the airport. There are people working at the seaport. Even this 100-percent inspection for drugs is still implied, but these [fakes] still come in even up till today. Who is deceiving who? It's a system thing, my sister."

Ikenna pointed out that fake drugs are rooted in a systemic network tied across regions, while Tony connected it to the unofficial drug markets. Both viewpoints are valid. Tony is right to indicate that the private drug markets are the primary site of distribution for fake drugs. But it is not simply that thousands of private wholesalers have helped to open up channels for fake drugs. Idumota is only one nodal point in a very complex transcontinental supply system.

The speculative marketplace has encouraged the search for lower manufacturing costs. As a result, new forms of abstracted manufacturing and multiple routes of distribution have provided a veiled means for fake drugs to be produced and distributed alongside "legitimate" products. Even before these drugs make it to Nigeria, the pricing and arbitrage strategies that tie drug manufacturing and distribution together ultimately make it difficult to tackle the enormous amount of fake and substandard drugs entering the country. Nonetheless, the

Nigerian public is hopeful that the drugs they purchase and consume will someday meet much higher standards.

Conclusion: Therapeutic Revolution or Monopoly Control?

The 1980s pharmaceutical industry revolution that set out to remake itself in the face of declining profits and ever-increasing generic industry competition was not simply about creating new therapies and transforming disease categories. These intentions were accompanied by dramatic changes in the global logics of speculative capital that reoriented capital from Northern-based multinational pharmaceutical firms back Northward, and opened Southern markets for very different kinds of investment and importation from India, China, and other locales.

The 1980s "therapeutic revolution" marked a turn to speculative capital as a way for the brand-name industry to survive a profitability crisis. Such survival strategies became possible in the context of African pharmaceutical market abandonment. The geopolitical architecture was important here: the drug industry relied upon fresh injections of capital from the investment community. That fresh capital coming into investment banks was harvested straight out of African debt repayments that had been orchestrated by structural adjustment programs and by other subsequent liberalization policies that forced dramatic debt repayments. Thus, the supposed therapeutic revolution of the 1980s relied upon the violent dispossession of the Nigerian brand-name drug market such that health indicators and outcomes became far worse as a result. In this context, the speculative turn put new investment-community pressures on the drug industry to increase profits by lowering costs, merging, acquiring, and dumping assets. These activities led to new logics of abstracted manufacturing processes and distribution arbitrage. As such, they leave little room for the actual development of a significant number of new drugs. In the process, structural adjustment programs that imposed massive economic disparity have essentially reduced the drug industry's earning power in foreign markets, and this has coincided with a dramatic inability for Nigerian consumers to purchase drugs.

Consolidation put some companies on top of the earning hierarchy. Indeed, a company such as Pfizer was making more money than it ever had in the past, and it became the highest-earning drug company in the world after the 1990s consolidations were somewhat exhausted. This is a question not entirely of greed as linked to Nigerian dispossession, but rather of the structural dynamics of an industry in which high earnings and staying on top are key to survival. Certainly a company such as Upjohn, which was the ninth highest earner in the industry at the same time, could not survive the frenzy—it was eaten by several companies, including Pfizer. This game of consolidation reduced the number of therapies in any given company's research and development pipeline: when companies merged, they tended to dump a number of low-margin drugs in order to focus marketing energy on the higher earners.[73]

The intensity of drug industry competition for survival continues today, as do the speculative practices found in a large West African wholesale drug market such as Idumota. Speculative and arbitrage practices are not just about increasing one's income (whether one is an Idumota trader, an international trading company, or a generic, fake, or brand-name company), but about increasing one's life or corporate chances. These speculative practices found in Idumota market and the pharmaceutical industry are two reverberations in a system of movements within a supply chain that must always anticipate market volatility. The offshoring of pharmaceutical production to India and China proliferates because investment industry expectations can be reached there—at least for now. As pharmaceutical capital is induced to move from one site to the next, the possibility of a new therapeutic revolution in even the wealthiest of drug markets may yet remain unrealized. Beyond a series of market abandonments made in the hopes of revolutionary therapies, what new pharmaceutical futures are possible? Until these market dynamics change, the idealized therapeutic revolution will likely continue to bypass West African markets.

NOTES

1. Meredeth Turshen, *Privatizing Health Services in Africa* (New Brunswick, NJ: Rutgers University Press, 1999); Adebayo O. Olukoshi, *The Politics of Structural Adjustment in Nigeria* (London: James Curry, 1993); Gloria Thomas-Emeagwali, *Women Pay the Price: Structural Adjustment in Africa and the Caribbean* (New Brunswick, NJ: Africa World Press, 1995).

2. Najmi Kanji, et al, *Drug Policies in Developing Countries* (London: Zed, 1992); Jeremy Greene, "Making Medicines Essential: The Evolving Role of Pharmaceuticals in Global Health" *BioSocieties* 6 (2011): 10–33.

3. Melinda Cooper, *Life as Surplus: Biotechnology and Capitalism in the Neoliberal Era* (Seattle: University of Washington Press, 2008); Kaushik Sunder Rajan, *Biocapital: The Constitution of Postgenomic Life* (Durham, NC: Duke University Press, 2006).

4. World Health Organization, *The World Drug Situation* (Geneva: World Health Organization, 1988), 47. The WHO compared the promise of biotech and the global disparity in access to safe, effective forms of essential medications.

5. More recently, the promise and future forecasts of an exponentially growing pharmaceutical market in Nigeria pay no attention to the history of pharmaceutical dispossession in the 1970s. I thank Javier Lezaum for pointing this out to me. See IMS Consulting Group, *Pharmerging Markets: Picking a Pathway to Success*, 2013.

6. Milton Silverman and Philip Randolph Lee, *Bad Medicine: The Prescription Drug Industry in the Third World* (Palo Alto, CA: Stanford University Press, 1992).

7. Fred Adenika, "Prospects for Pharmacists in the Nigerian Pharmaceutical Industry: The Next Twenty Years," *Nigerian Journal of Pharmacy* 13, no. 1 (1982): 6.

8. Fred Adenika, *Pharmacy in Nigeria* (Lagos, Nigeria: Panpharm, 1988), 84.

9. N. Mgbokwere, "The Pharmaceutical Wholesaler and His Supplier," *Pharmanews* 6, no. 4 (1984): 10–13.

10. S. E. Okereke, "Pharmacy Profession in Nigeria: Image on the Line," *Pharmanews* 10, no. 1 (January 1988): 12.

11. Akin Fadahunsi, "Devaluation: Implications for Employment, Inflation, Growth and Development," in *The Politics of Structural Adjustment in Nigeria*, ed. Adebayo O. Olukoshi (London: James Curry, 1993) 33–53. On corruption in Nigeria, see Daniel J. Smith, *A Culture of Corruption* (Princeton, NJ: Princeton University Press, 2008); Kristin Peterson, "Phantom Epistemologies," in *Fieldwork Isn't What It Used to Be*, ed. James D. Faubion and George E. Marcus (Ithaca, NY: Cornell University Press, 2009), 37–51.

12. "Shortfall in Drug Import Caused by Fake Companies," *Pharmanews* 7, no. 4 (1985): 1.

13. Vin Ujumadu, "Drug Importation and Import License," *Pharmanews* 8, no. 10 (October 1986): 6.

14. See Vin Ujumadu, "Drug Importation and Import License," arguing that no pharmacist was appointed to represent the Ministry of Health on the import licensing panel, and that the panel thus had no expert knowledge about how the industry would fare.

15. "Shortfall in Drug Import Caused by Fake Companies."

16. Ifeanyi Atueyi, "Good Riddance, 1984," *Pharmanews* 7, no. 1 (April 1985): 3.

17. Giovanni Arrighi, in "The African Crisis: World Systemic and Regional Aspects," *New Left Review* 15 (May-June): 22, states: "This was a reversal of historic proportions that reflected an extraordinary, absolute and relative capacity of the U.S. political economy to attract capital from all over the world. It is likely that this was the single most important determinant of the contemporaneous reversal in the economic fortunes of North America and of the bifurcation in the economic fortunes of Third World regions.

For the redirection of capital flows to the United States reflated both effective demand and investment in North America, while deflating it in the rest of the world. At the same time, this redirection enabled the United States to run large deficits in its balance of trade that created an expanding demand for imports of those goods that North American businesses no longer found profitable to produce."

18. In his work on the genomics company DeCode, Mike Fortun, in *Promising Genomics: Iceland and Decode Genetics in a World of Speculation* (Berkeley: University of California Press, 2008), describes the rise and fall of the company's public offerings on the stock market. Much of the volatility existed because investors' notions of the marketplace were in a constant state of flux—and the reason for that was the company's unfounded speculative promises of future earnings.

19. Mike Fortun, *Promising Genomics: Iceland and Decode Genetics in a World of Speculation*, 11.

20. Karen Ho, *Liquidated: An Ethnography of Wall Street* (Durham, NC: Duke University Press, 2009), 123–168.

21. Karen Ho, *Liquidated: An Ethnography of Wall Street*, 129.

22. Kaushik Sunder Rajan, "Pharmaceutical Crises and Questions of Value: Terrains and Logics of Global Therapeutic Politics," *South Atlantic Quarterly* 111, no. 2 (spring 2012): 321–346. As Sunder Rajan points out, companies do not produce enough new drugs to achieve these growth rates, and this amounts to an inherent structural contradiction. To meet the growth expectations of the investment community would require that a company bring to market multiple drugs each year—a highly difficult if not impossible goal to achieve because it takes up to fifteen years to develop a drug and bring it to market. This is why companies come to rely on blockbuster drugs that make over one billion dollars per year, as well as on producing "me-too" drugs—drugs that are chemically similar to existing marketed products—that are easy to develop, although not always easy to sell in a highly competitive environment.

23. Mike Fortun, *Promising Genomics: Iceland and Decode Genetics in a World of Speculation*; Sunder Rajan, *Biocapital: The Constitution of Postgenomic Life*.

24. Melinda Cooper, *Life as Surplus: Biotechnology and Capitalism in the Neoliberal Era*.

25. E. E. Obaseiki-Ebor et al., "Incidence of Penicillinase Producing Neisseria Gonorrhoeae (ppng) Strains and Susceptibility of Gonococcal Isolates to Antibiotics in Benin City, Nigeria," *Genitourin Medicine* 61, no. 6 (1985): 367–370.

26. G. M. Savage, "Spectinomycin Related to the Chemotherapy of Gonorrhea" *Infection* 1, no. 4 (1973): 227–233.

27. Funding Universe, "The Upjohn Company History." Accessed November 1, 2011. http://www.fundinguniverse.com/company-histories/the-upjohn-company-history/.

28. Funding Universe, "The Upjohn Company History."

29. Stella Okoli, *The Pharmaceutical Industry in Nigeria: Historical Review, Problems and Expectations* (Lagos: PGM-MAN, n.d.). By 1965, Pfizer was selling products in 100 countries, and its foreign sales amounted to $175 million; in 1989 it was doing business in 140 countries. See Corporate Watch, "Pfizer Inc."

30. The takeover of Warner-Lambert reportedly cost Pfizer $84 billion; the stock swapped as part of the acquisition was estimated to be worth $116 billion. See Reference for Business, "Pfizer Inc." Accessed November 10, 2010, at http://www.referenceforbusiness.com/history2/42/Pfizer-Inc.html.

31. Reference for Business, "Pfizer, Inc."

32. "Mrs. Okoli Pleads for Indigenous Producers," *Pharmanews* 12, no. 9 (1990): 1. As Susan O. Oremule, managing director of the Nigerian drug firm Associated Manufacturing Company, stated in an interview at the time, "Small scale manufacturing requires assistance of the government because it augments the services of the MNCs [multinational corporations], thereby forming a vital link to the entire pharmaceutical industry. So free markets work against the interests of MNC Pharma. Moreover, it works against the interest of public health because drugs that are needed are neglected. Only the cheap ones get produced."

33. G. D. Lahan, in "Drug Production in the Present Economic Situation," *Nigerian Journal of Pharmacy* 20, no. 1 (1989): 15–17, explains that in real terms this meant that within three years of SAP, manufacturers had to import more than 90 percent of raw materials, 30 percent of packaging materials, and all equipment and spare parts. The declining exchange rate meant that drug costs completely skyrocketed. "It calls for at least eight times the original capital base to operate at the same level as [before sap]" (15). These policy issues combined with increased competition with parallel imports, which undercut marketers working for the manufacturers. And, as the purchasing power of society dramatically decreased, industry garnered stockpiles, slowed down manufacturing momentum, and increased labor layoffs.

34. Julian Borger, "Industry That Stalks the US Corridors of Power," *Guardian* (13 February 2001). Accessed 13 July 2003 at http://www.guardian.co.uk/world/2001/feb/13/usa.

35. Understanding the reality of these dynamics remains difficult, as much of the new market-building and existing trading networks remain partially within shadow networks and economies. I have elsewhere (2009) described the way in which the public en masse attempts to account for such shadow activity in their daily lives, and I follow their lead by suggesting that social scientists need to reformulate what actually counts for the empirical within these social relations that are tied to massive economic activity. See Kristin Peterson, "Phantom Epistemologies," in *Fieldwork Isn't*

What It Used to Be, ed. James D. Faubion and George E. Marcus (Ithaca, NY: Cornell University Press, 2009), 37–51. See also Gernot Klantschnig, *Crime, Drugs and the State in Africa: The Nigerian Connection* (Dordrecht and Leiden, the Netherlands: Brill, 2013). Klantschnig interviewed imprisoned former narcotics dealers whose stories align much with the pharmaceutical stories told to me.

36. Kunle Okelola, *Pharmaceutical Manufacturers Group of the Manufacturers Association of Nigeria: An Overview of the Pharmaceutical Sector* (Abuja, Nigeria: PGM-MAN, 2009). See also Charles Wambebe and Nelson Ochekpe, *Pharmaceutical Sector Profile: Nigeria. Global Unido Project: Strengthening the Local Production of Essential Generic Drugs in Least Developed and Developing Countries* (Vienna: UNIDO, 2011), 11.

37. Charles Wambebe and Nelson Ochekpe, *Pharmaceutical Sector Profile: Nigeria. Global Unido Project: Strengthening the Local Production of Essential Generic Drugs in Least Developed and Developing Countries*, 1.

38. Ifeanyi Atueyi, *Fake Drugs in Nigeria: Topical Issues and Facts You Need to Know* (Lagos: Pharmanews, 2004), 38; Uwaezuoke, "Adulterated Drugs: Pharmacists Take Census," *Guardian Financial Weekly* (15 April 1991), 20.

39. Ifeanyi Atueyi, *Fake Drugs in Nigeria: Topical Issues and Facts You Need to Know*, 41.

40. UNODC, *Transnational Trafficking and the Rule of Law in West Africa: A Threat Assessment* (Vienna: UN Office on Drugs and Crime, 2009).

41. Sybil Ossei Agyeman Yeboah, "The Illicit Trafficking of Counterfeit Medicines," presentation made at the Technical Conference of Experts on the Trafficking in Fraudulent Medicines, Vienna (UN Office of Drugs and Crime, 14 February 2013), 6.

42. The literature produced by bench scientists in Nigeria indicates that malaria parasites and many species of bacteria are proving to be increasingly and highly resistant to older-generation drugs, and even to some newer ones. In the 1980s and early 1990s there were reports of extremely high resistance to tetracycline, ampicillin, chloramphenicol, and streptomycin. See F. O. Eko et al., "Antimicrobial Resistance Trends of Shigellae Isolates from Calabar, Nigeria," *Journal of Tropical Medicine and Hygiene* 94, no. 6 (December 1991): 407–410; A. N. Njoku- Obi, et al., "Resistance Patterns of Bacterial Isolates from Wound Infections in a University Teaching Hospital," *West African Journal of Medicine* 8, no. 1 (1989): 29–34; Iruka Okeke et al., "Antibiotic Resistance in *Escherichia coli* from Nigerian Students, 1986–1998," *Emerging Infectious Diseases* 6, no. 4 (July–August 2000); and Lamikanra et al., "Rapid Evolution of Fluoroquinolone-Resistant *Escherichia coli* in Nigeria Is Temporally Associated with Fluoroquinolone Use," *BMC Infectious Diseases* 11 (2011): 312. Lamikkanra et al. tracked microbial resistance to these drugs over time and found increasing resistance to strains of *E. coli*. By the 2000s and 2010s, researchers were recording extremely high resistance (90 to 100 percent) to older-generation drugs, as

well as existing and rising resistance to the newer generations of antibiotics known as cephalosporins and quinolones, in all parts of the country. See A. G. Habib et al., "Widespread Antibiotic Resistance in Savannah Nigeria," *African Journal of Medicine and Medical Sciences* 32, no. 3 (September 2003): 303–305; A. Oladipo Aboderin et al., "Antimicrobial Resistance in *Escherichia coli* Strains from Urinary Tract Infections," *Journal of the National Medical Association* 101, no. 12 (December 2009):1268–1273. More recently, others reported high levels of multiple drug resistance (even to cephalosporins and quinolones) in cases of bacterial infections including typhoid and tuberculosis; see Umolu P. Idia et al., "Antimicrobial Susceptibility and Plasmid Profiles of *Escherichia coli* Isolates Obtained from Different Human Clinical Specimens in Lagos, Nigeria," *Journal of American Science* 2, no. 4 (2006): 70–75; O. Olowe et al., "Antimicrobial Resistant Pattern of *Escherichia coli* from Human Clinical Samples in Osogbo, South Western Nigeria," *African Journal of Microbiology Research* 2, no. 1 (January 2008): 8–11.

43. Karen Politis Virk, "Biopharmaceutical Manufacturing and Offshoring in Puerto Rico: Linguistic and Cultural Challenges," *Language Connections* (June 2008, Boston). Accessed 15 June 2012 at www.languageconnections .com/portal/descargas/WhitePaper_BiopharmaManufacturing_Puerto Rico.pdf.

44. Hirokazu Miyazaki, *Arbitraging Japan: Dreams of Capitalism at the End of Finance* (Berkeley: University of California Press, 2013); Donald MacKenzie, *An Engine, Not a Camera: How Financial Models Shape Markets* (Cambridge, MA: MIT Press, 2006).

45. O. Victor Oparah, "NAFDAC: Sustaining the Grip on the Campaign against the Production, Importation, Exportation, Advertisement, and Distribution of Fake and Adulterated Food and Drug Products in Nigeria." Presented at the National Workshop on Regulation of ARV/OI Drugs and Related Biological Products, Lagos, Nigeria, 26 May 2005.

46. Adedotun Phillips, *The Nominalization of the Nigerian Economy* (Ibadan, Nigeria: Niser Institute for Social and Economic Research), 22.

47. Mridula Pore et al., "Offshoring in the Pharmaceutical Industry," in *The Offshoring of Engineering: Facts, Unknowns, and Potential Implications, Committee of the Offshoring of Engineering* (Washington: National Academy of Engineering, 2008), 103–124.

48. Since the 1990s, there has been a proliferation of lucrative "contract manufacturing organizations," which are industry brokers that facilitate offshore manufacturing at every level of the drug development process. A CMO industry report lists China as the site for outsourcing that is favored because manufacturing costs there can run 40 percent below typical US and European costs. See Oliver Mueller and Clifford Mintz, "CMOs and Final Dosage Manufacturing in China: The Far East Market Evolves," *Contract Pharma* (30 May 2012). Accessed 24 October

2013 at http://www.contractpharma.com/issues/2012-06/viewfeatures/
cmos-and-final-dosage-manufacturing-in-china/.

49. Pore et al., "Offshoring in the Pharmaceutical Industry," 103.

50. In addition to outsourcing and offshoring, the implementation of World
Trade Organization (WTO)–mandated intellectual property laws in both
India and China (as well as national changes in good manufacturing
practice standards) was very effective in reorganizing national industries.
These changes meant that medium-sized companies began to consolidate
and grow bigger while many, though not all, smaller companies were
squeezed out. Driving costs down even further, India and China have
become each other's preferred trading partners. For example, Indian drug
imports in China amounted to $58 billion in 2005, as opposed to US
imports, which were valued at less than $30 million. See Pore et al., "Off-
shoring in the Pharmaceutical Industry," 113.

51. Pore et al., "Offshoring in the Pharmaceutical Industry," 103.

52. Roger Bate, *The Deadly World of Falsified and Substandard Medicines* (Washington: AEI, 2012), 177.

53. Walt Bogdanich et al., "Chinese Chemicals Flow Unchecked to Market,"
New York Times (31 October 2007), A1.

54. Roger Bate, *The Deadly World of Falsified and Substandard Medicines*; US
Food and Drug Administration, *Global Engagement*, accessed 5 April 2013
at http://www.fda.gov/downloads/AboutFDA/ReportsManualsForms/
Reports/UCM298578.pdf.

55. Walt Bogdanich quotes congressional Representative John D. Dingell:
"China alone has more than 700 firms making drug products for the U.S.,
yet the FDA has resources to conduct only about 20 inspections a year in
China." Walt Bogdanich, "China Didn't Check Drug Supplier, Files Show,"
New York Times (16 February 2008), A1.

56. Gardiner Harris, "U.S. Identifies Tainted Heparin in 11 Countries," *New
York Times* (22 April 2008).

57. David Barboza and Walt Bogdanich, "Twists in Chain of Raw Supplies for
Blood Drug," *New York Times* (18 February 2008), A1.

58. Roger Bate, *The Deadly World of Falsified and Substandard Medicines*.

59. Katherin Eban, *Dangerous Doses: How Counterfeiters Are Contaminating
America's Drug Supply* (Orlando, FL: Harcourt, 2005).

60. Katherin Eban, *Dangerous Doses: How Counterfeiters Are Contaminating
America's Drug Supply*; Wyatt Yankus, "Counterfeit Drugs: Coming to a
Pharmacy Near You," *American Council on Science and Health*, 24 August
2006. Accessed 15 April 2011 at http://www.acsh.org/publications/pubID
.1379/pubdetail.asp.

61. Keith E. Maskus and Matthias Ganslandt, *Parallel Imports of Pharmaceutical
Products in the European Union* (Washington: World Bank, 1999). It might
help to briefly illustrate the drug distribution system in the United States,
where three major wholesalers—AmerisourceBergen, Cardinal Health, and

McKesson—procure drug stock directly from manufacturers. These three companies handle more than 80 percent of the drugs distributed in the United States. But in recent years, they have purchased from secondary wholesalers, who themselves often purchase from sources other than the manufacturers. The reason for these small diversions is, again, price. The secondary wholesalers search out discounted products because they do not necessarily have a price advantage over the larger wholesalers.

62. Donald deKieffer, "Trojan Drugs: Counterfeit and Mislabeled Pharmaceuticals in the Legitimate Market," *American Journal of Law & Medicine* 32, nos. 2/3 (2006): 329.

63. Roger Bate, *The Deadly World of Falsified and Substandard Medicines.*

64. Walt Bogdanich et al., "Chinese Chemicals Flow Unchecked to Market."

65. Walt Bogdanich et al., "Chinese Chemicals Flow Unchecked to Market"; UNODC, *Transnational Trafficking and the Rule of Law in West Africa: A Threat Assessment*; Matías Loewy, "Deadly Imitations." (Pan American Health Organization) 11, no. 1. Accessed April 16, 2010. http://www.paho.org/English/DD/PIN/Number23_article3.htm.

66. Walt Bogdanich, "Chinese Chemicals Flow Unchecked to Market."

67. India and China have both been stereotyped as centers for industrial piracy. In the case of China, Laikwan Pang, in "'China Who Makes and Fakes': A Semiotics of the Counterfeit," *Theory, Culture and Society* 25, no. 6 (2008):120, points out that the image of the pirate cannot be separated from the real or imagined industrial power that China possesses. China can produce any kind of product, and as such it "is tied to today's global capitalism in all senses." Because the bulk of both legitimate and illicit drug products comes from China and India, many diverse actors in Nigeria may encounter each other.

68. Yi-Chieh Jessica Lin, *Fake Stuff: China and the Rise of Counterfeit Goods* (New York: Routledge, 2011).

69. Lin, *Fake Stuff: China and the Rise of Counterfeit Goods.*

70. Carine Baxerres and Jean-Yves Le Hesran, "Where Do Pharmaceuticals on the Market Originate? An Analysis of the Informal Drug Supply in Cotonou, Benin," *Social Science and Medicine* 73 (2011): 1249–1256.

71. Ibid., 1252.

72. West African Health Organization, *Roundtable: A Joint Action against Fake Medicine in West Africa* (Bobo-Dioulasso, Burkina Faso: WAHO, 2011). Accessed 15 December 2011 at http://www.diplomatie.gouv.fr/fr/IMG/pdf/GeneralReportOuagadougou_Round_Table-En_cle4746d9.pdf 2011.

73. Kristin Peterson, "On the Monopoly: Speculation, Pharmaceuticals and Intellectual Property in Nigeria," *American Ethnologist* 41, no. 1 (2014a): 128–142.

TEN

Therapeutic Evolution or Revolution? Metaphors and Their Consequences

DAVID S. JONES

When the Cleveland surgeon René Favaloro published his description of coronary artery bypass grafting in 1968, he launched one of the most important surgical procedures in the United States.[1] Speaking at a conference in Houston in 1985, he described 1968 as the "year of revolution."[2] When he was interviewed a decade later, however, Favaloro used a different metaphor. As he described it, "The evolution took place in just a few months from patch graft to interposition graft to bypass graft."[3] So which was it: an evolution or a revolution? Debates about the meanings and merits of these two metaphors for historical change have been a fixture of the historiography of science and medicine for decades. Although historians do not argue as much about whether or not a particular development counted as a "scientific revolution" as they did when Thomas Kuhn's *Structure of Scientific Revolutions* was fresh, the choice of "evolution" or "revolution" remains important, especially in the history of medicine and therapeutics. The two metaphors carry very different connotations for our understandings of how and why medical practice changes over time.

Revolutions, as the chapters in this volume make clear, receive the lion's share of attention from historians.

Charles Rosenberg's classic essay on the "therapeutic revolution," revisited in chapter 11 at the end of this volume, has set the standard for therapeutic history for nearly forty years.[4] Yet Rosenberg's 1977 essay principally focused on a nosological revolution that only secondarily transformed therapeutics. Others have written about the bacteriological revolution, the antibiotic revolution that followed, and the broader pharmaceutical revolution in the 1950s. Geneticists have for decades been making promissory claims about a genetic revolution that will introduce a new epoch of personalized, precision medicine.[5] Historians of surgery have described the anesthetic and aseptic revolutions. One cardiologist, channeling Kuhn, has even described "the structure of cardiological revolutions."[6]

Evolution, however, is also ubiquitous in the medical literature. Consider the field of cardiology, once named "the youngest child of medical evolution."[7] Atherosclerotic plaques undergo "evolution,"[8] as do cardiac surgery procedures,[9] anesthetic techniques,[10] and the specialties of cardiology and cardiac surgery.[11] Doctors can use electrocardiograms to follow a heart attack's "electrocardiographic evolution."[12] New operations and instruments have been evolved.[13] When cardiac surgeons began to face competition from the new field of interventional cardiology, many realized that "only our ability to evolve will guarantee our survival."[14] Even patients joined the effort: "Patients undergoing coronary bypass grafting have undergone an evolution in recent years."[15] At times, physicians have explicitly debated about which metaphor—evolution or revolution—offers the more apt description for whatever therapeutic changes happen to interest them, whether they be heart-lung machines, statin therapy, or endovascular repair of abdominal aortic aneurysms.[16] A revolution itself, such as that produced by transesophageal echocardiography, can undergo evolution.[17]

The language of evolution has been entrenched in the history of medicine as well. In April 1913, for instance, William Osler gave lectures at Yale University entitled "Evolution of Medicine." He sought to tell the story of medical progress. Even though the path of that progress might not have been linear, medical theory and practice improved with evolution: "Like a living organism, truth grows, and its gradual evolution may be traced from the tiny germ to the mature product. Never springing, Minerva-like, to full stature at once, truth may suffer all the hazards incident to generation and gestation."[18] As the editors describe in their introduction to this volume, Fielding Garrison praised Osler's "panoramic survey" of the painful evolution of medicine from superstition to rationality. Garrison hoped that Osler's narrative of evolutionary progress would be an inspiration to students and other readers.[19]

Even though historians of medicine have since learned to be skeptical of positivism and Whiggish "just-so" stories, evolution remains widespread in historians' writing. Historians have published essays on the evolution of medical ideas, for instance of the term "chancre," of Darwin's concept of pangenesis, of clinical trials, or of Harvey Cushing's thoughts about specialization.[20] They have traced the evolution of medical techniques, including endotracheal anesthesia, prophylactic enucleation of the eye, bronchial casts, or frozen sections (and the impact of those on the evolution of surgical pathology).[21] And they have narrated the evolution of medical institutions, from the Mayo Clinic to health services in India.[22] Such articles rarely invoke anything more than the most superficial idea of evolution as a process of gradual, progressive change over time.[23]

What are we to make of the coexistence of evolution and revolution in medicine and its histories? Both words are often used casually in English without careful attention to their specific meanings or connotations. The meanings of "evolution" have themselves evolved over time, and many discordant meanings remain in use today.[24] From the Latin *evolver*, to rollout or unroll (as in unrolling a scroll), evolution first appeared in English in the mid-seventeenth century. It was used in different ways to describe the wheeling movement of dancers, the course of childbirth, or the working out of God's plan for creation. By the eighteenth century it increasingly implied a gradual change in a system from a simpler to a more complex state, as in embryological development. This meaning became generalized in biology to describe the transformation of organisms over time. "Revolution," as described elsewhere in this book, has followed an equally complex course, from a revolving movement in space or time to violent upheaval and the overthrow of an established social or political order.[25] By the nineteenth century, the sudden overthrow of revolution was contrasted against the gradual, organic reforms of evolution.[26] But this distinction was never perfect, with evolution in biology including ruthless struggles between species and dramatic extinctions. Do doctors and historians actually intend any of these specific meanings when they use "revolution" or "evolution" in their writing? Cardiological revolutions do not involve violent overthrow, and the evolution of cardiac surgery does not rely on surgeons' differential reproductive success.

There is meaning in the words nonetheless. Evolution and revolution are both models of change over time. It is easy to see the appeal of a claim of revolution for scientists, and for their historians: it pronounces a radical break from the past, confident and triumphant. Progress is implied by the decisiveness of the rupture. Such rhetoric is good for marketing, especially when contrasted against the cautious gradualism of evolution. But evolution has its own appeal, especially its reassuring connotation of progressive improvement. Roy Porter defined the stakes well in his 1986 essay on scientific revolutions.[27] He described the juxtaposition of evolution and revolution as a contrast between continuity and cataclysm. He argued that if historians would not stake a claim about this distinction, they put themselves "in danger of defaulting on the task of assessing overall patterns of science."[28] However, they had to proceed with caution. Porter advocated a narrow definition of scientific revolutions: they ought to involve a self-conscious process of challenge, resistance, and struggle, the deliberate "overthrow of an entrenched orthodoxy."[29] By this standard, the seventeenth century did bring some revolutionary changes to the sciences, but the changes in nineteenth-century medical theory that Rosenberg described were merely a crisis, not a revolution.[30] Even though he winnowed the list of scientific revolutions, Porter also warned against a "retreat into an *evolutionary* metaphor of science's development, on some specious analogy with the dictum *natura non facit saltum*."[31] What he wanted was deliberate, thoughtful, discussion of the pace and character of scientific change. His demand remains relevant today.

It is not enough simply to debate what counts, or not, as revolution or evolution. Instead, much can be gained through serious engagement with the theory and language of revolution and evolution in pursuit of the best possible accounts of scientific change. Porter did this with revolution, as do the authors of the chapters in this book. Something similar can be done with evolution. Relevant concepts and their components can be made into meaningful guides for historical analysis. Evolutionary biologists have developed an elaborate theoretical apparatus to understand the processes of organismic evolution, with analyses of niches, fitness, competition, the Red Queen hypothesis, extinction, taxonomy, island biogeography, and morphospace. Some of these ideas, such as that of the niche, have already been adapted by historians. Other aspects can be adapted to history as well—an exercise that can be thought-provoking and even productive.

It is of course important not to be cavalier when borrowing ideas across scholarly disciplines. Richard Lewontin, a noted evolutionary theorist, has warned scholars in other fields not to appropriate concepts of evolution, because evolutionary theory was developed to explain biological change, not social change, and its concepts cannot be casually applied across the latter domain.[32] Scholars have long contested efforts to apply evolution to psychology, sociology, and social policy.[33] Similar concerns exist with history. Applying biological theory to history risks naturalizing what are actually social, economic, and political processes. Moreover, theories of evolution, like those of revolution, carry connotations of progress. These can confound understandings of progress in medicine, something that has long been a vexing issue for historians. Used carefully, however, the language of evolution can be a valuable tool for historians to think with.

Niche

In basic ecological and evolutionary theory, a niche is the space or role in an environment occupied by a particular species. Bees pollinate flowers, bats eat mosquitoes, and so forth. Historians of medicine have taken up the niche concept in two different ways. In *Last Resort: Psychosurgery and the Limits of Medicine*, Jack Pressman explained why lobotomy worked in the 1940s but not forty years later.[34] He offered the niche as an intuitive, ecological metaphor. The efficacy of a treatment can only be understood in the context of the particular problem the treatment offers to solve: "The extent to which a treatment flourishes is directly dependent upon the specific features of the day's clinical landscape. In the long haul, viability is a matter of ecology, not virtue."[35] In the 1930s, asylums overflowed with patients, hopelessness, and horror. Psychiatrists desperately sought new treatments. Lobotomy, which could calm some patients (albeit at the cost of damaging their personality), offered "human salvage." It appealed to patients, their families, and psychiatrists. Pressman's metaphor was explicit: "From an ecological perspective, the treatment rapidly penetrated into a niche of almost limitless size that as yet had no competitors."[36]

Ian Hacking used niche models to explain the history of dissociative fugues and other diseases that appear in a society only to vanish at some future date: "I argue that one fruitful idea for understanding transient mental illness is the ecological niche, not just social, not just medical, not just coming from the patient, not just from the doctors, but from the concatenation of an extraordinarily large number of diverse types of elements which for a moment provide a stable home for certain manifestations of illness."[37] He argued that four "vectors" defined the extent of the niche: medical taxonomy (or nosology), cultural polarity, observability, and release. As these vectors change over time, so do the niches, and so do the diseases themselves: "To postulate a niche for an illness is to make two kinds of claims, one positive, one negative. In the presence of the relevant vectors, the illness flourishes; in their absence it does not."[38] For both Pressman and Hacking, the metaphor of the niche provided an analytic framework that accounted for changing diseases and treatments over time.

While the niche concept has clear value, it introduces some risks. As Lewontin has warned, invocation of a biological concept like "niche" in a historical analysis might reify the phenomena being studied. This is a risk, since existing scholarship on changing diagnostic categories and therapeutic practices has shown that there is little natural about these dynamics. Historians have described many cases in which interested groups have, in effect, created niches for diseases or treatments. Patient activists have pushed diseases onto the medical agenda. Pharmaceutical executives have publicized diseases to create new markets for their products. When diuretics and tricyclic antidepressants appeared in the 1950s, Merck and other companies distributed educational materials to popularize the diseases—hypertension and depression—that the drugs could treat.[39] This set the precedent for many diseases and their drugs, from social anxiety disorder to restless leg syndrome and erectile dysfunction.[40]

Historians have often analyzed these cases with an alternative metaphor, that of the market. While market analyses have obvious relevance and value, they focus on just one aspect of the phenomena: money. Niche models offer a broader approach that can incorporate other dynamics. Moreover, the risk of naturalization can be minimized by emphasizing the social factors that define the niche. Pressman described overflowing asylums, psychiatrists in search of respect, and legislatures concerned by growing mental health budgets. Hacking's vectors were intellectual and cultural, from medical theorizing about epilepsy to the new popularity of cycling. However, avoiding the biological baggage of niche can be tricky to do. Hacking, for instance, equivocates, suggesting that there had to be "an ecological niche in which the construction could thrive."[41] This just begs the question.

Tensions about whether a niche is natural or constructed are embedded deep in the origins of the word itself. "Niche" has been used since the eighteenth century to describe the lair of an animal or a suitable place for a person. This usage was borrowed from architecture.[42] "Niche" first appeared in English in 1610 to specify a space, often in a cathedral, built to house a statue or a relic; it replaced an older Latin term, *aedicula*, meaning a small house.[43] The derivation of "niche" itself remains contested. Some trace the word to a French source, also *niche*, meaning a kennel for a dog, or possibly *nichier*, meaning to make a nest. Others prefer an Italian source, *nicchio*, for seashell.[44] In either case, the architectural term "niche" has its roots, ironically, in nature. The ambiguity about whether a niche is natural or constructed simply recapitulates this etymology.

Recent developments in evolutionary theory offer a possible solution to this tension. When ecologists developed niche theories in the 1910s and the 1920s, they focused on characteristics of an organism's environment (e.g., availability of food and shelter, or competition and predation).[45] In 1957, however, G. Evelyn Hutchinson reconceptualized the niche as a property of the species in relation to its environment.[46] This definition introduced the distinction between the fundamental niche (i.e., one that was possibly achievable by a species) and the realized niche. Meanings of "niche" shifted again in the 1970s when Richard Lewontin popularized the idea of "niche construction."[47] Beavers build dams, grazers alter the species compositions of fields where they graze, and trees create myriad niches around themselves. As Lewontin later explained, organisms "are not simply *objects* of the laws of nature, altering themselves to bend to the inevitable, but active *subjects* transforming nature according to its laws."[48] By shifting the focus from adaptation to construction, evolution becomes a coupled process in which organisms are functions of their environment and environments are functions of their organisms.[49]

Understood in light of these modern formulations, the niche becomes a productive model for historians of medicine. It has ecological connotations, suggesting an opportunity within an environment, as well as architectural connotations, suggesting a built space (an idea that can encompass market strategies). In the simplest application, a therapeutic niche might simply be a disease or symptom in need of treatment. The rise of coronary artery disease in the twentieth century, for instance, opened a niche for a diverse assortment of pharmaceutical and surgical treatments. But the niche is not simply a phenomenon of the physical disease environment; it is also a social process. It might be recognition of the need to manage some aspect of the burden of disease. There was a lag of several decades, for instance, between the rise of coronary disease and the decisions of physicians and health officials to commit substantial resources against it. New disease concepts (e.g., atherosclerosis, coronary thrombosis), new technologies (e.g., the electrocardiogram), and new specialties (e.g., cardiology) all converged between the 1920s and 1950s to open the therapeutic niche for coronary artery disease.

Theories of niche construction suggest that a therapeutic niche will be altered by the treatments that attempt to fill it. Antibiotics have changed their niche by triggering the emergence (or evolution) of antibiotic resistant bacteria.[50] Chris Feudtner has described the transformation (or niche construction) of diabetes.[51] Before insulin, diabetes was an acute disease, with patients wasting away and then dying from ketoacidosis and hyperglycemic coma. After insulin, diabetes became a chronic disease, with patients developing diabetic retinopathy, nephropathy, neuropathy, and vascular disease. Each new complication opened a new therapeutic niche. The success of bypass surgery in the 1970s inspired cardiologists to develop angioplasty, which has now displaced bypass surgery from much of its niche. The complications of angioplasty, including restenosis and stent thrombosis, have created secondary niches, for platelet inhibitors and antiproliferative agents, that could not have been imagined in the 1950s. Used with attention to the subtleties that have been developed by evolutionary biologists, niche theory can be a valuable tool for historians of medicine.

Fitness

When doctors and patients think about therapeutics, they often focus on the most fundamental outcome: Did the treatment work? This can be surprisingly difficult to determine. Outcome can be assessed from the perspective of the physician or the patient; by changes in symptoms, laboratory values, imaging studies, or life expectancy; after short, medium, or long intervals; and with case series, cohort studies, randomized trials, and meta-analyses. Historians have also been extremely interested in efficacy. As Rosenberg explored in his classic essay on therapeutic revolutions, and revisits in the next chapter, one of the most interesting puzzles has been in understanding how and why the assessment of efficacy changes over time. Bloodletting, now dismissed by biomedical scientists, was popular in Western medicine for more than two thousand years. It must have worked. The crucial challenge is to understand what work it did.[52]

The concept of efficacy has productive parallels with the concept of fitness. Darwin used "fit" and "fitted" throughout *Origin*, but it was only in the 1866 edition, influenced by Alfred Russel Wallace and Herbert Spencer, that he began to use "survival of the fittest."[53] Population geneticists have defined fitness as differential reproductive success, something that is not an absolute attribute of an organism, but a measure of its success in a particular environment. Since reproductive success is sometimes random (e.g., an extremely "fit" organism could die in an accident), biologists have developed a "propensity" interpretation of fitness that distinguishes potential and realized fitness.[54]

It takes some tinkering to adapt evolutionary concepts of fitness to history of medicine. Treatments do not reproduce in any biological sense. Success is determined, instead, by the beneficial effect of a treatment on patients and the perception of that effect among physicians and patients. However, at an abstract level fitness can do useful work for historians. First, it actually is possible to think of fitness in terms of a treatment's ability to generate progeny.[55] As physicians work to improve treatments, whether pharmacological or procedural, they produce derivatives. Penicillin gave rise to methicillin, ampicillin, amoxicillin, and many other antibiotics. The first beta-blockers produced derivatives that diversified and filled other niches. Balloon angioplasty has inspired an ever-growing lineage of catheter-based interventions. If success at producing derivatives yields one with higher clinical efficacy, then the parent therapy dies out—a victim of its own reproductive success. Second, it is possible to think of therapeutic fitness in terms of a treatment's ability to expand a therapeutic niche. While sildenafil can produce erections, what really made it successful was its ability, through marketing, to transform the embarrassing problem of impotence into the profitable diagnosis of erectile dysfunction. In a similar way, it is possible for treatments to achieve success by creating subniches (segmenting the market?) for a series of treatments. The niche of hypertension now has space not just for one fittest antihypertensive, but for many fit diuretics, beta-blockers, and more.

The distinction between potential and realized fitness is useful as well. Doctors often think about both the optimal outcomes that can be achieved with a treatment and those realized in actual clinical practice. In this respect, randomized clinical trials measure potential fitness, while realized fitness is experienced by patients in routine clinical practice; this is the distinction between efficacy and effectiveness. The problem of noncompliance fits in here as well, as one of the many barriers that stand between potential and realized fitness.[56] Does a treatment work? That cannot be answered simply, just as a biologist cannot say whether or not an organism is fit. Like biologists who assess fitness in the context of a specific niche, physicians and historians must assess efficacy in the context of the problem being treated, the outcomes most valued by the patients and doctors, and the ability of the health care system to deliver the treatment.

Competition

Competition, one domain in which differential fitness reveals itself, has come to be seen as being nearly synonymous with natural selection. It plays a key role in evolutionary theory. Biologists define it specifically as "the simultaneous reliance of two individuals, or two species, on an essential resource that is in limited supply."[57] What is the limited resource in medicine? There are many possibilities. Illness episodes generate the need for treatment (and the opportunity for reimbursement). Patients host illness episodes. Health care resources are deployed to treat them. Competition for episodes, patients, and resources takes place between different treatments (e.g., medications or surgery for coronary disease), providers (e.g., cardiologists, cardiac surgeons, nutritionists), institutions (e.g., from neighborhood clinics to national referral centers), and insurers. While overt competition was once considered unseemly in medicine, it is now routine, and billions of dollars are spent each year on advertising to gain advantage. Each of these aspects of competition offers a productive target for historical analysis.

What determines the outcome of competition? Success in medicine is fickle. The best treatments, doctors, or health care systems do not necessarily outcompete the others. Doctors have sought to adjudicate competition between treatments with randomized clinical trials, but there have been many obstacles to the trials' power.[58] Success can come from better efficacy or from fewer side effects. Selective serotonin reuptake inhibitors, for instance, displaced tricyclic antidepressants not because of their superior efficacy, but because of their increased safety (especially in overdose). Marketing campaigns have pushed many blockbusters to prominence even when those blockbusters had no significant advantage over existing treatments.[59] Sometimes the cultural meanings of diseases and their treatments matter most. The science studies scholar Anne Pollock has shown how racial dynamics have influenced the popularity of treatments for hypertension (e.g., guidelines that once recommended diuretics for black patients and ACE inhibitors for white patients) and heart failure (e.g., the approval of BiDil for patients who self-identify as black).[60] The fittest might survive, but there are many ways for a treatment to be fit.

The Red Queen Hypothesis

In classic Darwinian theory, organisms struggle to adapt themselves to their environment. Biologists now recognize that niches change constantly over time, a result of both environmental change and the shifting of competitive landscapes as other species come and go. This has important consequences for adaptation and natural selection: organisms must adapt to something that is constantly changing. Invoking a scene from Lewis Carroll's *Through the Looking-Glass*, the evolutionary theorist Leigh van Valen in 1973 named this the Red Queen hypothesis.[61] As the Red Queen told Alice, in her world "it takes all the running you can do, to keep in the same place."[62] In biological terms, organisms might evolve constantly just to maintain a stable level of fitness in the changing environment. Subsequent theorists have introduced variants. One, restricting the Red Queen hypothesis to competitive interactions between species, coined a new phenomenon, the Court Jester effect, to analyze efforts by organisms to track random changes in their physical environments.[63] As a 2009 article explained, the "Red Queen model stems from Darwin, who viewed evolution as primarily a balance of biotic pressures, most notably competition." The Court Jester model, in contrast, argues "that evolution, speciation, and extinction rarely happen except in response to unpredictable changes in the physical environment, recalling the capricious behavior of the licensed fool of medieval times."[64]

The challenge of adapting to a changing niche provides a powerful intuitive model for understanding the fundamental task of medicine and public health: to provide relief from the diseases that afflict human populations. Physicians and public health officials seek to define and then eclipse the burden of disease.[65] The problem is that the burden of disease is never static. It changes constantly, in response to changing physical and social environments, the evolution of pathogenic microorganisms, the advent of new and dangerous technologies (e.g., cars, cigarettes), or the impact of decisive medical interventions (e.g., smallpox vaccination). Physicians and public health officials must struggle to keep up. Since innovation takes time, evolving medical therapies inevitably lag behind the changing burden of disease.

Physicians and medical researchers, for instance, set out to master bacterial disease in the 1880s. They studied patients, identified causative microorganisms, and then sought "magic bullets" that could cure the diseases, from immunizations and serotherapies early in the twentieth century to the "antibiotic revolution" in the 1950s, analyzed by Podolsky and Lie in chapter 1 of this volume. By that point, however, the burden of disease in the United States and other developed economies had shifted: cardiovascular disease and cancer had displaced infections as the leading causes of death.[66] Medical scientists took on these new challenges, supported by major investments in health care and research (e.g., the National Cancer Institute, the National Heart Institute). By the early 2000s, physicians could celebrate dramatic successes against coronary artery disease (e.g., diuretics, beta-blockers, ACE inhibitors, statins, bypass surgery, angioplasty, antismoking campaigns) and cancer (e.g., cytotoxic chemotherapy, surgery, radiation therapy, targeted chemotherapies). The burden of disease, however, continues to shift, with neuropsychiatric conditions rising to new prominence (e.g., depression, dementias, substance abuse). Medical science and public health will hopefully produce solutions to these conditions, but the burden of disease will surely shift once again.

A second Red Queen effect has played out in parallel. Just as medical and public health practitioners and institutions have struggled to keep pace with the changing burden of disease, clinical researchers have struggled to produce knowledge of therapeutic efficacy that keeps up with changing therapeutic practice. Definitive assessment of efficacy often requires long-term follow up (e.g., three- or five-year survival). Clinical trials that assess such outcomes necessarily last many years; design, patient recruitment, implementation, follow-up, and analysis all take significant time to complete. Trial outcomes often are not published until five to ten years after the design of the intervention protocol. Are the ensuing results relevant? It depends on assumptions about therapeutic evolution. If you believe, as many patients and doctors do, that treatments improve over time, then a trial's results are undermined before they are even published. They reflect treatment as it existed ten years earlier, in an ancestral—and more primitive—form.

Consider the trials of coronary angioplasty. By the mid-1990s angioplasty had become a routine treatment for stable coronary disease, even though there was little convincing evidence that it added value beyond optimal medical therapy. To produce decisive data, investigators from fifty sites designed the COURAGE trial to detect any incremental benefits provided by angioplasty.[67] They enrolled 2,287 patients between June 1999 and January 2004, and followed them through June 2006. Over a mean follow-up of 4.6 years, they found no significant differences in rates of death, heart attack, or hospitalization for acute coronary syndromes. This study, published in March 2007 in the *New England Journal of Medicine*, was trumpeted in the press as a "blockbuster." Shares of Boston Scientific, a leading stent manufacturer, fell, and stent use dropped 10 percent within a month.[68] Supporters of angioplasty rushed to the procedure's defense. Since enrollment began in 1999, most COURAGE patients (97.7 percent) received bare metal stents. In 2003, however, drug-eluting stents designed to prevent restenosis became available in the United States.[69] Most cardiologists assumed that the new stents would outperform the old ones. As a result, "one could very reasonably hypothesize" that the outcomes of COURAGE would have been better had drug-eluting stents been used.[70] And since drug-eluting stents had already come to dominate the marketplace, critics argued that COURAGE was obsolete on arrival. Its negative results need not diminish enthusiasm for the variants in current use. The evidence base, always running, can never catch up.

Extinction

Most species that have ever existed have gone extinct.[71] The same holds true in medicine. Many once-popular therapies have vanished, with competition probably the most common cause of extinction. When chlorpromazine appeared in the mid-1950s, lobotomy was made "redundant" and went extinct.[72] Chlorpromazine and other "typical" antipsychotics have since been driven close to extinction by newer (and heavily marketed) "atypical" antipsychotics. Sometimes a new competitor wipes out whole lineages. In the 1960s, surgeons used many different approaches to coronary revascularization; nearly all of those approaches disappeared with the emergence of bypass surgery in 1968.[73] Changes in the niche can be important as well. As Kehr and Condrau describe in chapter 5 of this volume, the decline of tuberculosis in the United States and Europe eliminated the need for rest cures, sanatoria, thoracoplasty, and a host of other once-popular interventions. Smallpox vaccine sowed the seeds of its own demise by eradicating its own niche. If enough individuals find ways to control coronary disease through lifestyle and prevention, then bypass surgery and countless other treatments might disappear as well.

While studies of the extinction of specific treatments can be productive, historians can also follow the lead of evolutionary biology and look at broader patterns in therapeutic evolution. Macroevolutionary theorists have examined the lineages of thousands of species to discern how rates of speciation and extinction have changed over time. Innovations in life forms—for instance, the development of multicellular organisms or the movement of plant and animal life from oceans to land—have led to the rapid emergence of new species. The astonishing diversity of Cambrian-era organisms found in the Burgess Shale provides perhaps the most important example, while the diversification of finches in the Galapagos Islands is the most famous. Extinctions often follow, as competition selects the best adapted organisms. Does something similar happen in medicine? Have periods of massive therapeutic proliferation, whether during the antibiotic revolution in the 1950s or the proliferation of angioplasty devices in the 1980s, been followed by periods of therapeutic mass extinction, as competition winnows out unfit therapies? It is necessary to organize the data of therapeutic evolution before one can see its patterns.

Taxonomy

Scholars in many fields, confronted with large data sets, have sought ways to organize them. In natural history this became the science of taxonomy. Taxonomy is not simply about description and sorting. It requires that arguments be made about affinity: Which things are most closely related? Taxonomists have long debated the merits of taxonomies based on morphology or genealogy.[74] This distinction is relevant in medicine as well. Doctors can classify diseases according to organ system or etiology, but ambiguities always persist. Does it make sense to define a category of pneumonia without regard to whether it is caused by staph or strep, or do strep infections form the "natural kind" regardless of whether they strike lung, throat, or skin?[75] The situation is different for classifying therapies. Many writers, especially in review articles and textbooks, offer typological classifications of medications. Psychiatric drugs can be divided into antidepressants, antipsychotics, mood stabilizers, and anxiolytics. Antihypertensives can be divided into diuretics, vasodilators, beta-blockers, calcium channel blockers, ACE inhibitors, angiotensin receptor blockers, and presumably others yet to come. But treatments, like species, have evolved over time. This makes it possible for physicians and historians to produce therapeutic genealogies. The different ways of classifying raise important questions for historian of medicine.

First, medical taxonomies, like biological taxonomies, have changed over time as medical knowledge has changed and as doctors have made new claims about affinity. Taxonomies of fever have changed with the rise of germ theory.[76] The classification of substance use has swung between vice and disease.[77] The shifts can be abrupt, especially when a bureaucratic power imposes a new taxonomic order. In 1892, for instance, the Department of the Interior issued new rules for physicians who worked on Indian reservations.[78] Consumption, which in 1891 had been a constitutional disease, along with cancer, anemia, dropsy, and rheumatism, now became tuberculosis, an infectious disease, like chicken pox, diphtheria, measles, and influenza. Theorists of cartography have long argued that maps are not simply descriptions of geographic space, but instead are arguments, the product of strategic decisions about what data to represent and how to represent them.[79] Taxonomies function similarly, making arguments about the affinity, etiology, or genealogy of diseases or therapeutics.

Second, the superimposition of genealogy on top of typological taxonomy reveals important boundary crossings in the history of therapeutics. The historian Walter Sneader, for instance, has used evolutionary taxonomy to organize knowledge of pharmacology and trace its history in his "genealogical approach to drug discovery."[80] Some lineages develop methodically, with all progeny staying within the same therapeutic class as the prototype. Penicillin gave rise to many generations of antibiotics, selected (designed) to be long-acting (e.g., procaine penicillin), resistant to penicillinases (e.g., methicillin), broad-spectrum (e.g., ampicillin), or orally absorbed (e.g., amoxicillin).[81] Other lineages are full of surprises. Consider the descendants of epinephrine. Analogs (i.e., adrenergic agonists such as albuterol) remain a mainstay of asthma therapy. Antagonists (i.e., beta-blockers such as propranolol), developed to protect the heart against adrenaline surges, proved useful not just for coronary artery disease but also for hypertension. Some researchers developed derivatives with less neurotoxicity (e.g., atenolol) to make hypertension regimens more tolerable. Other researchers, intrigued by the vivid dreams produced by lipophilic beta-blockers, sought more psychoactive derivatives, a pursuit that yielded the serotonin and norepinephrine reuptake inhibitors that have transformed the treatment of depression.[82] Many other pharmaceutical lineages have jumped across functional classes. Antimalarials produced antihistamines, and then antipsychotics.[83] B-vitamins gave rise to drugs for tuberculosis (e.g., isoniazid) and depression (e.g., iproniazid and other monoamine oxidase inhibitors).[84]

The ways in which drug lineages transgress therapeutic class reveal not just the complexity of pharmacology (e.g., the subtlety of drug-receptor interactions), but also the important role of serendipity. Researchers who develop derivatives for one purpose often stumble across drugs useful for another purpose. This resembles the processes of exaptation described by biologists. Just as feathers likely evolved as insulation before they enabled flight, drug derivatives often find unanticipated applications.

Similar processes take place in surgery. Between 1920 and 1970, surgeons developed a bewildering diversity of surgical procedures to treat coronary artery disease. Sometimes a lineage preserved its function even as its form changed completely. For instance, techniques used to slow the body's metabolism by reducing thyroid function evolved from surgical resection of the thyroid in the 1930s to destruction of thyroid tissue with radioactive iodine in the 1950s. Exaptation has been common, with techniques developed in one area of surgery (e.g., saphenous vein interposition grafts to repair renal artery stenosis) finding application elsewhere (e.g., for coronary artery disease). Once coronary artery bypass surgery achieved a foothold in its niche, it underwent adaptive radiation and gave rise to many variants, including recent attempts at minimally invasive procedures. The adaptive radiation of the angioplasty lineage has been even more dramatic (and profitable), with balloon techniques giving rise to atherectomy, laser ablation, stents, and many others.

Questions of lineage and taxonomy often become relevant for policy. How much change can accumulate in a therapeutic lineage while preserving functional identity? When is new evidence and regulatory oversight required to ensure that the treatment still works as its predecessors did? According to the 1976 Medical Device Amendment, a new device can be approved expeditiously if it is substantially equivalent to an existing device. This policy—called the 510(k) process—has since been extensively exploited by device manufacturers. One analysis of artificial hip implants included a branching tree diagram that traced the genealogical relations between sixty-three current implants and their ancestral forms.[85] The authors argued, despite serial claims of substantial equivalence, that significant changes had accumulated in the lineage over its many generations, and that these required new regulatory oversight. At what point has speciation, and thus the need for renewed regulatory scrutiny, taken place? It is not always clear. Generic drugs raise similar questions. What kinds of similarity produce sufficient taxonomic affinity such that a generic drug can be assumed to be therapeutically—and bureaucratically—interchangeable with the parent drug? As Jeremy A. Greene has shown, distinctions are made not just on the structure of the active ingredient, but also on the binders and fillers that might affect bioequivalence, and on the shapes, colors, and tastes that might affect pill-taking behavior.[86]

Island Biogeography

Taxonomies raise questions not just about change over time, but also about the distribution of diversity over space. For instance, evolutionary theorists have studied how variation emerges in geographically isolated populations ever since Darwin's famous voyage to the Galapagos Islands. As local varieties emerge, the isolated locales become sites for speciation. These intuitions were formalized in 1967 by the evolutionary biologists Robert MacArthur and E. O. Wilson in their analyses of how so many species can exist on islands. Subsequent work has examined the ways in which islands become sources of novelty (i.e., speciation). Sometimes a new species forms when an existing species expands to occupy an open niche, subsequently splitting into two. At other times, new species form when a geographic or behavioral barrier divides a group into two diversifying lineages.[87] The combination of isolation and small population size contributes to the rapid pace of change.

Medical geographers and historians have long wondered about the distribution of disease, and especially about the dynamics that influence the emergence of new pathogens in isolated regions and their potential dissemination.[88] The island biogeography of medical practice deserves similar attention. In chapter 6 of this volume, Jeremy A. Greene explores the significance of geographic variations in drug availability and pricing. A distinct literature exists about practice variation in surgery. From J. Alison Glover's 1938 description of a twenty-seven-fold disparity in tonsillectomy rates across London neighborhoods to the colorful maps of the *Dartmouth Atlas of Health Care* today, physicians have mapped striking disparities in medical practice between hospitals, cities, regions, and nations.[89] As John Wennberg and Alan Gittelsohn concluded in 1975, geographic variations in medical practice "are a rule for which there is yet no exception."[90] If practice variation simply reflected variation in the underlying burden of disease (i.e., if there were a perfect correlation between the biogeography of disease and the biogeography of medical practice), then it would not be interesting. However, an extensive body of research by physicians has concluded that much of the variation appears to be "unwarranted," reflecting not the application of evidence-based medicine to local burdens of disease, but rather the influence of physician supply, reimbursement practices, financial conflicts of interest, medical uncertainty, idiosyncratic differences in physicians' beliefs and practices, and myriad other influences on medical decision making. Health policy experts have long seen the existence of unwarranted variation as a problem. As Frederick Robbins, president of the Institute of Medicine, wrote in 1983, "It looks bad, and it looks bad because it is bad. It is not an appropriate way for a profession to behave."[91] Physicians and analysts have worked to identify the causes of unwarranted variation and to purge it from medicine.

Historians can offer different perspectives. The first is epistemological: Why did physicians become concerned about geographic variations when they did? The variations have existed for centuries.[92] When Glover identified them in 1938, his work triggered no interest in the problem. It was only in the 1960s and 1970s that the problem received attention in the United States, in the setting of two developments: concern about the skyrocketing costs of health care, and the emergence of evidence-based medicine.[93] It is not difficult to understand why the documentation of unwarranted variation has been an affront to the aspirations of evidence-based medicine. Advocates of this movement have sought to discipline medical practice and bring it into conformity with the dictates of clinical data. Historians can contribute to this endeavor, for instance, by helping to chart the forces that pull medical practice out of alignment with evidence-based medicine. They can also choose to complicate the endeavor. Is it plausible that medicine could ever be a fully rational science, isolated from social, economic, and political influences? Few historians think this likely. Their analyses of historical contingency and the importance of local context can reveal the inevitable limits of evidence-based medicine.

The second perspective turns the problem of geographic variation into an opportunity. Historians, informed by biologists' theories of island biogeography, could argue that local variation in medical practice is actually a good thing. Isolation and local variation have produced new traits and species in organismic evolution. Something similar has played out in the history of medicine. Different physicians and health care institutions have developed different approaches to particular clinical problems. Ideally, doctors share and compare practices and contribute to medical progress. Aseptic surgery first developed in a particular late-nineteenth-century German surgical culture, and then spread widely.[94] Directly observed therapy, developed to improve compliance with outpatient tuberculosis regimens in Madras in the 1950s, became a mainstay for treatment of many diseases in many places.[95] But these are the best-case scenarios. There has never been an efficient system that evaluates different local practices and determines whether one really is better than another. This is, of course, the nature of island biogeography. The barriers to exchange—physical, cultural, or otherwise—that foster local variation and innovation can also impede their dissemination.

Morphospace

One last concept is particularly thought-provoking. As Hutchinson formulated his niche theory in 1957, he realized that a niche was defined not just by two or three features of the environment and organism, but by innumerable factors. It was not simply a three-dimensional space, like an architectural niche, but an "n-dimensional hypervolume . . . every point in which corresponds to a state of the environment which would permit the species S1 to exist indefinitely."[96] This concept of the niche as a multidimensional hypervolume inspired a secondary idea, that of an n-dimensional trait space. As Steven Jay Gould wrote in 1991, "morphospace" represents the "full range of the abstract (and richly multivariate) space into which all organisms may fit."[97] Any creature, real or imagined, occupies just a small patch. Conceptualized this way, morphospace presented Gould and his fellow biologists with a challenge: "We need to measure density, range, clumping, and a host of other properties that determine differential filling of this totality; and we must be able to assess the variation in this differential filling through time."[98]

Morphospace provides evolutionary biologists with a teachable moment about contingency and developmental constraints. Large tracts of morphospace, once occupied, are now empty (e.g., trilobites, dinosaurs), the contingent result of meteor strikes and other causes of mass extinctions. But most morphospace has never been occupied. If you imagine every possible form a living creature could take (photosynthetic elephants! winged horses! dragons!), you quickly realize that most of these things have never existed. There are no six-limbed vertebrates. There are no talking horses. Instead, you find isolated clusters of creatures, with vertebrates in one region, crustaceans in another, trees someplace else, and an enormous—but still finite—cloud of bacteria. The lesson here is about constraint. Evolution works with a limited substrate: extant species. Since embryological development imposes constraints on how much one generation can vary from its parents, new species cluster near existing species and only slowly move into unfilled space. There is a wide gulf between realized and potential creatures.

Morphospace provides historians of medicine with two useful thought experiments. Thinking about disease space (pathospace?) is simple enough at first: it is the task of nosology and disease taxonomy. However, as you define the possible axes of disease space to capture every type of disease that does exist, and begin to wonder about every type of disease that might exist, it quickly becomes an exercise in morbid imagination, one pursued enthusiastically in horror films and science fiction. Zombie viruses are simply the most recent in a long line of appalling imagined diseases. Fiction aside, disease space raises an important question about the social determinants of disease: to what extent do we control which swathes of disease space are occupied? Many diseases exist now because of decisions people have made about how to structure their societies, from smoking-related illnesses to obesity, substance abuse, lead poisoning, and car accidents. Our hunter-gatherer ancestors were presumably spared these diseases. What about our descendants? It is possible to imagine a world free of lung cancer, bronchitis, and emphysema. If tobacco use ceased, those diseases would almost certainly slip back into the domain of diseases that could be imagined but do not actually exist.

The thought experiment is even more productive with therapeutics. Imagine an n-dimensional trait space for medical interventions; not just a pharmacospace or a surgerispace, but a therapospace, a remedispace—an iatrospace. The dimensions would allow the full range of conceivable interventions (pharmaceutical, surgical, interactional, natural, synthetic, magical, religious, specific, universal, etc.) for every possible disease. Within this iatrospace could be found actual treatments that do exist, abandoned treatments that once were popular, and ideal future treatments towards which medical research strives: magic bullets for cancer, drugs that reverse dementia, a vaccine for HIV, or an electromagnetic wand that dispels depression. As patients and doctors know too well, existing treatments occupy but a tiny fraction of potential iatrospace. The history of these shortcomings is in part a history of constraint. There are limits on what surgery can accomplish, and even though thousands of biologically active compounds have been tested, it has not been possible to find a perfect drug for every clinical problem. Furthermore, just as natural selection can only work with existing species, doctors largely use existing treatments to produce subsequent incremental derivatives.

But, unlike in biology, physicians can influence how iatrospace gets filled. They can consciously imagine the space of potential therapeutics, recognize gaps that exist, and work to fill them. Rational drug design, one of the many promissory sciences of contemporary biomedicine, demonstrates this well. As doctors characterize the mechanisms of disease in ever-increasing detail and improve the resolution of their map of the n-dimensional volume of disease space, they identify new destinations in iatrospace. Advances in cancer science have allowed doctors to move beyond surgical resection to cytotoxic chemotherapy, radiation, and now targeted kinase inhibitors. While there have been a few dramatic successes, many promising areas of iatrospace have not been reached. This model can help understand therapeutic failure as well. Psychiatrists, for instance, do not yet have a detailed enough map of psychiatric disease space to identify specific targets for therapeutic intervention. It might even be possible to construct a taxonomy of medical practice according to the barriers to a total eclipse of different segments of the burden of disease. In some areas, as in psychiatry, the problem is our understanding of disease space. In others, as is increasingly the case in oncology or infectious disease, the challenge is finding an actual molecule that performs a well-characterized function within iatrospace.

These abstractions of n-dimensional hypervolumes, of disease space or iatrospace, bring together different threads of evolutionary theory. They provide domains in which not only niches but also taxonomy, fitness, extinction, adaptive radiation, and many other processes play out. While evolutionary biology remains a distant analogy for the development of medical theory and practice, the theories of evolutionary biology can inspire productive theorizing within history of medicine.

The Problem of Progress

Historians of medicine can adapt theories and metaphors from evolutionary biology and develop new modes of description, new arguments about causation, and new perspectives on the dynamics of change over time. But they must think carefully if they do so. Is the analogy specific enough for evolutionary theory to add real value when applied to nonbiological systems? Can our understanding of efficacy really be enhanced by insights about fitness or the therapeutic niche? The rhetoric of evolution, like that of revolution, requires careful handling by historians of medicine. It is important to think not just about the potential creative insights it offers, but also about the potential downsides of evolutionary concepts. The most relevant dilemma with evolution for historians of medicine, as with revolution, is the problem of progress.

Progress has long been associated with the varied meanings of evolution. "Progress" entered English from Latin in the fifteenth century, to mean a step forward, as on a march or journey. The movement was not necessarily positive, as seen in the usage (which continues) of "the progress of a disease."[99] Through an association with "evolution," however, "progress" gradually gained the meaning of movement from worse to better, first as "an inherent principle of development of higher forms," and then more broadly to "an inherent process of social and historical improvement."[100] Most eighteenth- and nineteenth-century writers saw progress in idealist terms, though some became increasingly concerned about the costs of progress.

The association of evolution with progress has long been a bugaboo for biologists.[101] Traditional evolutionary thought assumed that evolution brought progress, as is seen in ubiquitous imagery of the great chain of being. It is true that there are creatures living today that are more complex than the most complex creatures two billion years ago, and it is unlikely that anyone living now would trade their human existence for that of a unicellular critter from eons past. Nonetheless, the scientific literature now takes a much more nuanced approach to progress. Phylogenetic lineages are full of dead ends. Some species lose functions over time (e.g., eyeless cave fish). A trait might satisfy a local selective pressure and proliferate, but decrease the fitness of a species in the long run (e.g., possibly the giant antlers of the Irish elk). Mass extinctions occurred repeatedly, with lineages vanishing sometimes for explicable causes and sometimes seemingly at random. At a global scale, evolution has actually maintained something of a status quo: if you plot complexity on the x-axis, and the number of species achieving that level of complexity on the y-axis, the median organism on earth for billions of years has always been, and still remains, a bacterium.[102] Nothing about natural selection or ecological dynamics, as now understood, necessitates progress.

Progress has been a similar problem in history of medicine, even among writers who would not self-identify as Whigs. Osler, Garrison, and many more recent historians have celebrated the progress of medicine. When doctors talk about treatments, practices, and institutions evolving, a sense of progress is part of this discourse. The assumption is that the new is better than the old, with evolution producing ever better understandings and interventions. In the 1960s, however, some historians of medicine turned away from these positivist assumptions and towards meta-narratives of relativism, skepticism, and critique. But progress is hard to set aside. Just as no one would want to live the life of an archaic bacterium, there are few who would choose to give up modern medical technology and live with medicine as it existed even fifty, let alone one or two hundred years ago. Historians have tried to find a balance by acknowledging the possibility of progress without accepting its inevitability.

Historians of medicine who are attuned to assumptions of evolutionary progress can offer perspective on progress in the medical literature. Physicians often deploy several different rhetorics of progress to generate faith and enthusiasm in new therapies—and to discount the need for scholarly or regulatory scrutiny. In some cases, they accentuate the merits of a break from the past. For instance, when coronary artery bypass grafting was launched in the late 1960s, it was the latest in a long series of surgical attempts to treat coronary artery disease. Since prior techniques had ended in disillusionment, skeptics often assumed that the new operation would be no different. They demanded that bypass surgery be subjected to rigorous trials. Surgeons did not deny this history; rather, they denied its relevance. They argued that past surgical treatments had failed because they had relied on inadequate diagnostic technology. The advent of coronary angiography in the 1960s, however, allowed surgeons to visualize the coronary arteries before making a decision about surgical intervention—a "leap forward in our ability to read coronary disease that can be fairly likened to the impact of the invention of the printing press on the written word."[103] This diagnostic revolution ruptured any kind of historical continuity. As the surgeon Donald Effler explained, "Whatever surgical efforts were expended before are of historical interest only, and it does little good to dwell on past failures."[104]

In other cases, doctors place their emphasis on gradual progress. A physician might develop a variant on an existing treatment and make a claim of incremental, evolutionary progress: the new is similar enough to the old, but improved, so that it should be trusted at the outset. This strategy allows doctors to tweak the dose of an approved regimen or adjust an operation in an attempt to make it safer, quicker, cheaper, or more effective. As long as everyone assumes that the tweak is positive, then there is no need for new clinical trials or regulatory review. For instance, just as the Food and Drug Administration allows expedited approval if a new device is substantially equivalent to an existing device, it also grants the benefit of the doubt if the device involves "incremental innovations" of an existing device.[105] Is this wise? It depends on assumptions of progress. Device manufacturers argue that if the first device was safe and effective, then their slightly improved devices should be safe and effective as well, and hopefully more so. This intuition has worked well in many instances: incremental change has allowed for the safe flourishing of numerous medical devices and operative procedures. But device companies have now spawned so many generations of derivatives that some new devices bear little resemblance to the distant ancestors on which their approval relied, and many have been approved without specific clinical evidence. Consider the implantable devices that are used to control cardiac arrhythmias. Between 1979 and 2012, the Food and Drug Administration granted 77 formal premarket approvals and an additional 5,829 supplements, 37 percent of which involved a change in design.[106] Several of these devices failed—a consequence of unfulfilled assumptions of progress.

The challenge for historians is to use the language and theories of evolution skillfully. Evolutionary language can certainly imbue historical writing with assumptions of progress, just as assumptions of progress still pervade popular understandings of organismic evolution. However, biologists have learned to disentangle evolution and progress and tell stories about the multiple possible outcomes of evolution. Historians should also be able to invoke medical evolution (or revolution) and simultaneously subject the question of progress to the scrutiny it requires.

Evolution or Revolution?

Physicians, patients, and historians share an interest in the dynamics of medical change. Physicians and patients want rapid progress. Historians want to understand the dynamics and causes of change (and, when they get sick, most hope that medical science has progressed). The rhetoric of revolution holds much appeal for physicians celebrating an innovation or for historians drawing attention to the importance of their object of study. A claim of revolution is a demand for attention. However, as Roy Porter warned, historians must take care not to be drawn into the drama and overstate the claim. The chapters in this volume provide a nuanced view of the subtleties and stakes of revolutionary claims. What about the opposing metaphor, of evolution? The rhetoric of evolution also looks to progressive improvement, but with reassuring gradualism in place of frightening rupture. If revolutionary change satisfies those who are dissatisfied with existing practice and want something fundamentally new, then evolutionary change reassures those who want gradual improvement of existing practice.

Historians need not adjudicate whether evolution or revolution is better. Instead, they can make two important contributions. First, they can mine scholarship on revolution and evolution, whether from political science or biology, to develop tools to refine our understanding of the past. Porter defined strict standards for revolution (i.e., a self-conscious overthrow of an existing scientific orthodoxy) and used them to characterize purported scientific revolutions. Historians can adapt concepts of evolution to analyze and understand change over time. Second, they can attend closely to language and its connotations. Whether the model is evolution or revolution, one core consequence seems to be the same: the expectation of a better future. However, there is nothing inherent in the theory of either evolution or revolution that ensures progress. In fact, there is much in the dynamics of evolution—whether of niches, competition, Red Queen effects, or morphospace—that argues against progress. While progress is a possible outcome of organismic evolution, it is not an inevitable one. When it takes place, it requires specific explanation. The same holds true for medicine and its history.

NOTES

1. René G. Favaloro, "Saphenous Vein Autograft Replacement of Severe Segmental Coronary Artery Occlusion," *Annals of Thoracic Surgery* 5 (1968): 334–339.
2. Favoloro, quoted in *THI Today* (December 1985): 2, in John P. McGovern Historical Collections and Research Center (Houston Academy of Medicine), Institutional Collection #43 (Texas Heart Institute), box 2, folder "THI Today, 1985."
3. Favaloro, "Oral History" (3 March 1997), in William S. Stoney, ed., *Pioneers of Cardiac Surgery* (Nashville: Vanderbilt University Press, 2008), 357–368, quote on 364.
4. Charles Rosenberg, "The Therapeutic Revolution: Medicine, Meaning, and Social Change in Nineteenth-Century America," *Perspectives in Biology and Medicine* 20 (1977): 485–506.
5. David S. Jones, "The Prospects of Personalized Medicine," in Sheldon Krimsky and Jeremy Gruber, eds., *Genetic Explanation: Sense and Nonsense* (Cambridge, MA: Harvard University Press, 2013), 147–170; Reza Mirnezami, Jeremy Nicholson, and Ara Darzi, "Preparing for Precision Medicine," *New England Journal of Medicine* 366 (2012): 489–491.

6. Burton E. Sobel, "The Structure of Cardiological Revolutions," *Circulation* 87 (1993): 2047–2054. Sobel described three revolutions: a "social revolution" that led to declining prestige of the profession, the revolution of interventional cardiology, and the revolution of molecular and cellular biology.

7. Louis Faugères Bishop and John Neilson, *History of Cardiology* (New York: Medical Life Press, 1927), 71.

8. Peter Libby, Paul M. Ridker, and Attilio Maseri, "Inflammation and Atherosclerosis," *Circulation* 105 (2002): 1135–1143.

9. Floyd D. Loop, Delos M. Cosgrove, Bruce W. Lytle, Robert L. Thurer, Conrad Simpfendorfer, Paul C. Taylor, and William L. Proudfit, "An 11-Year Evolution of Coronary Arterial Surgery," *Annals of Surgery* 190 (October 1979): 444–454.

10. A. S. Keats, "Evolution of Anesthesia for Cardiac Surgery," *Cleveland Clinic Quarterly* 48 (1981): 75–79.

11. Ellis J. Jones, William S. Weintraub, Joseph M. Carver, Robert A. Guyton, and Caryn L. Cohen, "Coronary Bypass Surgery: Is the Operation Different Today?" *Journal of Thoracic and Cardiovascular Surgery* 101 (1991): 108–115.

12. Anthony P. Fletcher, Sol Sherry, Norma Alkjaersig, Fotios E. Smyrniotis, and Sidney Jick, "The Maintenance of a Sustained Thrombolytic State in Man," *Journal of Clinical Investigation* 38 (1959): 1111–1119.

13. Michael E. DeBakey, "Changing Concepts in Thoracic and Vascular Surgery," *Journal of Thoracic and Cardiovascular Surgery* 38 (1959): 145–165.

14. Michael Mack, Ralph Damiano, Robert Matheny, Hermann Reichenspurner, and Alain Carpentier, "Inertia of Success: A Response to Minimally Invasive Coronary Bypass: A Dissenting Opinion," *Circulation* 99 (1999): 1404–1406.

15. Jones et al., "Coronary Bypass Surgery."

16. F. Trojette, A. Benamar, S. Beloucif, D. Foure, H.J. Poulain, "Clinical Experience with the Mini-extracorporeal Circulation System: An Evolution or a Revolution?" *Annals of Thoracic Surgery* 77 (2004): 2172–2176; P. Rehfield, C. Kopes-Kerr, and M. Clearfield, "The Evolution or Revolution of Statin Therapy in Primary Prevention: Where Do We Go from Here?" *Current Atherosclerosis Reports* 15 (2013): 298; C. Lee, C. S. Barroso, and P. J. Troped, "Endovascular Aneurysm Sealing for Abdominal Aortic Aneurysm Repair: Evolution or Revolution?" *Cardiovascular and Interventional Radiology* 37 (October 2014): 1129–1136.

17. J. R. T. C. Roelandt, I. R. Thomson, W. B. Vletter, P. Brommersma, N. Bom, and D. T. Linker, "Multiplane Transesophageal Echocardiography: Latest Evolution in an Imaging Revolution," *Journal of the American Society of Echocardiography* 5 (1992): 361–367.

18. William Osler, *The Evolution of Medicine* (New Haven: Yale University Press, 1922), 219.

THERAPEUTIC EVOLUTION OR REVOLUTION?

19. Fielding H. Garrison, preface to Osler, *The Evolution of Medicine*, xiii.
20. Harry Keil, "The Evolution of the Term Chancre and Its Relation to the History of Syphilis," *Journal of the History of Medicine and Allied Sciences* 4 (1949): 407–416; Gerald L. Geison, "Darwin and Heredity: The Evolution of His Hypothesis of Pangenesis," *Journal of the History of Medicine and Allied Sciences* 4 (1969): 375–411; Abraham M. Lilienfeld, "Ceteris Paribus: The Evolution of the Clinical Trial," *Bulletin of the History of Medicine* 56 (1982): 1–18; Samuel H. Greenblatt, "Harvey Cushing's Paradigmatic Contribution to Neurosurgery and the Evolution of His Thoughts about Specialization," *Bulletin of the History of Medicine* 77 (2003): 789–822.
21. Noel Gillespie, "The Evolution of Endotracheal Anaesthesia," *Journal of the History of Medicine and Allied Sciences* 1 (1946): 583–594; R. K. Blach, "Prophylactic Enucleation in Sympathetic Ophthalmitis: The Evolution of an Heroic Form of Treatment," *Medical History* 15 (1971): 190–192; Andrew Davies, "The Evolution of Bronchial Casts," *Medical History* 17 (1973): 386–391; James R. Wright, "The Development of the Frozen Section Technique, the Evolution of Surgical Biopsy, and the Origins of Surgical Pathology," *Bulletin of the History of Medicine* 59 (1985): 295–326.
22. H. P. Tait, "Health Services in India and Burma: Their Evolution and Present Status," *Medical History* 16 (1972): 184–193; W. Bruce Fye, "The Origins and Evolution of the Mayo Clinic from 1864 to 1939: A Minnesota Family Practice Becomes an International 'Medical Mecca,'" *Bulletin of the History of Medicine* 84 (2010): 323–357.
23. One essay, for instance, on the evolution of the concept of febrile seizures describes how "beginning in the mid-nineteenth century and continuing as a gradual process to the present, this thinking has changed dramatically." See John W. Gardner and Robert C. Dinsmore, "Evolution of the Concept of the Febrile Seizure as It Developed in the American Medical Literature, 1800–1980," *Journal of the History of Medicine and Allied Sciences* 50 (1995): 340–363, quote on 341.
24. "Evolution," *Oxford English Dictionary,* 3rd edition (March 2008), available at www.oed.com; Raymond Williams, "Evolution," in *Keywords: A Vocabulary of Culture and Society*, revised edition (New York: Oxford University Press, 1983), 120–123; Robert J. Richards, "Evolution," in *Keywords in Evolutionary Biology*, ed. Keller and Elisabeth A. Lloyd (Cambridge, MA: Harvard University Press, 1992), 95–105, on 95; Richard C. Lewontin, "Organism and Environment," in *Learning, Development, and Culture*, ed. H. C. Plotkin (New York: John Wiley & Sons, 1982), 151–170, especially 152–156.
25. "Revolution," *Oxford English Dictionary,* 3rd edition (March 2010), available at www.oed.com; Raymond Williams, "Revolution," in *Keywords*, 270–274.
26. Williams, "Evolution," 122; Williams, "Revolution," 273.
27. Roy Porter, "The Scientific Revolution: A Spoke in the Wheel?" in *Revolution in History*, ed. Porter and Mikuláš Teich (Cambridge: Cambridge University Press, 1986), 290–316.

295

28. Porter, "The Scientific Revolution," 300. He writes: "Is it helpful to picture the course of the history of science as revolutionary? Or might it not make better sense to stress its 'evolutionary' aspects, its continuities and accommodation to the wider socio-intellectual environment? These large questions matter, not least because, with the irresistible rise of specialization, scholarship becomes myopic and fragmented" (300).

29. Ibid., 300.

30. Ibid., 308.

31. Ibid., 309.

32. Richard Lewontin, conversation with the author, 23 May 2011.

33. For one revealing exchange, see Stephen J. Gould, "Darwinian Fundamentalism," *New York Review of Books* 44 (12 June 1997): 34–37; Steven Pinker, with a reply by Gould, "Evolutionary Psychology: An Exchange," *New York Review of Books* 44 (9 October 1997).

34. Jack Pressman, *Last Resort: Psychosurgery and the Limits of Medicine* (Cambridge: Cambridge University Press, 1998).

35. Ibid., 14; see also 160.

36. Ibid., 190.

37. Ian Hacking, *Mad Travelers: Reflections on the Reality of Transient Mental Illness* (Charlottesville, VA: University of Virginia Press, 1998), 13.

38. Ibid., 82.

39. David Healy, *The Antidepressant Era* (Cambridge, MA: Harvard University Press, 1997); Jeremy A. Greene, *Prescribing by Numbers: Drugs and the Definition of Disease* (Baltimore: Johns Hopkins University Press, 2007).

40. Ray Moynihan and David Healy, "The Fight against Disease Mongering: Generating Knowledge for Action," *PLoS Medicine* 3 (2006): e191.

41. Hacking, *Mad Travelers*, 101.

42. James R. Griesemer, "Niche: Historical Perspectives," in *Keywords in Evolutionary Biology*, 230–240. Caroline Jones, at the Massachusetts Institute of Technology, alerted me to this history. Mark Jarzombek to Caroline Jones, 24 June 2009, e-mail shared with author.

43. "Niche," *Oxford English Dictionary*, available at www.oed.com. For the evolution of aedicula in medieval cathedral architecture, see John Summerson, "Heavenly Mansions: An Interpretation of Gothic," in *Heavenly Mansions and Other Essays on Architecture* (London: Cresset Press, 1949): 1–28.

44. *Online Etymology Dictionary*, available at http://www.etymonline.com.

45. Griesemer, "Niche."

46. Griesemer, "Niche," 238–239; Robert K. Colwell, "Niche: A Bifurcation in the Conceptual Lineage of the Term," in *Keywords in Evolutionary Biology*, 241–248.

47. Richard Lewontin, "Adaptation," *Scientific American* 239 (September 1978): 212–230, quote on 215. See also F. John Odling-Smee, Kevin N. Laland, and Marcus W. Feldman, "Niche Construction," *American Naturalist* 147

(1996): 641–648. As they explain, "the idea here, in retrospect, is obvious: Organisms, through their metabolism, their activities, and their choices, define, partly create, and partly destroy their own niches" (641).

48. Richard C. Lewontin, "Organism and Environment," in *Learning, Development, and Culture*, ed. H. C. Plotkin (New York: John Wiley & Sons, 1982), 151–170, quote on 163.

49. Richard Lewontin, "Gene, Organism, Environment," in *Evolution from Molecules to Men*, ed. D. S. Bendall (Cambridge: Cambridge University Press, 1983), 273–285; Clive G. Jones, John H. Lawton, and Moshe Shachak, "Positive and Negative Effects of Organisms as Physical Ecosystem Engineers," *Ecology* 78 (1997): 1946–1957; F. John Odling-Smee, Kevin N. Laland, and Marcus W. Feldman, "Niche Construction," *American Naturalist* 147 (1996): 641–648; K. N. Laland, F. J. Odling-Smee, and M. W. Feldman, "The Evolutionary Consequences of Niche Construction: A Theoretical Investigation Using Two-Locus Theory," *Journal of Evolutionary Biology* 9 (1996): 293–316. For a discussion of how niche construction can be applied to the human social sciences (e.g., human niche construction, social learning, cultural inheritance, etc.) and a vigorous debate about that approach, see Kevin N. Laland, John Odling-Smee, Marcus W. Feldman, and commentators, "Niche Construction, Biological Evolution, and Cultural Change," *Behavioral and Brain Sciences* 23 (2000): 131–175.

50. Robert Bud, *Penicillin: Triumph and Tragedy* (New York: Oxford University Press, 2007); Scott Podolsky, *The Antibiotic Era: Reform, Resistance, and the Pursuit of a Rational Therapeutics* (Baltimore: Johns Hopkins University Press, 2014).

51. Chris Feudtner, *Bittersweet: Diabetes, Insulin, and the Transformation of Illness* (Chapel Hill: University of North Carolina Press, 2003).

52. Rosenberg, "The Therapeutic Revolution," 1977.

53. Diane Paul, "Fitness: Historical Perspectives," in *Keywords in Evolutionary Biology*, 112–114.

54. John Beatty, "Fitness: Theoretical Contexts," in *Keywords in Evolutionary Biology*, 115–119.

55. One example is provided by Walter Sneader, discussed in more detail below.

56. David S. Jones, "Technologies of Compliance: Surveillance of Self-Administration of Tuberculosis Treatment, 1956–1966," *History and Technology* 17 (Winter 2001): 279–318; Jeremy A. Greene, "Therapeutic Infidelities: 'Noncompliance' Enters the Medical Literature, 1955–1975," *Social History of Medicine* 17 (2004): 327–343.

57. Evelynn Fox Keller, "Competition: Current Usages," in *Keywords in Evolutionary Biology*, 68–73, quote on 68.

58. Harry Marks, *The Progress of Experiment: Science and Therapeutic Reform in the United States, 1900–1990* (Cambridge: Cambridge University Press, 1997).

59. Greene, *Prescribing by Numbers*; Joseph Dumit, *Drugs for Life: How Pharmaceutical Companies Define Our Health* (Durham, NC: Duke University Press, 2012); Healy, *Antidepressant Era*.

60. Anne Pollock, *Medicating Race: Heart Disease and Durable Preoccupations with Difference* (Durham, NC: Duke University Press, 2012).

61. Leigh Van Valen, "A New Evolutionary Law," *Evolutionary Theory* 1 (1973): 1–30, especially 17–22. See also Lewontin, "Adaptation," 215.

62. Lewis Carroll, *Through the Looking-Glass, and What Alice Found There* (London: Macmillan and Co., 1872).

63. Anthony D. Barnosky, "Distinguishing the Effects of the Red Queen and Court Jester on Miocene Mammal Evolution in the Northern Rocky Mountains," *Journal of Vertebrate Paleontology* 21 (2001): 172–185. A recent analysis invokes another piece of fiction (J. R. R. Tolkien's *Lord of the Rings*) to mark a different distinction. Organism-environment mismatches, whether from Red Queen or Court Jester effects, can cause both increased extinction rates and decreased origination rates of new species, a phenomenon of "evolutionary sterility that we call the Entwives effect." See Tiago B. Quental and Charles R. Marshall, "How the Red Queen Drives Terrestrial Mammals to Extinction," *Science* 290 (2013): 290–292, quote on 291.

64. Michael J. Benton, "The Red Queen and the Court Jester: Species Diversity and the Role of Biotic and Abiotic Factors through Time," *Science* 323 (2009): 728–732, quote on 728.

65. David S. Jones, Scott H. Podolsky, and Jeremy A. Greene, "The Burden of Disease and the Changing Task of Medicine," *New England Journal of Medicine 366* (2012): 2333–2338.

66. Thomas McKeown, *The Role of Medicine: Dream, Mirage, or Nemesis?* (Princeton, NJ: Princeton University Press, 1979).

67. William E. Boden, Robert A. O'Rourke, Koon K. Teo, et al., "Design and Rationale of the Clinical Outcomes Utilizing Revascularization and Aggressive Drug Evaluation (COURAGE) Trial," *American Heart Journal* 151 (2006): 1173–1179; Boden, O'Rourke, Teo, et al., "Optimal Medical Therapy with or without PCI for Stable Coronary Disease," *New England Journal of Medicine* 356 (2007): 1503–1516.

68. Keith J. Winstein, "Stent Implants Declined in April: Doctors Attribute Drop to Study Showing Drugs May Have Similar Benefits," *Wall Street Journal*, 17 May 2007; Winstein, "A Simple Health-Care Fix Fizzles Out," *Wall Street Journal*, 11 February 2010.

69. Dean J. Kereiakes, "PCI Is No Better Than Medical Therapy for Stable Angina? Seeing Is Not Believing," *Cleveland Clinic Journal of Medicine* 74 (2007): 637–642. See also Barnaby J. Feder, "First, a New Artery Stent Study; Now, Questions about What It All Means," *New York Times*, 28 March 2007.

70. Kereiakes, "PCI Is No Better," 640.

71. John Damuth, "Extinction," in *Keywords in Evolutionary Biology*, 106–111.

72. Pressman, *Last Resort*, 401.

73. Jones, "Surgical Practice and the Reconstruction of the Therapeutic Niche: The Case of Myocardial Revascularization," in Thomas Schlich and Chris Crenner, *Innovation in Surgery* (Rochester, NY: University of Rochester Press, forthcoming).

74. David C. Hull, *Science as a Process: An Evolutionary Account of the Social and Conceptual Development of Science* (Chicago: University of Chicago Press, 1990).

75. Ian Hacking, *The Social Construction of What?* (Cambridge, MA: Harvard University Press, 1999); Geoffrey C. Bowker and Susan Leigh Star, *Sorting Things Out: Classification and Its Consequences* (Cambridge, MA: MIT Press, 2000).

76. Christopher Hamlin, *More Than Hot: A Short History of Fever* (Baltimore: Johns Hopkins University Press, 2014).

77. Sarah W. Tracy, *Alcoholism in America: From Reconstruction to Prohibition* (Baltimore: Johns Hopkins University Press, 2005).

78. For the old nomenclature, see *Annual Report of the Commissioner of Indian Affairs to the Secretary of Interior* (Washington: Government Printing Office, 1891), 806–812. For the new nomenclature, see *Annual Report of the Commissioner of Indian Affairs to the Secretary of Interior* (Washington: Government Printing Office, 1892), 919–925.

79. Dennis Wood, *The Power of Maps* (New York: Guilford Press, 1992); Dennis Wood and John Fels, *The Natures of Maps: Cartographic Constructions of the Natural World* (Chicago: University of Chicago Press, 2009).

80. Walter Sneader, *Drug Prototypes and Their Exploitation* (New York: John Wiley & Sons, 1996), ix. In the preface to Sneader's work, historian John Swann describes how Sneader, by "walking us step-by-step through the evolution," is able to "document the modern revolution of real therapeutics" (vii).

81. Ibid., 463–482.

82. Ibid., 264–281.

83. Ibid., 661–679.

84. Ibid., 410–415.

85. Brent M. Ardaugh, Stephen E. Graves, and Rita F. Redberg, "The 510(K) Ancestry of a Metal-on-Metal Hip Implant," *New England Journal of Medicine* 368 (2013): 97–100.

86. Jeremy A. Greene, *Generic: The Unbranding of Modern Medicine* (Baltimore: Johns Hopkins University Press, 2014).

87. Robert J. Whittaker and José María Fernández-Palacios, *Island Biography: Ecology, Evolution, and Conservation*, 2nd ed. (New York: Oxford University Press, 2007).

88. Tom Koch, *Cartographies of Disease: Maps, Mapping, and Medicine* (Redlands, CA: ESRI Press, 2005).

89. Glover, J. Alison. "The Incidence of Tonsillectomy in School Children,"

Proceedings of the Royal Society of Medicine 31 (1938): 1219–1236; Center for the Evaluative Clinical Sciences, *Dartmouth Atlas of Health Care* (Chicago: American Hospital Publishing, 1996). For a discussion, see David S. Jones, *Broken Hearts: The Tangled History of Cardiac Care* (Baltimore: Johns Hopkins University Press, 2013). For a critique of the Dartmouth approach, see Michael E. Chernew, Lindsay M. Sabik, Amitabh Chandra, Teresa B. Gibson, and Joseph P. Newhouse, "Geographic Correlation between Large-Firm Commercial Spending and Medicare Spending," *American Journal of Managed Care* 16 (2010): 131–138; Joseph P. Newhouse and Alan Garber, "Geographic Variation in Health Care Spending in the United States," *JAMA* 310 (2013): 1227–1228.

90. John Wennberg and Alan Gittelsohn, "Health Care Delivery in Maine: I. Patterns of Use of Common Surgical Procedures," *Journal of the Maine Medical Association* 66 (May 1975): 123–130, 149, quote on 127.

91. Frederick Robbins, quoted in John K. Iglehart, "From the Editor," *Health Affairs* 3 (Summer 1984).

92. For most of this history, doctors argued that geographic variations in medical practice were an appropriate response to variations in environmental conditions. See, for instance, John H. Warner, *The Therapeutic Perspective: Medical Practice, Knowledge, and Identity in America 1820–1885* (Cambridge, MA: Harvard University Press, 1986).

93. Jones, *Broken Hearts*. The research, when it did emerge, focused almost exclusively on surgery. Greene, in chapter 6 of this volume, describes how practice variation in drug prescriptions also existed and had become a subject of intense interest among pharmaceutical companies. Yet the literatures about surgical variation and pharmaceutical variation have remained distinct, having been produced by different kinds of analysts and published (or not) in different venues, and having motivated different kinds of policy interventions.

94. Thomas Schlich, "Asepsis and Bacteriology: A Realignment of Surgery and Laboratory Science," *Medical History* 56 (2012): 308–334.

95. Ronald Bayer and David Wilkinson, "Directly Observed Therapy for Tuberculosis: History of an Idea," *Lancet* 345 (1995): 1545–1548.

96. G. Evelynn Hutchinson, "Concluding Remarks," *Cold Spring Harbor Symposium on Quantitative Biology* 22 (1957): 415–427.

97. Stephen Jay Gould, "The Disparity of the Burgess Shale Arthropod Fauna and the Limits of Cladistic Analysis: Why We Must Strive to Quantify Morphospace," *Paleobiology* 17 (1991): 411–423, quote on 420.

98. Gould, "The Disparity," 420. Gould admitted that this task was "dauntingly difficult" (420) at best, and possibly "logically intractable" (421). See also Benjamin Blonder, Christine Lamanna, Cyrille Violle, and Brian J. Enquist, "The N-Dimensional Hypervolume," *Global Ecology and Biogeography* 23 (2014): 595–609, especially 603.

99. Raymond Williams, "Progressive," in *Keywords*, 243–245, quote on 244.

100. Williams, "Progressive," 244, 245.

101. Richard Dawkins, "Progress," in *Keywords in Evolutionary Biology*, 263–272, especially 263.

102. Variation around this mean, however, has increased. This accounts for the appearance of more complex life forms over time.

103. Donald Effler, "Surgery for Coronary Disease," *Scientific American* 210 (1968): 36–43, quote on 38.

104. Effler, "Myocardial Revascularization at the Community Hospital Level," *American Journal of Cardiology* 32 (1973): 240–242, quote on 240.

105. Benjamin N. Rome, Daniel B. Kramer, and Aaron S. Kesselheim, "FDA Approval of Cardiac Implantable Electronic Devices via Original and Supplement Premarket Approval Pathways, 1979–2012," *JAMA* 311 (2014): 385–391, quote on 388.

106. Rome, Kramer, and Kesselheim, "FDA Approval," 390.

ELEVEN

A Therapeutic Revolution Revisited

CHARLES ROSENBERG

"Therapeutic revolution" is a familiar term. To most of us, physicians and laypersons alike, it is a shorthand label for a period in the mid-twentieth century when for the first time medicine was able to intervene in the trajectory of many diseases. One thinks of insulin, of antibiotics, of steroids. The term implies a transformative efficacy and a reassuring history, a decisive inflection point in a narrative of clinical progress, of the laboratory's power to inform the practice of medicine. And the term is not entirely misleading. It describes and encapsulates significant events and changed attitudes. Although they may not always label it a "therapeutic revolution," the great majority of physicians and educated laypersons assume that a fundamental shift took place in the last three-quarters of the twentieth century—a change that gave medicine new and powerful tools in the diagnosis and treatment of disease. Most assume as well that this revolution was not only one *in* therapeutics, but one in morbidity patterns and life expectancy that was wrought *by* therapeutics. Even the wide cultural acceptance of a *thing* called "therapeutic revolution" has played a role in the subsequent history of medical care, helping shape expectations and social policy, legitimating roles, and defining norms of practice.

But the concept of a twentieth-century therapeutic revolution obscures as well as illuminates. It obscures the incremental and multidimensional change that created the world in which this particular twentieth-century revolution could have taken place. It also implies a perhaps too all-encompassing role for medical and surgical intervention in changing patterns of morbidity and mortality. Even more fundamentally, the casual invocation of a "therapeutic revolution" obscures the ways in which clinical practice is necessarily a component in a complex time- and place-specific system of ideas and social practices that cannot be adequately understood outside that larger context. The origins and ultimate social fate of particular technological innovations are neither inevitable nor entirely predictable.

This may seem no more than a litany of truisms, but these are ideas that were not clearly discernible in the canon of medical history when I first began to study the field as a young man. Everyday practice before the mid-nineteenth century was, in fact, not a subject of serious interest to medical historians academic or amateur; it was a source of quaint anecdote, more an occasion for embarrassment than for pride. There were a few exceptions. The introduction of vaccination, for example, or the uses of digitalis and quinine seemed worth discussing as atypical points of physiologically rational light in a darkness of traditional practice. Otherwise, historians found it hard to make sense of traditional therapeutics. If they thought about it at all, it seemed a timeless dead end of placebolike ritual allied with folk practice and the healing power of nature—entirely unrelated to the nineteenth- and twentieth-century development of a "rational," laboratory-derived, and increasingly objective science-based medicine.

But over a period of several decades I became fascinated by a nagging reality. Why had traditional interventions changed so little over so many centuries in Western medicine?[1] Some practices, such as a concern with diet and moderation in stimulants, made a kind of intuitive sense. But why had bleeding, cupping, and the administration of emetics and diuretics lasted so long? Why had physicians and lay people shared and accepted this array of—in retrospect—ineffective and unpleasant incursions into their bodies? If they persisted for so long, such measures and medicaments must have been in some sense effective, even if in a way inconsistent with modern notions of physiological efficacy.

My ultimate conclusion was simple enough. Traditional therapeutics persisted, and in that sense must necessarily have worked, because such practices were part of a social and cultural system and made functional sense within that system. It was a system in which understandings of health and disease were widely shared, in which practice took place often in a face-to-face domestic setting (not in a hospital or clinic or even a doctor's office), in which the physician or practitioner was known to family members, and in which the physiological effects of a prescribed cathartic or emetic, of bleeding or cupping could be seen—witnessed—by all concerned. It was an era of individual and nonspecific ideas of disease, in which disease was labile, cumulative, and individual. A cold might evolve into pleurisy or pneumonia, and then into tuberculosis. Constitution and circumstance interacted over time to create idiosyncratic outcomes. An individual's life-sustaining intake and outgo constituted and reconstituted the vital essence of life; balance preserved health, imbalance brought illness. And individual choices—diet, exercise, rest, sexual behavior—could shape that ultimate balance; thus, volition as well as constitution played a role in an individual's path to health or disease. Etiology was as much biography as specific pathological mechanism. This system was coherent in terms of shared ideas and social setting, and it was responsive to the human need for intervention, admonition, and explanation in time of sickness. The seemingly quaint medical practices of 1800 should not, I concluded, be dismissed or ignored, but should be understood in terms of the worldviews and social practices—the choices available—to those past actors.

We now inhabit a world very different from the one I have just described, and it is the fundamental and multidimensional change between the "then" of 1800 and the "now" of the mid-twentieth century that constituted what I referred to as a "therapeutic revolution." We live in a world of metrics and molecules, of aggregate truths, of evidence-based reality. It is a world in which professional understandings of health and disease separate rather than link patient and physician, a world that has assumed and assimilated a revolution in conceptions of disease, a hospital and laboratory revolution, and revolutions in the economics of health and in policy and bureaucracy. And it is a world characterized by an altered incidence of morbidity and life expectancy, of pervasive chronic disease and long-term care delivered by credentialed strangers in institutional settings. It is a world in which disease is understood as specific and definable, reduced to and in part constituted by disease categories and treatment protocols. It is a world of corporate strategies and government policy. This complex of ideas, roles, institutions, and practices is the current endpoint of the therapeutic revolution that I delineated four decades ago when trying to express the distance between medical care in the late eighteenth and late twentieth centuries. The therapeutic revolution I had in mind at once mirrored, incorporated, and in part constituted larger structural changes in society.

There has never been a time or place without modes of curing; we have always had therapeutics with us. But it is a characteristic of our particular system that we assume that modern therapeutic practices are categorically different—the result of a cumulative understanding of the natural world and a capacity to intervene that somehow removes Western therapeutic practices of the past century from the contingency that is culture, from the very constructedness and interconnectedness of medicine that is so apparent to the historian or ethnographer of other times and places. This is a powerful and culturally dominant narrative, appealing to our faith in science and the inevitability of progress, to the hope that sickness will ultimately be vanquished through the inevitable accumulation of "breakthroughs" and "insights." We have not cured cancer yet, for example, but few doubt our capacity to ultimately find effective treatments for this family of intractable ills. For most of our contemporaries, lay and medical, this is the basic narrative, and everything else is contingent, arbitrary, and somehow less. This faith in progress and rationality is in fact a significant aspect of our collective worldview, but one that impairs our ability to understand the very complexity of that medical culture in which we bestow such faith—and in which we live.

This narrative of technological triumphalism coexists with a parallel and oppositional yet logically consistent twin. That parallel narrative is one of declension, of the individual patient abstracted and alienated, a mirror of larger forces of modernity in which the individual patient is an imperfect example of a more fundamental reality, a variety of corporeal background noise from which the physician discerns an actionable signal, a diagnostic choice among aggregate pictures of disease arrayed in an agreed-upon taxonomy (such as the most recent revision of the *International Classification of Diseases*, or the *Diagnostic and Statistical Manual* of the American Psychiatric Association). It is a narrative of impersonality and abstraction, of the individual patient as epiphenomenon. How many of us have not heard a patient's plaintive complaint that the physician was looking at a screen and not at her? And that physician was often unfamiliar, if highly credentialed. Such complaints echo and reinforce a long familiar body of theory that traces the development of Western society from a face-to-face communal world to our impersonal world of bureaucracy and credentialed status.

Fault Lines: Anticipating Revolutions

This complexity of attitude and expectation illustrated by these asymmetrical narratives of progress is not simply anomalous but instructive. It makes clear that therapeutic change cannot be understood in terms of the creation of new drugs and procedures alone. Differences in social location, perception, and interest help shape the complex negotiations surrounding the adoption of new clinical technologies. Such tensions are hard to ignore. As I write these words, the world faces an Ebola epidemic, the media speak of a post-antibiotic age, and intractable chronic disease in an aging population makes us question the very definition and boundaries of therapy. Pharmaceutical companies fret about empty pipelines and intellectual property, mergers and tax strategies. Governments concern themselves with regulatory issues and research support. Worries about toxic substances in the environment and climate change only intensify public sensitivity to the broadly ecological dimensions of the variables that determine disease incidence. Issues relating to therapeutics appear not only in the science section of newspapers, but also in the business section, and even on the front page. Ordinary men and women are, to cite a conspicuous example, faced with confusing messages about screening for breast, prostate, and thyroid cancer. The moral seems clear. Thinking about therapeutics means thinking not only about the physiological activity of particular drugs and devices, but about societies in the whole, their values and structures.

There are five areas aside from the world of biomedical research in which one can sense tensions and anticipate change. One is the role of the physician. A second is the conflict already mentioned between the individual and the structure of knowledge and bureaucracy that sees him or her in the aggregate; when teaching, I refer to it as the "*n* of one problem." Third is the fact that the world is global in a variety of ways that implicate therapeutic options. A fourth area of tension is the relationship between public and private sectors in a world of regulatory policy and practice and research strategies prioritized or neglected. Finally, every developed and developing country faces a shifting patient population, one that is older and increasingly burdened by chronic disease and incapacity. All these areas are inextricably related, but for the sake of analysis I think it helpful to tease them apart.

In a world of randomized clinical trials, guideline committees, and restrictive formularies, it is clear that the physician's autonomy is being constrained. Bureaucratic protocols, as opposed to judgment and idiosyncrasy, play a larger and larger role in clinical practice. Physicians echo their patients' complaints that they are forced to spend too much time looking at screens, and thus have that much less time to interact with the individuals they seek to evaluate. An alarmist might describe this as a de facto deskilling of the profession, or, perhaps more accurately, a de facto constriction of the individual practitioner's range of choice. In addition, the boundaries between physicians and other healers promise to become murkier and murkier. In the United States, for example, brick-and-mortar retail chains are rushing to establish walk-in clinics staffed by physician assistants and nurse practitioners. Assisted-living facilities exist in a quasimedical space staffed by nurses, aides, and social workers, with the occasional visit by a physician. Home hospice care is increasingly managed by teams of nurses, nurse practitioners, social workers, and aides, with the physician an occasional electronic voice. Such phenomena are indicators of greater change to come—and are already significant aspects of day-to-day therapeutics in its realistically comprehensive sense.

Related to all these issues is a question I have alluded to previously: the issue of human idiosyncrasy—the atypical—in a world ordered by standard aggregate pictures. This has, in some sense, always been a part of medical history. Even Hippocratic doctors judged a particular intermittent fever against a general, historically derived picture of the prognosis and treatment of previous cases exhibiting a similar course and pattern of symptoms. But in our world of bureaucracy and quality control, of treating numbers and images, this process has become even more intrusive and controlling. It is not necessarily a thing to be categorically dismissed, but it is a thing to be acknowledged and understood. The growing dominance of electronic data bases and widespread screening only exacerbates and intensifies a critique already a century old: that physicians treat diseases, not patients, and that reality is increasingly a chart and a record of laboratory findings, and not the particular man or woman in a particular bed, an individual with her or his personal and physiological needs and idiosyncrasies—and family. Increasing constraints on therapeutic choice only mirrors and exacerbates this asymmetry. One solution, already entertained widely, is the ultimate reductionism of genetic or personalized medicine—a final solution of sorts to the problem of biological, if not social, diversity.

Developments in therapeutics are necessarily global in a variety of ways. Perhaps most obvious are the ways in which standards of best practice have become widely disseminated—if not followed—throughout the world's medical community. Faith in the cogency and reliability of universal standards (randomized clinical trials, for example) creates not only a tactical but a moral agenda. If a drug, immunization technique, or surgical practice is accepted as efficacious, then failure to provide it to vulnerable populations implies a motivating critique of current practices, even if poverty, culture, and isolation make such interventions unrealistic in the short term, and "local standards of care" continue to prevail. No procedure or drug functions in the abstract; similar procedures can have different ecological niches and constituencies in different cultures. One thinks of the American-trained surgeons replacing hips or heart valves in India for an international constituency of patients. The globalization of travel and trade also helps shape the availability of drugs and devices, just as such spatial mobility increases the hazard of epidemic disease. Corporate strategies and market realities are global as well, and necessarily play a role in the selective availability of particular drugs and devices. But the global is necessarily local. In many regions, for example, traditional medical practices and beliefs continue to exist and even thrive alongside the practices and presumptions of Western medicine. And, finally, of course, our increasing awareness of the epidemiological consequences of climate change makes us anticipate a shifting disease burden in future years—and thus new therapeutic needs and priorities (including, of course, the weighing of preventive as opposed to therapeutic interventions). Of course, such global trends have a variety of local manifestations: each state responds in its own fashion to these pervasive tensions.

In most countries, the role of the state in the provision and management of health and welfare has grown dramatically over the past two centuries. It has added layers of decision making and resource allocation, along with regulation, to the practices and relationships we call medical care. How drugs are approved, how research goals are defined, and how basic research is paid for and articulated with translational research and clinical practice all play a role in shaping the everyday therapeutic realities of how and where patients are treated. Medicine has always been clothed with a special moral weight—I have compared it elsewhere to national defense—that makes health-related decisions reflect a special kind of rationality, one that marries economic motives and perspectives to what might be called transcendent considerations.[2] The great majority of us assume that access to at least a minimal level of health care—like some modest level of education—is a de facto human right, and the policies of most governments have in some measure reflected this special relationship. Medicine is always a hybrid enterprise. And the narrative I have already described, of progress through a succession of therapeutic revolutions, only intensifies this moral argument. Progress and efficacy imply at least some access to care. Such considerations have and will shape health policy and thus the therapeutic environment in most countries.

Finally, all these shifting variables need to be seen in a demographic and epidemiological context. Men and women are living longer in most parts of the world, and, as scores of commentators have underlined, chronic ills ranging from diabetes to circulatory ailments have become increasingly common in what used to be called the developing world. The provision of care in such environments necessarily reflects particular realities of class, culture, and region. This vast new burden of chronic disease constitutes in itself a revolution of a sort. In this world of complexity and change, therapeutic practice is indicator as well as substance, a window onto the societies in which it is provided.

This world of medical care and its history is far more complex and ambiguous than the widely assumed tale of laudable progress embedded in a narrative of technical accomplishment—of inevitable therapeutic revolutions emerging from the ever-maturing world of biomedicine. One needs to think of individual and social efficacy as well as the more narrowly and operationally defined efficacy of the randomized clinical trial and meta-analysis.[3] The concept of social efficacy underlines the need to think about the impact and nature of particular practices and policies at a variety of social and spatial locations and points in time, and thus, necessarily, on particular women and men. It makes us ask about questions of access, of patient experience, even of the definition and boundaries of what counts as therapeutics. How does hospital or nursing home routine relate to therapeutics? Or mass screening and its consequences? Or self-medication with over-the-counter drugs? Are we living through a new sort of therapeutic revolution—or evolution? It is time to think with as well as about therapeutics.

NOTES

1.   Charles E. Rosenberg, "The Therapeutic Revolution: Medicine, Meaning, and Social Change in Nineteenth-Century America," *Perspectives in Biology and Medicine* 20 (1977): 485–506. A somewhat different and extended version appears in Rosenberg, *Explaining Epidemics and Other Studies in the History of Medicine* (Cambridge: Cambridge University Press, 1992), 9–31.

2. Charles Rosenberg, *The Care of Strangers: The Rise of America's Hospital System* (New York: Basic Books, 1987), 350.

3. I employ parallel language in "Introduction: The History of Our Present Complaint," in *Our Present Complaint: American Medicine, Then and Now* (Baltimore: Johns Hopkins University Press, 2007), 9–10.

Contributors

MATTHEW BASILICO is a medical student at Harvard Medical School and a PhD candidate in economics at Harvard University. He was a Fulbright Scholar in Malawi and is coeditor of *Reimagining Global Health* (Berkeley: University of California Press, 2013).

CHRISTIAN BONAH is professor of the history of medical and health sciences at the University of Strasbourg. He has worked on comparative history of medical education, history of medicaments, and the history of human experimentation. His recent work includes research on risk perception and management in drug scandals and courtroom trials, as well as studies on medical film.

FLURIN CONDRAU has been professor of the history of medicine at the University of Zurich, Switzerland, since 2011. He has published widely on the history of infection, with particular focus on cholera, tuberculosis, and, more recently, hospital infections. His publications include the coedited volume *Tuberculosis Then and Now* (Montreal: McGill-Queen's University Press, 2010).

PAUL FARMER is Kolokotrones University Professor and chair of the Department of Global Health and Social Medicine at Harvard Medical School, chief of the Division of Global Health Equity at Brigham and Women's Hospital in Boston, and cofounder of Partners in Health. He has written extensively on health, human rights, and the consequences of social inequality.

JEREMY A. GREENE is professor of medicine and the history of medicine at the Johns Hopkins University School of Medicine, the author of *Prescribing by Numbers: Drugs and the Definition of Disease* (Baltimore: Johns Hopkins University Press, 2007) and *Generic: The Unbranding of Modern Medicine* (Baltimore: Johns

Hopkins University Press, 2014), and a practicing internist at a community health center in East Baltimore.

NICOLAS HENCKES is a sociologist and associate researcher at the Center for Research in Medicine, Science, Health, Mental Health, and Society at the National Center for Scientific Research in Paris. He works on the changing landscape of mental health in France during the twentieth century. His latest project looks at the current status of psychosis risk in Europe from both ethnographic and historical perspectives.

JANINA KEHR is a medical anthropologist and lecturer in the History of Medicine Section at the University of Zurich, Switzerland. She holds a PhD from the Ecole des hautes études en sciences sociales Paris, and works on infectious disease control and the biopolitics of austerity in contemporary Europe.

NILS KESSEL is a postdoctoral researcher in science studies at the Institute for Research and Innovation in Society in Paris. He studied modern and contemporary history, history of medicine, and French before undertaking his PhD thesis on the history of drug consumption in West Germany. He has also authored a book on the history of West German emergency services since 1945: *Geschichte des Rettungsdienstes 1945–1990: Vom Volk von Lebensrettern zum Berufsbild Rettungsassistent* (Pieterland, Switzerland: Peter Lang AG, 2008).

DAVID S. JONES is the Ackerman Professor of the Culture of Medicine at Harvard University. Trained as a psychiatrist and historian, he has studied health disparities, human subjects research, and decision making in clinical medicine, especially in cardiac therapeutics. His current research explores the evolution of coronary artery bypass surgery.

ANNE KVEIM LIE is associate professor in medical history and head of studies in community medicine at the University of Oslo. Among her research interests are temporalities in health and medicine, the history of antibiotic resistance, and disease categories of the eighteenth century.

JULIE LIVINGSTON is professor of history and social and cultural analysis at New York University. Her books include *Improvising Medicine: An African Oncology Ward in an Emerging Cancer Epidemic* (Durham, NC: Duke University Press, 2012), and *Debility and the Moral Imagination in Botswana* (Bloomington: Indiana University Press, 2005). In 2013 she was named a MacArthur fellow.

LUKE MESSAC is an MD/PhD student in the Department of History and Sociology of Science at the University of Pennsylvania. His dissertation is a history of the role of medical care in development planning and popular politics in Malawi. His publications include articles in *Social Science & Medicine* and the *Journal of Policy History*.

KRISTIN PETERSON is associate professor of anthropology at the University of California, Irvine. Her research interests include science and technology studies, African studies, and postcolonial iterations of political economy. She is the author of *Speculative Markets: Drug Circuits and Derivative Life in Nigeria* (Durham, NC: Duke University Press, 2014).

SCOTT H. PODOLSKY is associate professor of global health and social medicine at Harvard Medical School, and director of the Center for the History of Medicine at the Francis A. Countway Medical Library. His most recent book is *The Antibiotic Era: Reform, Resistance, and the Pursuit of a Rational Therapeutics* (Baltimore: Johns Hopkins University Press, 2015).

CHARLES ROSENBERG is the Ernest E. Monrad Professor Emeritus in Harvard University's Department of the History of Science. He has written widely on medicine and science in America; his most recent book is *Our Present Complaint. American Medicine, Then and Now* (Baltimore: Johns Hopkins University Press, 2007). He is currently at work on a study of changing disease concepts in the past two centuries.

ELIZABETH SIEGEL WATKINS is dean of the graduate division, vice chancellor of student academic affairs, and professor of history of health sciences at the University of California, San Francisco. She is the author of *On the Pill: A Social History of Oral Contraceptives* (Baltimore: Johns Hopkins University Press, 1998) and *The Estrogen Elixir: A History of Hormone Replacement Therapy in America* (Baltimore: Johns Hopkins University Press, 2007).

Index

The letter *f* following a page number denotes a figure.

bacteriological revolution, 5, 6, 270
barbituric acids, 113–14
Barboza, David, 254
Basaglia, Franco, 85
Batten, Lindsey, 29
Baxerres, Carine, 257
Bean, William, 25–26, 34
Bernard, Claude, 6
beta-blockers, 115
Bibile, Senaka, 159–62, 165, 173
Biehl, Joao, 223
biomedicalization, process of, 138–39
birth control pill: end of revolution of,
    55–58; invention of, 44–47; legacies of,
    as therapeutic revolution, 58–61; new
    uses for, 57–58; rebranding of, 57; revo-
    lution bred by, 48–53; safety concerns
    of, 53–55; tactics used by manufactur-
    ers to promote use of, 57; as therapeutic
    revolution, 43–44, 50–53
Bogdanich, Walt, 254
Bonds, Matthew, 204
Botswana: antiretrovirals in, 220–21;
    cancer in, 225–33; chemotherapy in,
    223–24, 227–33; therapeutic revolution
    in, 219–20. *See also* southern Africa
Bradford Hill, Austin, 134–35
Brill, Henry, 77
Bud, Robert, 22, 33

cardiovascular drugs, IMS data and,
    105–11
Carson, Rachel, 30
chemical revolution, 7; in psychiatry, 11
chemotherapy, 21, 68; in Botswana, 223–
    24; skepticism of, 78–79
Chloromycetin, 24
chlorpromazine, 90n13; as cause in reduc-
    tion of psychiatric hospitalizations, 77–
    78; discovery of, 68; historical account
    of development of, 90n12; marketing
    of, 68–69; psychiatry and, 72; psycho-
    logical effects of, 70–71; spread of use
    of, 68–69. *See also* neuroleptics
Ciba-Geigy, 75
Clarke, Adele, 138
clinical trials, 9, 11, 29, 33, 47, 74, 79, 134–
    38, 252, 271, 277–82, 291, 306, 308
competition: as an evolutionary concept,
    278, 282, 293; as a market concept, 57,
    238

Cooper, Melinda, 243
Cornet, George, 133
Croften, John, 135
Cushing, Harvey, 271
Cutler, David, 190

Darwin, Charles, 276
Davies, Dame Sally, 18
Deaton, Angus, 194
deinstitutionalization: emergence of, 76;
    role of neuroleptics in, 77–78
de Kieffer, Donald, 255
de Kruiff, Paul, 5, 21
Delay, Jean, 68, 70–71
delivery, science of, 207
Deniker, Pierre, 68, 70–71
dependency theory, 164–65
DESI (Drug Efficacy Study and Implemen-
    tation) process, 27–33
Dettweiler, Peter, 132
diagnosis, standardization of, 73–75; battle
    over, 76
Diagnostic and Statistical Manual, third edi-
    tion (DSM-III), 73–75
Domagk, Gerhard, 133–34
dopamine hypothesis, 71
DOTS (directly observed treatment, short
    course) approach, 136–37, 142
Dowling, Harry, 24, 25, 26, 27, 32, 34
Drug Efficacy Study and Implementation
    (DESI) process, 27–33
drug industry. *See* pharmaceutical industry
Drug Reform Act of 1978, 166
DSM-III (*Diagnostic and Statistical Manual*,
    third edition), 73
Dubos, René, 29, 34
Duramed oral contraceptive, 57
dystopias, 34–35; collision of, with antibi-
    otic resistance, 29–31

Eban, Katherine, 254
Edgerton, David, 99, 117
efficacy, 1, 2, 11–12, 27, 33, 50, 52, 58, 68,
    87, 97–99, 109, 130–40, 154, 158, 162,
    166, 169, 188–98, 219–20, 222–24,
    229–33, 276–77. *See also* Drug Efficacy
    Study and Implementation (DESI)
    process
Effler, Donald, 291
Eisenstadt v. Baird, 49
Elkeles, Barbara, 131

INDEX

multinational pharmaceutical industry, 161. *See also* American pharmaceutical industry; pharmaceutical industry
Myrdal, Gunnar, 191

n-dimensional trait space, 288–89
Nelson, Gaylord, 53–54, 154, 156, 157
neuroleptic revolution, 65–66, 87–88. *See also* chlorpromazine
neuroleptics: debate over, 78; defined, 89n3; early appraisals of, 71–72; efforts to understand, 71; history of development of, 75–76; role of, of deinstitutionalization, 77–78
niche concept, 273–76
Nigeria: effects of oil bust at end of 1970s in, 241–42; fake and substandard drugs in, 256–60; histories of market making in, 240–46; impact of multinational pharmaceuticals in, 240–41; Pfizer in, 245–46; pharmaceutical companies in, 241–44; in pharmaceutical market, 238–39; restructuring of pharmaceutical industry in, 238, 243; SAPs in, 238; therapeutic revolution in, 260–61. *See also* Idumota neighborhood, Lagos, Nigeria

Ogden, Jessica, 137
Okelola, Kunle, 246–47
Omran, Abdel, 128–29, 160, 192, 195
Oppenheimer, Gerald, 193
oral contraceptives. *See* birth control pill
oral rehydration therapy (ORT), 198
Ortho Tri-Cyclen, 57
Osler, William, 4, 270, 290
Osseo-Asare, 163

Packard, Randall, 219
Packard, Vance, 157
Paris Clinic, 5
Pasteur, Louis, 5, 6
"past future" concept, 19
patents, pharmaceutical availability and, 170–71
Paterson, Marcus, 132
penicillin, 22, 24
Pernick, Martin, 6
Peterson, Olster, 153, 156–57
Pettenkofer, Max von, 130

Pfizer, in Nigeria, 245–46
pharmaceutical access, 155; patents, promotion, and pricing of, 170–72; problems of, 162–63. *See also* access to medicines
pharmaceutical geography, 150–86
pharmaceutical industry: Senator Kefauver's hearings on, 27; multinational, 161; role of, in public health, 167–70; role of, revolution in psychiatry and, 72–73; symbolic power of, 7–10. *See also* American pharmaceutical industry; Idumota neighborhood, Lagos, Nigeria; Nigeria
pharmaceutical marketing, 154; of chlorpromazine, 68–70; critics of, 157; as diffusion mechanism, 156–57; Gaylord Nelson's hearings into practices of, 160
pharmaceuticals, historical accounts of, 98–99
pharmacotherapeutic revolution, 97–98; drugs of, 116–17; narrative of, 98
Pieters, Toine, 70
pill, the. *See* birth control pill
Pincus, 45, 47
Pinel, Philippe, 83
Pison, Gilles, 202
Planned Parenthood Federation of America, 45, 48
Pollock, Anne, 278
Porter, Roy, 3, 6, 272, 292
Pressman, Jack, 68, 273–74
Preston, H., 201
Primary Health Care (PHC) movement, 192
Pritchett, Lant, 202
progress, problem of, 289–92
psychiatric revolution: bitter fruit of, 83–87; third, 83
psychiatry: chemical revolution in, 11; chlorpromazine and, 72; diagnostic standards and, 73–75; French, 74, 80–81; revolutionary rhetoric and, 77–80; revolution in, and role of pharmaceutical industry, 72–73; standardization of, 72–73
psychometrics, 74
psychopharmacology, 67, 72; psychiatric standards and, 72–73; revolutionary character of, 117–18. *See also* neuroleptic revolution

319

public health, role of pharmaceuticals in, 167–70
public health nihilism, 193–94

Rajkumar, Andrew Sunil, 202
Rasmussen, Nicolas, 67–68
Red Queen hypothesis, 279–81
Reed, James, 60
Reinfeld, Fred, 21
revolution(s), 6; anatomo-pathological, 5; anesthetic, 5–6; antibiotic, 20–25, 33–36; antiseptic, 5–6; bacteriological, 5; birth control pill and, 43–44, 48–53, 55–58; chemical, 7, 11; concept of, 23; defined, 271–72; vs. evolution, 292–93; laboratory, 5; psychiatric, 81–87; in psychopharmacology, 117–18; scientific, 3, 21. *See also* therapeutic revolutions
revolutionary narratives, 3–7
Rhône-Poulenc, 68–69
Ribicoff, Abraham, 154
Rieff, Phillip, 82
Robbins, Frederick, 286
Rock, John, 45, 47
Rosa, Hartmut, 35
Rosenberg, Charles E., 270, 272; concept of therapeutic revolution, 2
Rostow, Walt, 191
rotavirus, 198
Rwanda, mortality declines in, 203–6, 204f

safety, drug, 27, 33, 50–56, 253, 278
Sandbu, Martin, 202
Sanger, Margaret, 43, 45, 46
sanitariums, 132–33
SAPs (structural adjustment programs), 238
Sarett, Lewis H., 169
Sargant, William, 79
schizophrenia, 71, 75–76
Schweiker, Richard, 166
science, role of, and modern medicine, 4–5
scientific revolution, 3, 21
Seaman, Barbara, 53–54
Seasonale oral contraceptive, 57, 58
sedatives, 112, 113
Semmelweis, Ignaz, 5
Shapin, Stephen, 3
shock therapy, 68
Shyrock, Richard, 6

Sigmamycin, 27
Silverman, Milton, 165
Simmons, Henry, 28
Smith, Austin, 155
southern Africa: antiretrovirals in, 220–21; therapeutic revolution in, 218–20. *See also* Botswana
Soviet pharmaceutical access, American pharmaceutical industry and, 155
Speaker, Andrew, 126–27
Specia, 69–70
Sri Lanka, 159–60
Starr, Paul, 5
Stetler, C. Joseph, 156
Stolley, Paul, 28
streptomycin, 21, 24, 128, 133
streptomycin trials, 135
structural adjustment programs (SAPs), 238
sulfa drugs, 8, 20, 21, 34
sulfonamides, 133–34
superbugs, 29
Swaroop, Vinaya, 202
Swazey, Judith, 69
Szasz, Thomas, 85
Szreter, Simon, 190, 195

tardive dyskinesia, 86
taxonomy, science of, 282–85
TB. *See* tuberculosis (TB)
TB Alliance, 142
technologies, doctrine of appropriate, 164
Teeling-Smith, George, 161
Terramycin, 24
therapeutic modernity, 162–64
therapeutic revolutions, 5, 9–10; anticipating, 306–9; beneficiaries of, 153–58; birth control pill and, 43–44, 50–51; concept of twentieth-century, 302–5; IMS data and, 115–18; invention of term, 8–9; Koselleck's definition of, 3; narratives of, 1–2, 7–8; negative side of, 154; in Nigeria, 260–61; Rosenberg's concept of, 2. *See also* revolution(s)
Thomas, Lewis, 7–8
Tiefenbacher, Max, 170–71
Tilley, Helen, 163
Tomes, Nancy, 7
Toulouse, Edouard, 79
Toynbee, Arnold, 3
tranquilizers, 116–17